# CONDUCTION HEAT TRANSFER

D. POULIKAKOS

*Mechanical Engineering Department*
*University of Illinois at Chicago*

PRENTICE HALL, Englewood Cliffs, New Jersey 07632

**Library of Congress Cataloging-in-Publication Data**

Poulikakos, D.
Conduction heat transfer / D. Poulikakos.
p. cm.
Includes bibliographical references and index.
ISBN 0-13-175845-4
1. Heat—Conduction. I. Title.
TJ260.P634 1994
621.402'2—dc20 93-1504
CIP

Acquisitions editor: Bill Zobrist
Production editor: Bayani Mendoza de Leon
Copy editor: Zeiders & Associates
Cover designer: Violet Lake Design
Manufacturing buyer: Linda Behrens
Editorial assistant: Susan Handy

Printed in the United States of America
10 9 8 7 6 5 4 3 2 1

ISBN 0-13-175845-4

Prentice-Hall International (UK) Limited, *London*
Prentice-Hall of Australia Pty. Limited, *Sydney*
Prentice-Hall Canada, Inc., *Toronto*
Prentice-Hall Hispanoamericana, S.A., *Mexico*
Prentice-Hall of India Private Limited, *New Delhi*
Prentice-Hall of Japan, Inc., *Tokyo*
Simon & Schuster Asia Pte. Ltd., *Singapore*
Editora Prentice-Hall do Brasil, Ltda., *Rio de Janeiro*

To my parents, Voula and Yiorgos Poulikakos,
and to my wife, Lily,
for being a continuous source of support,
encouragement, and motivation.

# CONTENTS

# PREFACE

Heat conduction constitutes the topic of a semester or quarter course at the beginning graduate level in the great majority of mechanical engineering departments in the United States. After teaching this course numerous times in the past decade (both on campus and in industrial sites in the Chicago area) it became abundantly clear to me that a serious need existed for a new conduction heat transfer textbook. This textbook should contain an appropriate topic selection sufficient for the needs of the course while blending the necessary mathematics with tangible engineering applications. To this end, modern application of heat conduction should be emphasized to inform the student of the current state of a mature field.

In undertaking the task of writing such a book, I decided to include material appropriate for a semester-long course for graduate and advanced senior students. Both the level of presentation of the material and the selection of the topics covered obey this principle. In selecting what to include (and what to omit) out of a large number of topics in heat conduction, I used the experience gathered from interactions with colleagues, students, and practing engineers over the past decade, as well as, I must confess, an amount of personal bias toward topics such as heat conduction from sources and sinks or heat conduction during the freezing of pure substances and alloys that are timely and appropriate for classroom presentation. In what follows the defining features of the book are listed.

- The solution of heat conduction problems calls for a good knowledge of partial (and ordinary) differential equations that beginning graduate students do not always possess. To aid the student with the purely mathematical intricacies, a good number of intermediate steps have been included in the description of the solution methods throughout. It is hoped that with this approach the student will absorb the material better and will be able to appreciate the physical and engineering aspects of heat conduction problems.

- The examples and the problems are, almost without exception, drawn from engineering applications. Instead of solving, for example, heat conduction in a square, rectangle, or cylinder, an engineering problem is presented and it is shown how this problem can be modeled as heat conduction in a square, rectangle, or cylinder.
- Whenever appropriate, solution of the examples and the derivations are carried out one step further—to calculation of the heat flux or heat transfer rate from a surface of interest. Hence the student can recognize the usefulness of the solution for temperature distribution.
- While some examples are drawn from traditional heat transfer areas such as heat exchangers and furnaces, most are drawn from areas of current interest and appeal, such as heat transfer in manufacturing processes (crystal growth, casting, coating, welding, etc.), cooling of electronic equipment, and heat transfer in biomedical, environmental, food processing, and cooking applications. Through these examples the student will have the opportunity to get a flavor of topics of current research and everyday life in heat conduction without having to encounter more involved, research-level material.
- Chapter 10 is, to the best of my knowledge, a unique feature of the book. It deals with the timely topic of freezing of binary alloys in the presence of a mixed-phase region (mushy zone). The level of the material is the same as that of Chapter 9, and it is meant to initiate the student in phase-change heat transfer in alloys, which constitutes a contemporary research area in thermal sciences.
- Another special feature is Chapter 12, a collection of recommended term projects in heat conduction that require computer usage. I have consistently assigned such projects in my heat conduction classes and have gathered excellent feedback from students. In addition, the solution of conduction problems in the industry often requires familiarity with computer usage. Either a single project can be assigned to the entire class, or the class can be divided into groups of students with different projects assigned to each group.
- A serious effort was made to connect the appendixes with the material in the various chapters. To this end, examples from the various chapters are used in the appendixes to illustrate how the material in the appendixes can be used to complement the material in the text. In addition, wherever possible, the presentation style in the appendixes follows that of the main chapters. This approach maximizes the usability and usefulness of the appendixes.
- The book is accompanied by an Instructor's Manual containing solutions to the chapter problems.

The instructor may cover the material in the text in any sequence that he or she considers appropriate. My personal preference is to teach the chapters sequentially except for Chapter 11, which I introduce (at least partially) after the completion of Chapter 6. Note that Chapter 11 deals with numerical methods in conduction. It was placed at the end of the book based on organization considerations alone, since Chapters 2 to 10 deal largely with analytical methods. Regarding the use of this chapter, however, I feel that after Chapter 6 is completed, the student has acquired a reasonably good knowledge of analytical methods and can benefit from comparisons between analytical and numerical techniques. I also feel that the term project should be assigned about this time in the semester so that the student has enough time to complete it. The fact that the completion of the term project requires knowledge of numerical methods in heat conduction is an additional reason why Chapter 11 may be introduced after Chapter 6.

A number of people provided me with help, advice, and constructive commentary during the development of this book. Professor R. A. Gardner of Washington University in St. Louis, Professor C. Grigoropoulos of the University of California at Berkeley, Professor J.W. Sheffield of the University of Missouri at Rolla, and Professor E. M. Sparrow of the University of Minnesota reviewed the manuscript for Prentice Hall and provided me with valuable comments and suggestions. Professor A. Bejan of Duke University, Professor M. Kazmierczak of the University of Cincinnati, and Professor K. J. Renken of the University of Wisconsin at Milwaukee also reviewed the manuscript and provided me with constructive comments and ideas from which the manuscript has benefited. Professor W. J. Minkowycz of the University of Illinois at Chicago shared with me his viewpoint that a textbook for a one-semester course should not be all-inclusive. Instead, it should contain a topic selection appropriate for a sequence of lectures spanning over the period of a semester.

Dr. W.-Z. Cao, Dr. S. K. Rastogi, Dr. T. L. Spatz, Mr. T. Bennet, Mr. B. Kang, Mr. Y. Shen, Mr. J. Waldvogel, Mr. C.-H. Yang, and Mr. Z. Zhao, my graduate students during the time that I was working on this project, helped me in various ways, such as checking some of my algebra and preparing many of the preliminary illustrations.

I would like to take this opportunity to thank my wife, Lily, who has been a great companion and a continuous source of support and motivation over the years, including the time during which I was working on this project in addition to my usual professional activities and obligations. She gave me her valuable input and ideas in various aspects of the book and is responsible for the utilization of color graphics on the cover.

The manuscript was typed very efficiently by Ms. Denise Burt, who in addition to the actual task of typing, had to "decipher" my handwriting. As usual, she did an excellent job.

*D. Poulikakos*
*Chicago, Illinois*

# ABOUT THE AUTHOR

Dimos Poulikakos received his Diploma in Mechanical Engineering (1978) from the National Technical University of Athens, Greece, and his M.S. in Mechanical Engineering (1980) and Ph.D. in Mechanical Engineering (1983) both from the University of Colorado at Boulder. He joined the faculty of the Mechanical Engineering Department of the University of Illinois at Chicago as an assistant professor in 1983. He was promoted to the rank of associate professor with tenure in 1987 and to the rank of professor in 1993. His past research activities cover a variety of areas, exemplified by convective heat and mass transfer in fluids and in fluid-saturated porous media, energy conservation, electronic equipment cooling, film boiling, and film condensation. Currently, his research is in the areas of transport phenomena in materials processing (foundry and rapid solidification of pure substances and alloys) and fluid mechanics of atomized high-speed jets. He is also involved in the development and utilization of state-of-the-art optical techniques (holography) for the accurate, nonintrusive measurement of temperature, concentration, and flow fields in engineering applications.

Among the awards that he has received in recognition of his contributions in research and education in mechanical engineering are the Presidential Young Investigator Award, the Pi Tau Sigma Gold Medal, the SAE Ralph R. Teetor Award, and the University of Illinois Scholar Award. He has published over 85 research articles in peer-reviewed journals as well as numerous articles in proceedings of professional conferences.

CHAPTER 1

# INTRODUCTION

*Heat conduction* is a mode of heat transfer in which energy is transported from regions of higher temperature within a medium to regions of lower temperature, through the interaction and energy exchange of elementary particles. To this end, elementary particles (such as atoms or molecules) in higher-temperature regions are at higher energy levels. As a result, they oscillate rapidly and collide and set into motion (by exchange of kinetic energy) neighboring particles that are at lower energy levels. The process is repeated and propagates through the medium. The energy transported by this process is termed *heat*.

The other two modes of heat transfer are *heat convection* (where fluid motion is involved in the transport of heat) and *thermal radiation* (where heat is transferred by the action of electromagnetic waves). Both of these heat transfer modes are outside the scope of the present book.

It is not easy to measure directly heat transfer in a medium. Instead, the temperature distribution of a body is measured with devices such as thermocouples, thermistors, and thermometers, or by optical techniques. The flow of heat in a medium is then determined by utilizing the measured temperature field as well as the fundamental law of heat conduction, commonly known as *Fourier's law*. This law relates the heat flux ($W/m^2$) at a specific location inside the medium of interest, to the temperature gradient at the same location. Although the law is named after the famous French physicist and mathematician Joseph Fourier, its origin, as well as the origin of the mathematical theory of heat conduction, is not without controversy. A detailed historical account of the relevant events is contained in Ref. [1]. Here we present only a brief summary.

The first theoretical modeling of heat conduction problems was published by Fourier in 1804 [2]. In this partly erroneous work, Fourier refers to an 1804 paper by J. B. Biot [3] on the topic of propagation of heat in a medium. It is speculated by science historians [1] that the contents of Biot's paper had perhaps been communicated to Fourier prior to the publication of the paper, prompting Fourier to turn his attention to the problem of heat propagation within a continuum. Perhaps, then, Biot's name should be mentioned alongside Fourier's name when credit is to be given for the origination of the mathematical theory of heat conduction.

Fourier's original work was expanded and published in his 1807 memoir [4,5]. A further expanded and corrected theory of heat propagation in a continuum was put forth by Fourier in his 1811 prize essay [6,7]. This essay was Fourier's submission for the grand prize of the Institute of Mathematics in 1811. The award committee that judged the work consisted of Lagrange, Laplace, Malus, Haüy, and Legendre. Despite the severe criticism of Fourier's essay by the committee and in particular by Laplace and Lagrange, the grand prize was awarded to Fourier. However, the Institute of Mathematics did not publish Fourier's award-winning paper. It was after Fourier became the secretary of the French Academy of Sciences several years later that his work was published in two parts [6,7].

In 1822, Fourier published an improved account of his earlier work on the propagation of heat in his analytical theory of heat [8], which expands and generalizes previous mathematical concepts on the expansion of a function into a series of trigonometric (sine and cosine) functions. References [6–8] are considered to be the first well-documented cornerstones of today's theory of heat conduction.

With the aforementioned historical perspective in mind, Fourier's law of heat conduction for an isotropic material reads

$$\dot{\mathbf{q}}'' = -k \, \nabla T \tag{1.1}$$

In eq. (1.1), $\dot{\mathbf{q}}''$ is the heat flux vector (W/m$^2$), $T$ the temperature (K), and $k$ the thermal conductivity of the material (W/mK). The thermal conductivity is a thermophysical property of the material and is, generally speaking, temperature dependent. This is particularly true for low-temperature applications. In many everyday engineering applications of heat conduction, however, the dependence of the thermal conductivity on temperature is negligible.

Since the heat flux is proportional to the thermal conductivity, it is reasonable to expect that materials with high thermal conductivities result in high heat fluxes and facilitate the heat transfer for prescribed temperature gradients. These materials are called *thermal conductors*. Similarly, materials with low thermal conductivities hinder heat transfer and are called *thermal insulators*.

While thermal conductors, such as most metals, feature thermal conductivities of the order $k \approx O(10^2)$ W/mK, the thermal conductivity of many insulators is three to four orders of magnitude smaller, $O(10^{-2})$ W/mK $< k < O(10^{-1})$ W/mK. On the other hand, gases at atmospheric pressure possess even smaller thermal conductivities $O(10^{-3})$ W/mK $< k < O(10^{-1})$ W/mK. The thermal conductivity of a sample of typical substances is contained in Table 1.1. Extensive data on thermal conductivities are contained in Refs. [9–11].

**TABLE 1.1** THERMAL CONDUCTIVITY OF SELECTED SUBSTANCES AT ATMOSPHERIC PRESSURE

| Material | Temperature (K) | $k$(W/mK) |
|---|---|---|
| | Metals | |
| Aluminum | 297 | 190 |
| Copper | 297 | 390 |
| Iron | 297 | 78 |
| Lead | 297 | 38 |
| Nickel | 297 | 90 |
| Silver | 297 | 418 |
| Steel | 297 | 46 |
| Tin | 297 | 64 |
| | Nonmetals | |
| Asbestos | 273 | 0.151 |
| Cork | 297 | 0.042 |
| Glass (Pyrex) | 297 | 1.177 |
| Ice | 273 | 2.2 |
| Limestone | 297 | 1.7 |
| Quartz sand | 297 | 0.26 |
| Silicone | 297 | 85 |
| Water | 297 | 0.6 |
| | Gases | |
| Air | 297 | 0.026 |
| Ammonia | 273 | 0.022 |
| Argon | 311 | 0.018 |
| Carbon dioxide | 297 | 0.016 |
| Hydrogen | 297 | 0.182 |
| Nitrogen | 297 | 0.026 |
| Oxygen | 297 | 0.027 |

Returning to eq. (1.1), we apply this equation with respect to a Cartesian coordinate system to obtain an expression for the heat flux at the outer surface of the body shown in Fig. 1.1a.

$$\dot{\mathbf{q}}''_A = \left(-k\frac{\partial T}{\partial x}\mathbf{i} - k\frac{\partial T}{\partial y}\mathbf{j} - k\frac{\partial T}{\partial z}\mathbf{k}\right)_A \tag{1.2}$$

where the subscript $A$ denotes the fact that the relevant quantities are evaluated at the outer surface of the body. Clearly, the three components of the heat flux vector at surface $A$ are

$$\dot{q}''_{Ax} = -\left(k\frac{\partial T}{\partial x}\right)_A \tag{1.3}$$

$$\dot{q}''_{Ay} = -\left(k\frac{\partial T}{\partial y}\right)_A \tag{1.4}$$

$$\dot{q}''_{Az} = -\left(k\frac{\partial T}{\partial z}\right)_A \tag{1.5}$$

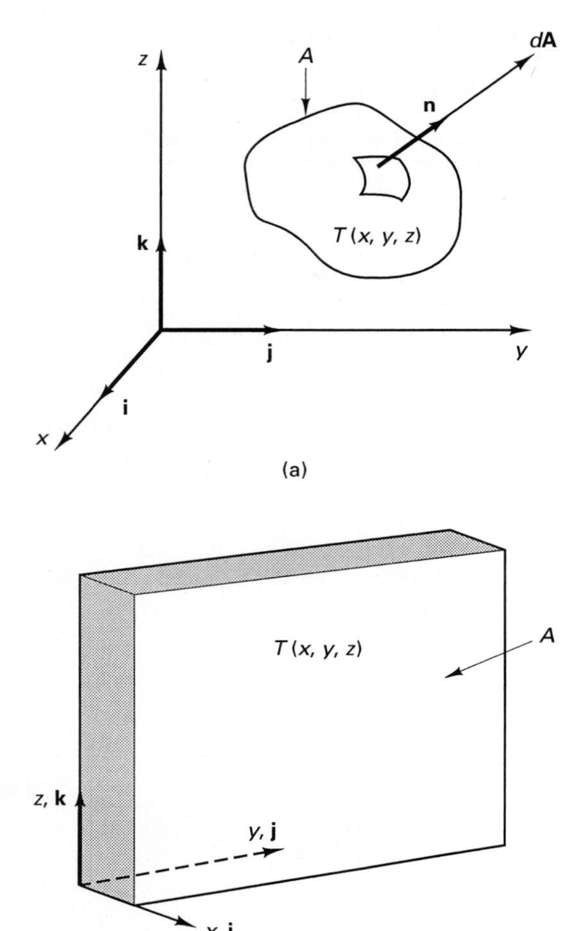


**Figure 1.1** (a) Schematic of a body of arbitrary shape with respect to the Cartesian coordinate system. (b) Schematic of a flat wall.

To obtain the net heat transfer rate ($d\dot{q}_A$) to or from a differential surface area $d\mathbf{A} = dA\,\mathbf{n}$, where $\mathbf{n}$ is the unit area vector, the following equation needs to be utilized:

$$d\dot{q}_A = \dot{\mathbf{q}}''_A \cdot dA\,\mathbf{n} \tag{1.6}$$

Combining eqs. (1.1) and (1.6) yields

$$d\dot{q}_A = -(k\,\nabla T)_A \cdot dA\,\mathbf{n} \tag{1.7}$$

The net heat transfer rate through the entire surface A results from the integration of the elementary heat transfer rate of eq. (1.7) over the surface $A$:

$$\dot{q}_A = -\int_A (k\,\nabla T)_A \cdot dA\,\mathbf{n} \tag{1.8}$$

In the special case of the rectangular slab shown in Fig 1.1b, the total heat flow through the right side (of area $A$) of the slab is, by application of eq. (1.8),

$$\dot{q}_A = -\int_A \left(k \frac{\partial T}{\partial x}\right)_A dA \tag{1.9}$$

Further, if the temperature and the thermal conductivity of the slab are dependent only on the $x$-direction, eq. (1.9) simplifies to

$$\dot{q}_A = -k_A A \left(\frac{\partial T}{\partial x}\right)_A \tag{1.10}$$

Finally, the presence of the minus sign on the right-hand side of Fourier's law [eq. (1.1)] is worth discussing. Since heat flows from regions of higher temperature to regions of lower temperature, the temperature gradient in the direction of the heat flux vector is negative. To obtain a positive and physically meaningful result for the heat flux in the direction of the heat flux vector, a minus sign was placed by convention on the right-hand side of eq. (1.1). To exemplify, let us examine eq. (1.10). If heat flows in the positive $x$-direction (the temperature is decreasing in the positive $x$-direction), $\partial T/\partial x$ is negative. It is then because of the presence of the minus sign on the right-hand side of eq. (1.10) that the result for $\dot{q}_A$ becomes positive and physically meaningful.

### Fourier's Law in Anisotropic Materials

In the case of isotropic materials, Fourier's law, expressed by eq. (1.1), relates the heat flux to the temperature gradients at any point inside the isotropic material, as discussed earlier. The thermal conductivity at any point of an isotropic material, appearing in eq. (1.1), is independent of direction.

In the case of anisotropic materials, on the other hand, the thermal conductivity is directionally dependent [12,13]. Therefore, Fourier's law needs to be generalized to account for the directional dependence of the thermal conductivity. The generalized Fourier's law for heat conduction in anisotropic materials in a Cartesian coordinate system reads [12,14]

$$q_i = -\sum_{j=x,y,z} k_{ij} \frac{\partial T}{\partial x_j} \qquad i = x, y, z \tag{1.11}$$

Applying eq. (1.11) to the coordinate system of Fig. 1.1a yields three equations for the heat flux, each corresponding to one Cartesian direction:

$$q_x = -\left(k_{xx} \frac{\partial T}{\partial x} + k_{xy} \frac{\partial T}{\partial y} + k_{xz} \frac{\partial T}{\partial z}\right) \tag{1.12}$$

$$q_y = -\left(k_{yx} \frac{\partial T}{\partial x} + k_{yy} \frac{\partial T}{\partial y} + k_{yz} \frac{\partial T}{\partial z}\right) \tag{1.13}$$

$$q_z = -\left(k_{zx} \frac{\partial T}{\partial x} + k_{zy} \frac{\partial T}{\partial y} + k_{zz} \frac{\partial T}{\partial z}\right) \tag{1.14}$$

Hence the thermal conductivity, $k_{ij}$, of an isotropic material is a second-order tensor. Equations (1.12–1.14) can be written in the following matrix form:

$$\begin{bmatrix} q_x \\ q_y \\ q_z \end{bmatrix} = \begin{bmatrix} k_{xx} & k_{xy} & k_{xz} \\ k_{yx} & k_{yy} & k_{yz} \\ k_{zx} & k_{zy} & k_{zz} \end{bmatrix} \begin{bmatrix} \partial T/\partial x \\ \partial T/\partial y \\ \partial T/\partial z \end{bmatrix} \tag{1.15}$$

The meaning of the thermal conductivity coefficients in the matrix of eq. (1.15) is similar to that of the stress coefficients in continuum mechanics. For example, $k_{xx}$ is the conductivity in the $x$-direction at a point located on a plane perpendicular to the $x$-axis. Similarly, $k_{xy}$ is the conductivity in the $y$-direction at a point located on a plane perpendicular to the $x$-axis, and $k_{xz}$ is the conductivity in the $z$-direction at a point located on a plane perpendicular to the $x$-axis. In general, the first subscript in the thermal conductivity coefficient refers to the axis perpendicular to which a plane passing through the point of interest is located. The second index refers to the direction of the thermal conductivity. The various coefficients in the thermal conductivity matrix are not independent. Several relations between coefficients exist [15,16]. For example, $k_{ij} = k_{ji}$, $k_{ii} > 0$, $k_{ii}k_{jj} - k_{ij}^2 > 0$ for $i \neq j$.

Expressions similar to eqs. (1.11)–(1.15) can be obtained in other coordinate systems [14]. Since the study of heat conduction in anisotropic materials is outside the main focus of this book, these expressions are not presented. The principles of heat conduction in anisotropic media are often used to study heat transfer in several classes of composite materials and constitute a topic of contemporary research in the area of thermal sciences.

## REFERENCES


1. J. Herivel, *Joseph Fourier, the Man and the Physicist,* Clarendon Press, Oxford, 1975.
2. J. Fourier, Draft paper, *Bibliothèque nationale,* BN MS.ff.22525, pp. 107–149, 1804.
3. J. B. Biot, Mémoire sur la propagation de la chaleur, *Bibl. Br.*, Vol. 37, pp. 310–329, 1804.
4. J. Fourier, *Mémoire sur la propagation de la chaleur,* MS.1851, Ecole Nationale des Ponts et Chaussee's, Paris, 1807.
5. I. Grattan-Guinness and J. R. Ravetz, *Joseph Fourier 1768–1830,* MIT Press, Cambridge, MA, 1972.
6. J. Fourier, Théorie du mouvement de la chaleur dans les corps solides, *Mem. Acad. R. Sci.,* Vol. 4, pp. 185–555, 1819–20.
7. J. Fourier, Théorie du mouvement de la chaleur dans les corps solides, *Mem. Acad. R. Sci.,* Vol. 5, pp. 153–246, 1821–22.
8. J. Fourier, *Théorie analytique de la chaleur* (Analytical Theory of Heat), Paris, 1822.
9. R. W. Powell, C. Y. Ho, and P. E. Liley, *Thermal Conductivity of Selected Materials,* NSRDS-NBS 8, U.S. Department Commerce, NBS, Washington, DC, 1966.
10. *Thermophysical Properties of Matter,* Vols. 1–3, 1F1, Plenum Data Corp., New York, 1969.

11. W. M. ROHSENOW, J. P. HARTNETT, and E. N. GANIC, Eds., *Handbook of Heat Transfer Fundamentals,* 2nd ed. pp. 3-1 to 3-135, 1985.
12. L. ONSAGAR, *Phys. Rev.,* Vol. 37, pp. 1405–1426, 1931.
13. L. ONSAGAR, *Phys. Rev.,* Vol. 38, pp. 2265–2279, 1931.
14. M. N. ÖZISIK, *Heat Conduction,* Wiley, New York, 1980.
15. I. PRIGOGINE, *Thermodynamics of Irreversible Processes,* Wiley, New York, 1961.
16. H. B. G. CASIMIR, *Rev. Mod. Phys.,* Vol. 17, pp. 343–350, 1945.

CHAPTER 2

# THE ENERGY EQUATION IN HEAT CONDUCTION AND COMMON TYPES OF BOUNDARY CONDITIONS

To obtain the theoretical temperature field in a system (body) in which heat transfer in the mode of conduction takes place, one needs to derive a model that governs the heat transfer process, such that temperature is the primary variable describing the heat transport. The central equation of this model is the energy equation, which, together with a set of boundary and initial conditions, can be solved to yield the temperature field in the system. The main reason for the choice of the temperature as the primary variable descriptive of the conduction process lies in the fact that from a practical standpoint, temperature is easily measured with a variety of methods, ranging from simple thermometers, thermocouples, and thermistors to more sophisticated optical techniques [1, 2]. Following this logic, we first derive the energy equation for conduction and then cast it in a form in which temperature is the unknown variable to be determined. Two alternatives are presented: a differential formulation and an integral formulation.

## 2.1 ENERGY EQUATION, DIFFERENTIAL FORMULATION

Consider the elementary *fixed* control volume shown in Fig. 2.1a. This control volume is located inside a large (by comparison) system (medium) and is chosen arbitrarily. For the sake of generality, we allow for the system to move with *constant* velocity $\mathbf{V}(U, V, W)$. If no work transfer occurs across the boundary of the control

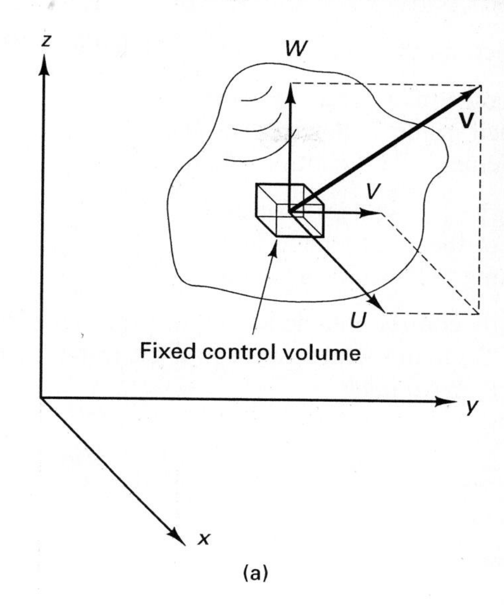


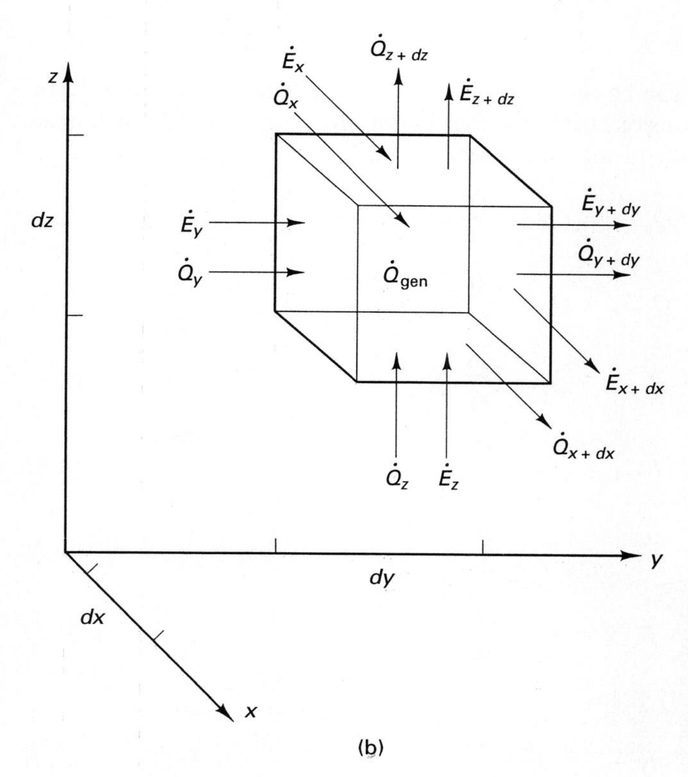


**Figure 2.1** (a) Schematic of a medium moving with constant velocity through a stationary differential control volume. (b) Schematic indicating the flow of energy through the differential control volume.

volume, an energy balance performed on the control volume is stated as follows:

$$\begin{array}{l} \text{time rate of change of} \\ \text{energy contained inside} \\ \text{the control volume} \end{array} = \begin{array}{l} \text{rate of energy} \\ \text{flowing into the} \\ \text{control volume} \end{array}$$

$$- \begin{array}{l} \text{rate of energy} \\ \text{flowing out of the} \\ \text{control volume} \end{array} + \begin{array}{l} \text{rate of energy} \\ \text{generated in the} \\ \text{control volume} \end{array}$$

Figure 2.1b is an enlargement of the control volume shown in Fig. 2.1a. With the help of the notation shown in Fig. 2.1b and with respect to a Cartesian coordinate system $(x, y, z)$, the energy balance above reads

$$\frac{\partial}{\partial t}(E) = \sum_{i=x,y,z} \dot{Q}_i - \sum_{\substack{i=x+dx,\\ y+dy,\\ z+dz}} \dot{Q}_i + \sum_{i=x,y,z} \dot{E}_i - \sum_{\substack{i=x+dx,\\ y+dy,\\ z+dz}} \dot{E}_i + \dot{Q}_{\text{gen}} \tag{2.1}$$

Equation (2.1) is expanded as follows:

$$\frac{\partial}{\partial t}(E) = \dot{Q}_x - \dot{Q}_{x+dx} + \dot{Q}_y - \dot{Q}_{y+dy} + \dot{Q}_z - \dot{Q}_{z+dz}$$
$$+ \dot{E}_x - \dot{E}_{x+dx} + \dot{E}_y - \dot{E}_{y+dy} + \dot{E}_z - \dot{E}_{z+dz} + \dot{Q}_{\text{gen}} \tag{2.2}$$

The heat transfer rate at $x$ can be related to the heat transfer rate at $x + dx$ with the help of a Taylor series expansion, using the first two terms of the series, which is an approximation with accuracy of the first order.

$$\dot{Q}_{x+dx} = \dot{Q}_x + \frac{\partial \dot{Q}_x}{\partial x} dx + \cdots \tag{2.3}$$

$$\dot{Q}_{y+dy} = \dot{Q}_y + \frac{\partial \dot{Q}_y}{\partial y} dy + \cdots \tag{2.4}$$

$$\dot{Q}_{z+dz} = \dot{Q}_z + \frac{\partial \dot{Q}_z}{\partial z} dz + \cdots \tag{2.5}$$

$$\dot{E}_{x+dx} = \dot{E}_x + \frac{\partial \dot{E}}{\partial x} dx + \cdots \tag{2.6}$$

$$\dot{E}_{y+dy} = \dot{E}_y + \frac{\partial \dot{E}}{\partial y} dy + \cdots \tag{2.7}$$

$$\dot{E}_{z+dz} = \dot{E}_z + \frac{\partial \dot{E}}{\partial z} dz + \cdots \tag{2.8}$$

Combining eqs. (2.2)–(2.8) yields

$$\frac{\partial E}{\partial t} = -\frac{\partial \dot{Q}_x}{\partial x} dx - \frac{\partial \dot{Q}_y}{\partial y} dy - \frac{\partial \dot{Q}_z}{\partial z} dz - \frac{\partial \dot{E}}{\partial x} dx - \frac{\partial \dot{E}}{\partial y} dy - \frac{\partial \dot{E}}{\partial z} dz + \dot{Q}_{\text{gen}} \tag{2.9}$$

To introduce the easily measured temperature in eq. (2.9), we invoke Fourier's law [eqs. (1.3)–(1.5)]:

$$\dot{Q}_x = -k \frac{\partial T}{\partial x} dy\, dz \tag{2.10}$$

$$\dot{Q}_y = -k \frac{\partial T}{\partial y} dx\, dz \tag{2.11}$$

$$\dot{Q}_z = -k \frac{\partial T}{\partial z} dx\, dy \tag{2.12}$$

On the other hand, the terms involving the energy flowing through the control volume in eq. (2.0) can be cast in the following form:

$$\frac{\partial \dot{E}}{\partial x} dx = \frac{\partial E}{\partial x} U \tag{2.13}$$

$$\frac{\partial \dot{E}}{\partial y} dy = \frac{\partial E}{\partial y} V \tag{2.14}$$

$$\frac{\partial \dot{E}}{\partial z} dz = \frac{\partial E}{\partial z} W \tag{2.15}$$

In addition, we express the energy content of the control volume in terms of the internal energy per unit mass (specific internal energy, $e$):

$$e = \frac{E}{\rho\, dx\, dy\, dz} \qquad \frac{\partial E}{\partial t} = dx\, dy\, dz \frac{\partial(\rho e)}{\partial t} \tag{2.16, 2.17}$$

Combining eqs. (2.9)–(2.17) yields

$$\frac{\partial}{\partial t}(\rho e) = \frac{\partial}{\partial x}\left(k \frac{\partial T}{\partial x}\right) + \frac{\partial}{\partial y}\left(k \frac{\partial T}{\partial y}\right) + \frac{\partial}{\partial z}\left(k \frac{\partial T}{\partial z}\right)$$
$$-U \frac{\partial(\rho e)}{\partial x} - V \frac{\partial(\rho e)}{\partial y} - W \frac{\partial(\rho e)}{\partial z} + \dot{Q}'''_{\text{gen}} \tag{2.18}$$

In eq. (2.18), $\dot{Q}'''_{\text{gen}}$ is the volumetric heat generation rate (W/m$^3$). Next, we relate the specific internal energy, $e$, to the specific enthalpy, $h$, by using from thermodynamics the definition

$$h = e + \frac{P}{\rho} \tag{2.19}$$

where $P$ is the pressure. Substituting eq. (2.19) in eq. (2.18), performing the algebra, and invoking vector notation, we obtain (see Problem 2.4)

$$h\left(\frac{\partial \rho}{\partial t} + \mathbf{V} \cdot \nabla \rho\right) + \rho\left(\frac{\partial h}{\partial t} + \mathbf{V} \cdot \nabla h\right) - \frac{\partial P}{\partial t} - \mathbf{V} \cdot \nabla P$$
$$= \nabla \cdot (k\, \nabla T) + \dot{Q}'''_{\text{gen}} \tag{2.20}$$

Note that owing to the utilization of vector notation, eq. (2.20) is general and it holds in other coordinate systems, not simply the Cartesian coordinate system based on which it was derived.

The mass conservation equation (continuity equation) for a moving continuum reads [3]

$$\frac{\partial \rho}{\partial t} + \mathbf{V} \cdot \nabla \rho + \rho \nabla \cdot \mathbf{V} = 0 \tag{2.21}$$

In the present derivation of the conduction equation, the velocity of the body is assumed to be constant. Based on this observation, we utilize the continuity equation (2.21) to conclude that the first term in brackets on the left-hand side of eq. (2.20) is equal to zero. Hence eq. (2.20) simplifies to

$$\rho\left(\frac{\partial h}{\partial t} + \mathbf{V} \cdot \nabla h\right) - \frac{\partial P}{\partial t} - \mathbf{V} \cdot \nabla P = \nabla \cdot (k\, \nabla T) + \dot{Q}'''_{\text{gen}} \tag{2.22}$$

The next step is to express the specific enthalpy, $h$, in terms of temperature. We invoke the canonical relation for the specific enthalpy [4,5]:

$$dh = T\,ds + \frac{1}{\rho}\,dP \tag{2.23}$$

where the specific entropy change $ds$ is

$$ds = \left(\frac{\partial s}{\partial T}\right)_P dT + \left(\frac{\partial s}{\partial P}\right)_T dP \tag{2.24}$$

It can easily be shown from thermodynamic relationships that [4,5]

$$\left(\frac{\partial s}{\partial P}\right)_T = -\frac{\beta}{\rho} \tag{2.25}$$

$$\left(\frac{\partial s}{\partial T}\right)_P = \frac{c_p}{T} \tag{2.26}$$

where $c_p$ is the specific heat at constant pressure and $\beta$ is the coefficient of thermal expansion defined as follows:

$$\beta = -\frac{1}{\rho}\left(\frac{\partial \rho}{\partial T}\right)_P \tag{2.27}$$

Combining eqs. (2.23)–(2.26) gives us

$$dh = c_p\,dT + \frac{1}{\rho}(1 - \beta T)\,dP \tag{2.28}$$

Substituting eq. (2.28) into eq. (2.22) and performing the algebra yields

$$\rho c_p\left(\frac{\partial T}{\partial t} + \mathbf{V} \cdot \nabla T\right) - \beta T\left(\frac{\partial P}{\partial t} + \mathbf{V} \cdot \nabla P\right) = \nabla \cdot (k\, \nabla T) + \dot{Q}'''_{\text{gen}} \tag{2.29}$$

Neglecting compressibility effects, since they are indeed negligible in the vast majority of conduction applications ($\rho$ = const, $\beta = 0$, $c_p = c$), eq. (2.29) simplifies to

$$\rho c\left(\frac{\partial T}{\partial t} + \mathbf{V} \cdot \nabla T\right) = \nabla \cdot (k \nabla T) + \dot{Q}'''_{\text{gen}} \tag{2.30}$$

If the thermal conductivity of the medium, $k$, is assumed constant, the conduction equation becomes

$$\rho c\left(\frac{\partial T}{\partial t} + \mathbf{V} \cdot \nabla T\right) = k \nabla^2 T + \dot{Q}'''_{\text{gen}} \tag{2.31}$$

In the majority of heat conduction applications, the body of interest is motionless with respect to the differential control volume of Fig. 2.1 (i.e., $\mathbf{V} = 0$). In this case, eqs. (2.30) and (2.31), respectively, simplify to

$$\rho c \frac{\partial T}{\partial t} = \nabla \cdot (k \nabla T) + \dot{Q}'''_{\text{gen}} \tag{2.32}$$

$$\rho c \frac{\partial T}{\partial t} = k \nabla^2 T + \dot{Q}'''_{\text{gen}} \tag{2.33}$$

Since eqs. (2.30)–(2.33) are written in vector form, they hold in other coordinate systems as well (i.e., not only in the Cartesian coordinate system with respect to which they were derived). It is, therefore, straightforward to obtain the energy equation for conduction in a cylindrical and a spherical coordinate system (the other two coordinate systems that are used commonly in heat transfer studies, in addition to the Cartesian system). All one has to do is substitute in eqs. (2.30)–(2.33) the expressions for the vector operators corresponding to the coordinate system of interest. The results for the cylindrical and spherical coordinate systems of Fig. 2.2 for the most

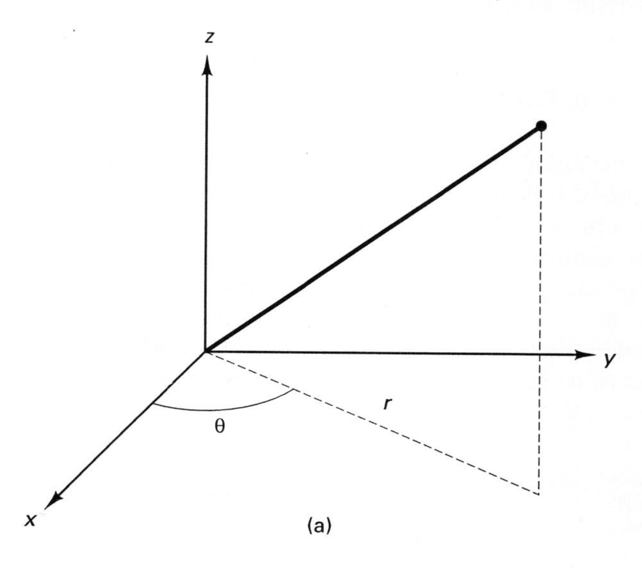


**Figure 2.2** (a) Cylindrical coordinate system ($r$, $\theta$, $z$). (b) Spherical coordinate system ($r$, $\phi$, $\theta$).

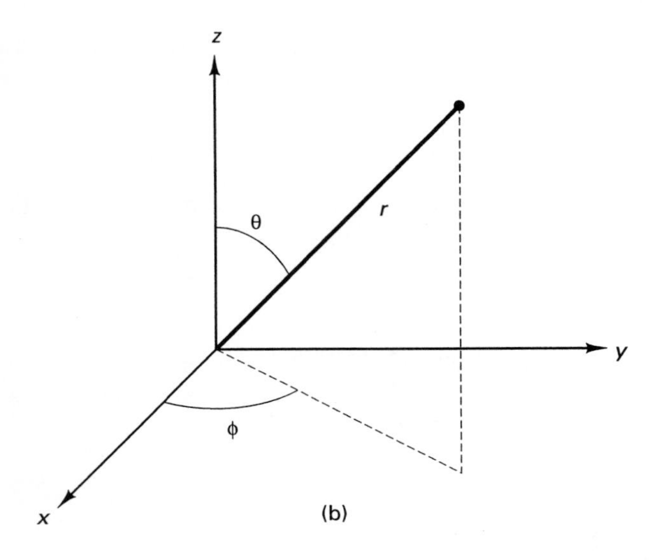


**Figure 2.2** (*cont.*)

commonly used equation, (2.33), are:

*Conduction equation in cylindrical coordinates (Fig. 2.2a):*

$$\rho c \frac{\partial T}{\partial t} = k\left[\frac{1}{r}\frac{\partial}{\partial r}\left(r\frac{\partial T}{\partial r}\right) + \frac{1}{r^2}\frac{\partial^2 T}{\partial \theta^2} + \frac{\partial^2 T}{\partial z^2}\right] + \dot{Q}'''_{\text{gen}} \tag{2.34}$$

*Conduction equation in spherical coordinates (Fig. 2.2b):*

$$\rho c \frac{\partial T}{\partial t} = k\left[\frac{1}{r^2}\frac{\partial}{\partial r}\left(r^2\frac{\partial T}{\partial r}\right) + \frac{1}{r^2}\frac{1}{\sin\theta}\frac{\partial}{\partial \theta}\left(\sin\theta\frac{\partial T}{\partial \theta}\right) + \frac{1}{r^2\sin^2\theta}\frac{\partial^2 T}{\partial \varphi^2}\right] + \dot{Q}'''_{\text{gen}} \tag{2.35}$$

## 2.2 ENERGY EQUATION, INTEGRAL FORMULATION

To begin the derivation of the energy equation following an integral formulation, let us consider the system of interest shown in Fig. 2.3. The volume of the system is denoted by $V$, its surface area by $S$, and the outward unit normal to the surface by $\mathbf{n}$. The system is a *finite, fixed* control volume inside a very large medium moving with a constant velocity $\mathbf{V}$. The energy equation states a balance between the rate at which energy is stored in the system, on the one hand, and the net rate of energy, either incoming to the system through its boundary or generated within the system, on the other hand. The terms mentioned above can be expressed as follows:

*Rate of energy stored in the system:*

$$\iiint_V \frac{\partial}{\partial t}(\rho e)\, dV \tag{2.36}$$

*Net rate of energy generation in the system:*

$$\iiint_V \dot{Q}'''_{\text{gen}}\, dV \tag{2.37}$$

*Net rate of energy through the surface S:*

$$-\iint_S \dot{\mathbf{Q}}'' \cdot \mathbf{n}\, dS \tag{2.38}$$

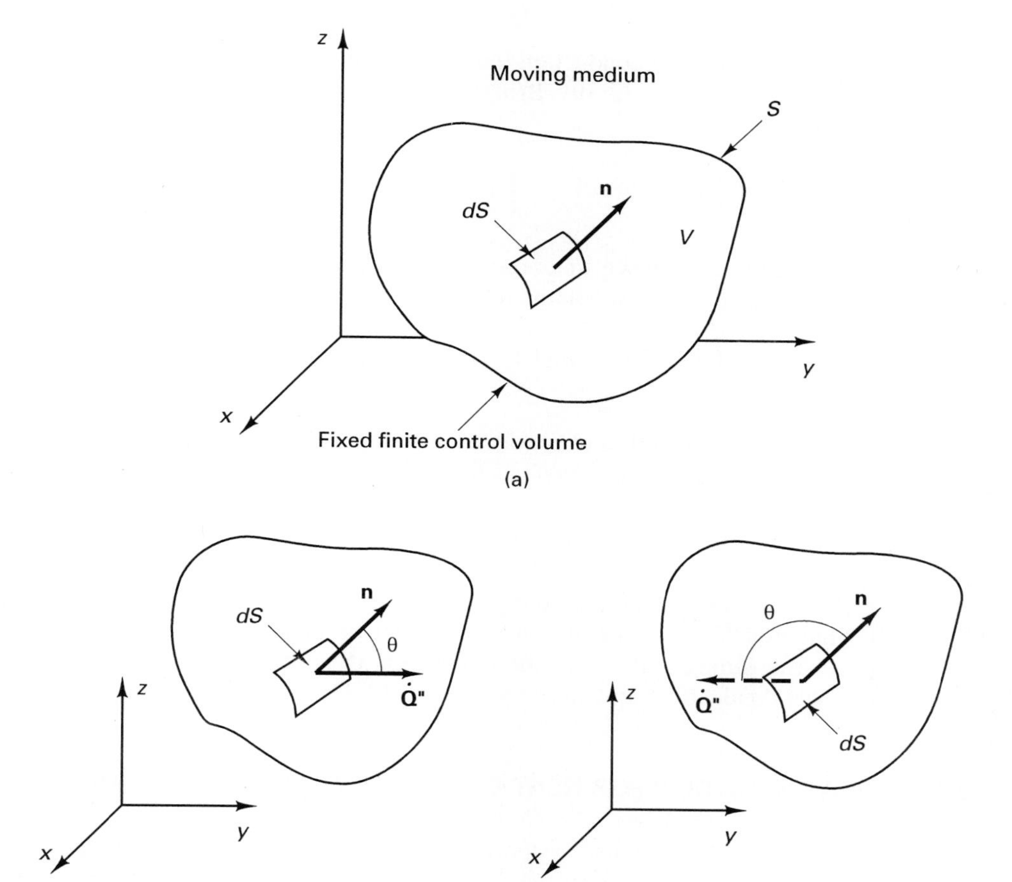


**Figure 2.3** (a) Basic system. (b) $0° < \theta < 90°$ (energy flux exiting the control volume) $\dot{\mathbf{Q}}'' \cdot \mathbf{n} = |\dot{\mathbf{Q}}''|\,|\mathbf{n}| \cos\theta > 0$. (c) $90° < \theta < 180°$ (energy flux entering the control volume) $\dot{\mathbf{Q}}'' \cdot \mathbf{n} = |\dot{\mathbf{Q}}''|\,|\mathbf{n}| \cos\theta < 0$.

The minus sign in statement (2.38) accounts for the convention that *energy flux is positive when entering the system and negative when exiting the system*. As shown in Fig. 2.3b and c the quantity $\dot{\mathbf{Q}}'' \cdot \mathbf{n}$ is positive when $0° < \theta < 90°$, in which case the energy flux is exiting the control volume and negative when $90° < \theta < 180°$, in which case the energy flux is entering the control volume. Hence the sign of the quantity $\dot{\mathbf{Q}}'' \cdot \mathbf{n}$ is opposite to that required by the aforementioned convention, and the "minus" sign in eq. (2.38) is necessary to satisfy this convention.

The net rate of energy flux through the control surface consists of two parts: energy conducted through the surface and energy carried through the surface by the moving medium, that is,

$$\dot{\mathbf{Q}}'' = -k\,\nabla T + \rho e \mathbf{V} \tag{2.39}$$

Fourier's law of heat conduction [eq. (1.1)] was accounted for in the first term of the right-hand side of eq. (2.39). Bringing together eqs. (2.36)–(2.39) yields the energy equation

$$\iiint_V \frac{\partial}{\partial t}(\rho e)\,dV = \iint_S (k\,\nabla T - \rho e \mathbf{V}) \cdot \mathbf{n}\,dS + \iiint_V \dot{Q}'''_{\text{gen}}\,dV \tag{2.40}$$

Applying the divergence theorem, the surface integral on the right-hand side of eq. (2.40) is converted to a volume integral:

$$\iint_S (k\,\nabla T - \rho e \mathbf{V}) \cdot \mathbf{n}\,dS = \iiint_V \nabla \cdot (k\,\nabla T - \rho e \mathbf{V})\,dV \tag{2.41}$$

Combining eqs. (2.40) and (2.41), accounting for the constancy of the velocity vector and equating the integrands of the right- and left-hand sides, yields

$$\frac{\partial}{\partial t}(\rho e) = \nabla \cdot (k\,\nabla T) - \mathbf{V} \cdot \nabla(\rho e) + \dot{Q}'''_{\text{gen}} \tag{2.42}$$

Equation (2.42) is identical to eq. (2.18), which was derived based on a differential formulation. From this point on, one can follow identically the steps outlined earlier under the differential formulation approach and obtain the more common forms of the heat conduction equation, eqs. (2.30)–(2.33).

## 2.3 BIOHEAT EQUATION FOR HEAT CONDUCTION IN LIVING TISSUES

Heat transfer in biological systems is an area of current interest and appeal with obvious direct applications to the welfare of human beings, animals, and other living organisms. A comprehensive review paper of the basic heat transfer models utilized to study heat transfer in biological systems was published by Chato [6].

An important mode of heat transfer in biological systems is heat conduction. To this end, Pennes [7] first proposed an approximate bioheat equation that models heat conduction in living tissues. Pennes's bioheat equation can account for the presence of blood perfusion, metabolic heat generation, and radiation absorption. The blood perfusion heat transfer is modeled as a cooling (sink) mechanism that removes heat from the tissue at a rate $\dot{m}_b''' c_{pb}(T - T_A)$ per unit volume, where $\dot{m}_b'''$ is the volumetric blood flow rate in the tissue (kg/m$^3$s), $c_{pb}$ the specific heat of the blood, $T$ the tissue temperature, and $T_A$ the arterial blood temperature, which usually is approximated to be constant and equal to the core temperature of the body (37°C). The metabolic heat generation ($\dot{q}'''$) can be modeled as a heat source. The same is true for the radiation absorption ($\dot{q}_r'''$), which is usually zero unless the tissue is subjected to some type of radiation treatment. Based on the above, it can be shown [6,8] that the heat conduction in the tissue, for constant blood properties, is governed by the following bioheat equation:

$$\rho_t c_{pt} \frac{\partial T}{\partial t} + \dot{m}_b''' c_{pb}(T - T_A) = \nabla \cdot (k_t \nabla T) + \dot{q}''' + \dot{q}_r''' \tag{2.43}$$

where subscript $t$ denotes the tissue. The derivation of this equation can be obtained utilizing the differential or the integral formulation outlined earlier in this chapter and is the subject of Problem 2.6. Note that eq. (2.43) is a generalization of eq. (2.32) accounting for the internal convection mechanism in the tissue obtained by blood perfusion. The metabolic heat generation and radiation absorption terms in eq. (2.43) could be handled by the heat generation term of eq. (2.32) as well.

The bioheat equation (2.43) has served well workers in the area of biological heat transfer and is still in use today. Researchers have attempted to improve this equation over the years. Perhaps the most complete model for heat transfer in a perfused tissue was proposed by Chen and Holmes [9]. In this model, a convection term, $\rho_b c_{pb} \mathbf{V}_b \cdot \nabla T$, where $\mathbf{V}_b$ is the blood perfusion velocity vector, was added to the left-hand side of eq. (2.43). Next, the tissue thermal conductivity ($k_t$) in this equation was replaced with an effective thermal conductivity, $k_{\text{eff}} = k_t + k_b$, where $k_b$ is an apparent thermal conductivity accounting for blood perfusion in the small blood vessels. According to the model of Chen and Holmes [9], the blood flow rate $\dot{m}_b$ and the arterial temperature should be evaluated at a specific size level. The effective thermal conductivity in the bioheat equation of Chen and Holmes [9] is, in general, temperature dependent. To this end, Gautherie [10] has proposed graphically such a temperature-dependent effective thermal conductivity in the region $25°\text{C} < T < 50°\text{C}$. Hurlburt and Chato [11] accurately approximated the effective thermal conductivity of Gautherie [10] with the following expression:

$$k_{\text{eff}} = k_t + k_b = 4.82 - 4.44833[1.00075^{-1.575^{T-25}}] \frac{W}{\text{m} \cdot °\text{C}} \tag{2.44}$$

Additional models for the effective thermal conductivity of living tissues have been developed and can be found in the open literature [6,12]. For the purposes of the classroom treatment of heat conduction in living tissues, the bioheat equation (2.43) is sufficient and will be used whenever necessary throughout this book.

## 2.4 DISCUSSION OF COMMON BOUNDARY CONDITIONS

There exist three common types of boundary conditions in conduction heat transfer and they are summarized in Figs. 2.4 to 2.6.

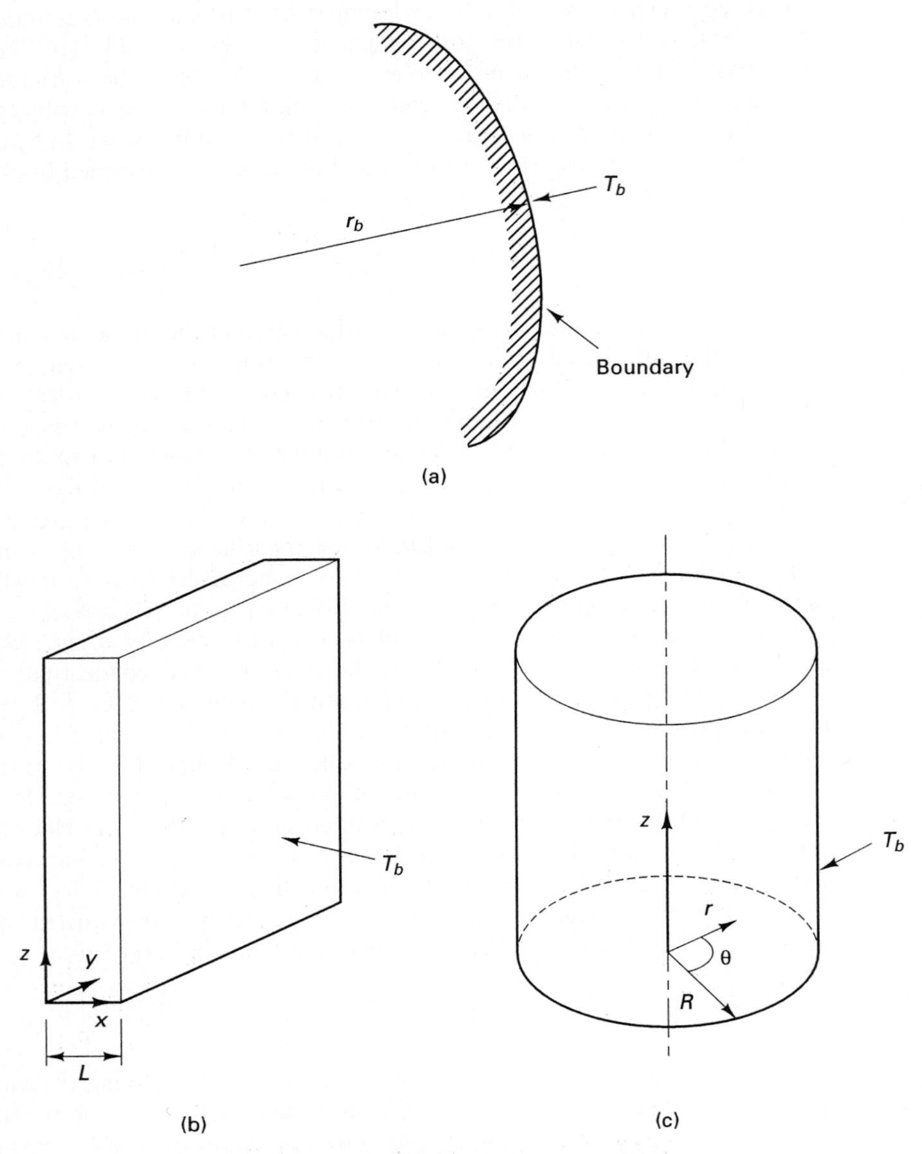


**Figure 2.4** Boundary condition of the first kind.

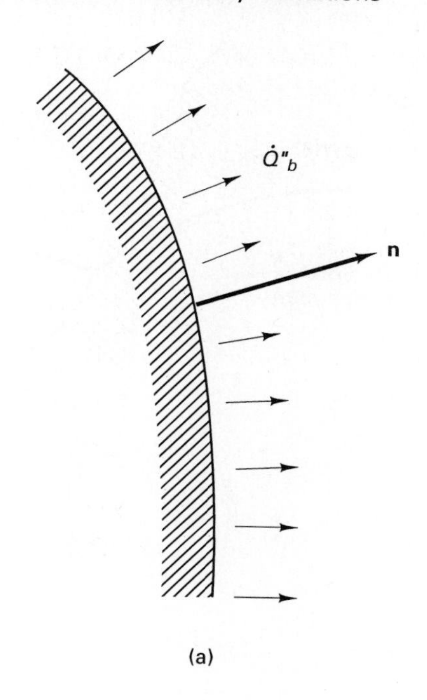


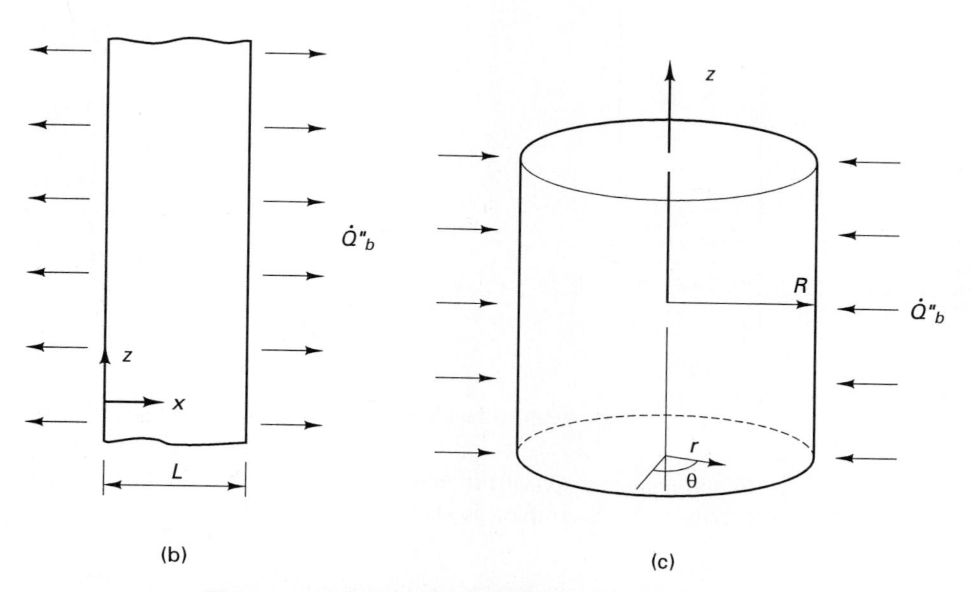


**Figure 2.5** Boundary condition of the second kind.

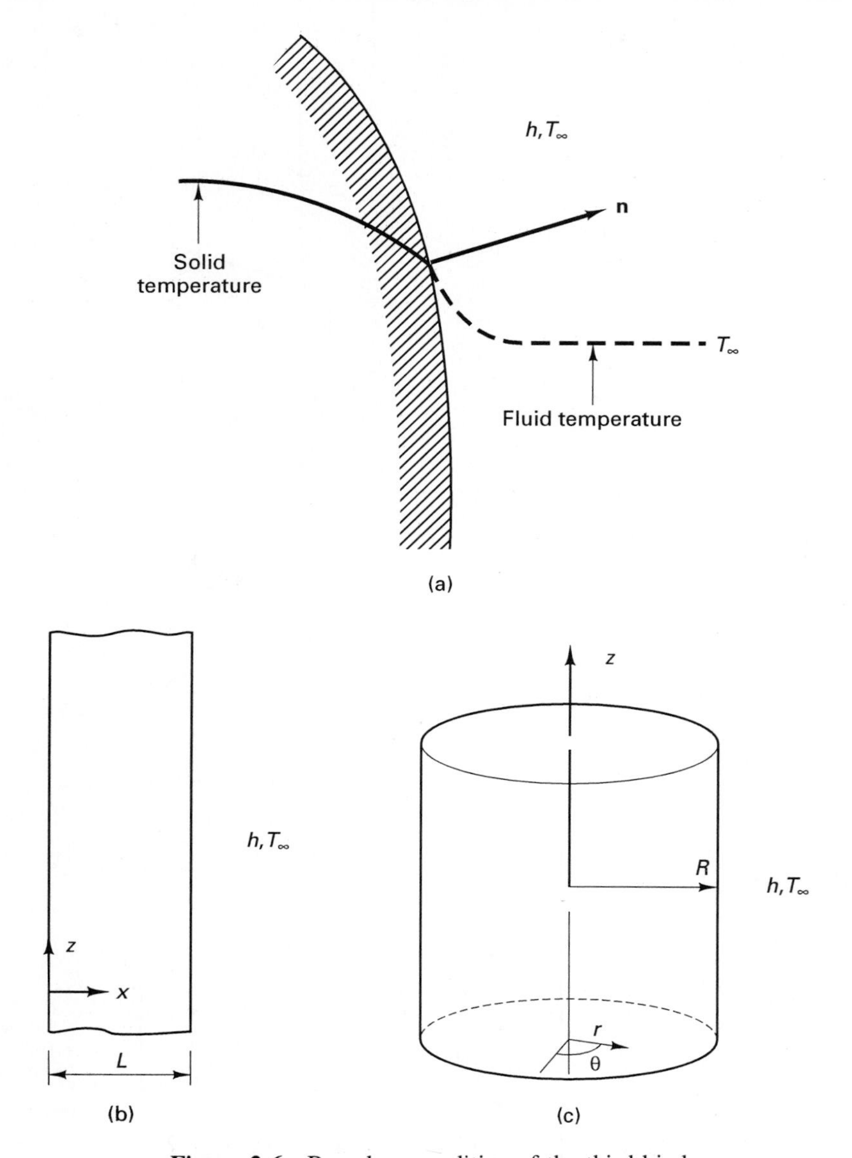


**Figure 2.6** Boundary condition of the third kind.

**1.** *Boundary condition of the first kind.* In this case the boundary *temperature* ($T_b$) is prescribed. For example, as shown in Fig. 2.4a,

$$T(r_b, t) = T_b \tag{2.45}$$

More specifically, if one is dealing with a Cartesian or cylindrical coordinate system (Fig. 2.4b and c) the boundary conditions of the first kind at $x = L$ and $r = R$, respectively, read

$$T(L,y,z,t) = T_b \tag{2.46}$$

$$T(R,\theta,z,t) = T_b \tag{2.47}$$

**2.** *Boundary condition of the second kind.* A boundary condition is of the second kind when it states that the *heat flux* at a boundary is prescribed. As shown in Fig. 2.5a,

$$\dot{Q}_b'' = -k_b\left(\frac{\partial T}{\partial n}\right)_b \tag{2.48}$$

In eq. (2.48), $n$ denotes the outward unit normal to the wall. If the differentiation is along the outward unit normal, the minus sign of the right-hand side of eq. (2.48) stays as is. If the differentiation is along the inward unit normal, the sign of the right-hand side needs to be changed from a minus to a plus. $\dot{Q}_b''$ is the prescribed heat flux at the boundary. This heat flux may be created by, say, electrical or radiative heating. The sign of $\dot{Q}_b''$ is positive [as shown in eq. (2.48)] if it denotes heat flux from the boundary and negative [opposite to what is shown in eq. (2.48)] if it denotes heat flux toward the boundary.

In the special case of a Cartesian coordinate system like the one used in Fig. 2.5b, the boundary conditions of second kind at $x = L$ and $x = 0$, respectively, are

$$\dot{Q}_b'' = -k_b\left(\frac{\partial T}{\partial x}\right)_{x=L} \tag{2.49}$$

$$\dot{Q}_b'' = k_b\left(\frac{\partial T}{\partial x}\right)_{x=0} \tag{2.50}$$

Similarly, for the cylindrical geometry of Fig. 2.5c,

$$-\dot{Q}_b'' = -k_b\left(\frac{\partial T}{\partial r}\right)_{r=R} \tag{2.51}$$

Note that in the special case of an insulated wall ($\dot{Q}_b'' = 0$) the temperature gradient on the right-hand side of eqs. (2.48)–(2.50) is equal to zero.

**3.** *Boundary condition of the third kind.* This type of boundary condition is commonly termed a *convection* boundary condition. It simply states that if a body is in contact with a moving convecting fluid, the heat conduction in the body equals the heat convected by the fluid. A typical temperature distribution in the vicinity of a boundary between a solid body and a fluid is shown in Fig. 2.6a. For the configuration of Fig. 2.6a,

$$-k\left(\frac{\partial T}{\partial n}\right)_b = h(T_b - T_\infty) \tag{2.52}$$

In eq. (2.52), $T_\infty$ is the temperature of the moving fluid far away from and unaffected by the boundary and $h$ is the heat transfer coefficient [13,14]. The minus sign on the left-hand side of eq. (2.52) denotes differentiation along the outward normal direction to the surface. If the differentiation is along the inward normal direction, the sign needs to be changed from minus to plus. For the Cartesian system of Fig. 2.6b, the boundary conditions at $x = 0$ and $L$, respectively, are

$$k\left(\frac{\partial T}{\partial x}\right)_{x=0} = h(T - T_\infty)_{x=0} \tag{2.53}$$

$$-k\left(\frac{\partial T}{\partial x}\right)_{x=L} = h(T - T_\infty)_{x=L} \tag{2.54}$$

For the cylinder of Fig. 2.6c, the boundary condition at the outer surface is

$$-k\left(\frac{\partial T}{\partial r}\right)_{r=R} = h(T - T_\infty)_{r=R} \tag{2.55}$$

In addition to the three common types of boundary conditions discussed above, different boundary conditions, more complex than the above, may be encountered in the solution of heat conduction problems. For example, simultaneous convection and radiation heating may occur at a boundary, or heat release (absorption) because of freezing (melting) at a moving phase-change interface may take place. Boundary conditions not falling into any of the three categories presented in this section are discussed in later chapters in connection with the conduction problems in which these boundary conditions need to be met.

### Homogeneous Boundary Conditions

Before closing this section, it is worth identifying a special category of boundary conditions that is of particular importance to the solution of heat conduction problems with the popular method of separation of variables (this method is discussed in detail in Chapter 4). This category of boundary conditions consists of the various types of *homogeneous* boundary conditions, defined as follows: *A boundary condition is homogeneous if all of its terms other than zero itself are of the first degree in the unknown function (temperature) and its derivatives.*

To clarify the meaning of the terminology *first degree* in this definition, we offer the following examples. A term of the type $T(\partial T/\partial x)$ or $T(\partial^2 T/\partial x^2)$ (the product of the temperature and one of its derivatives) is of the second degree. A term of the type $T$ or $\partial T/\partial x$ or $\partial^2 T/\partial x^2$ is of the first degree. A constant term $T_\infty$, $T_b$, or $\dot{Q}_b''$ (if these quantities are constant) is of the zeroth degree.

According to the discussion above, if $T_b$ is other than zero, boundary conditions (2.45)–(2.47) are *not* homogeneous. If $T_b$ is zero, these boundary conditions are homogeneous. Similarly, boundary conditons (2.48)–(2.51) are *not* homogeneous. They become homogeneous in the special case of an insulated wall. Finally, boundary conditions (2.52)–(2.55) are homogeneous only in the special case where $T_\infty = 0$. If $T_\infty \neq 0$, boundary conditions (2.52)–(2.55) are not homogeneous. As mentioned above, the concept of homogeneous boundary conditions will be used extensively in Chapter 4, where the method of separation of variables is introduced. The presence of a number of homogeneous boundary conditions in the heat conduction model is necessary for this method to work.

## PROBLEMS

**2.1.** Show that the heat conduction equation for a solid medium moving in the $x$-direction with a constant velocity $U$ is given by

$$\rho c\left(\frac{\partial T}{\partial t} + U\frac{\partial T}{\partial x}\right) = \nabla \cdot (k\,\nabla T) + \dot{Q}'''_{\text{gen}}$$

Next, show that for one-dimensional steady-state conduction (in the $x$-direction of an $x,y,z$ Cartesian system) in a body with constant properties and without heat generation the equation above reduces to

$$U\frac{\partial T}{\partial x} = \alpha\frac{\partial^2 T}{\partial x^2}$$

where $\alpha = k/\rho c$ is the thermal diffusivity of the medium. Finally, show that for radial steady-state conduction in cylindrical coordinates $(r, \theta, x)$ in a body with constant properties this equation becomes

$$U\frac{\partial T}{\partial x} = \alpha\frac{1}{r}\frac{\partial}{\partial r}\left(r\frac{\partial T}{\partial r}\right)$$

**2.2.** Verify that the heat conduction equation for a body with constant thermophysical properties in cylindrical coordinates is given by eq. (2.34).

**2.3.** Verify that the heat conduction equation for a body with constant thermophysical properties in spherical coordinates is given by eq. (2.35).

**2.4.** Prove that by substituting eq. (2.19) into eq. (2.18), we obtain eq. (2.20).

**2.5.** Prove that substituting eq. (2.28) into eq. (2.22) yields eq. (2.29).

**2.6.** Heat conduction takes place in living tissues. Consider the case where temperature variations exist in a living tissue in the presence of blood perfusion, metabolic heat generation, and radiation absorption. The blood perfusion effect can be modeled as a cooling mechanism that removes heat from the tissue at a rate $\dot{m}_b''' c_{pb}(T - T_A)(W/m^3)$, where $\dot{m}_b'''$ is the volumetric blood flow rate in the tissue, $c_{pb}$ the specific heat of the blood, $T$ the tissue temperature, and $T_A$ the arterial blood supply temperature assumed constant. The metabolic heat generation is denoted by $\dot{q}'''$. The radiation absorption in the tissue is denoted by $\dot{q}_r'''(W/m^3)$ and is nonzero only when the tissue is subjected to radiation diathermy. Using the differential control volume formulation as well as the integral formulation as demonstrated in this chapter and assuming that the thermophysical properties of the tissue and of the blood are constant, show that the heat conduction process in the tissue is governed by the bioheat equation [6–8]

$$\rho_t c_{pt}\frac{\partial T}{\partial t} + \dot{m}_b''' c_{pb}(T - T_A) = k_t\,\nabla^2 T + \dot{q}''' + \dot{q}_r'''$$

where the subscript $t$ denotes the tissue.

**2.7.** If the human leg is approximated as a long cylinder of the cross section shown in Fig. P2.7, construct (but do not solve) a mathematical model for the study of heat conduction in the human leg. Allow for dependence of the thermal conductivity on location [$k = k(r)$].

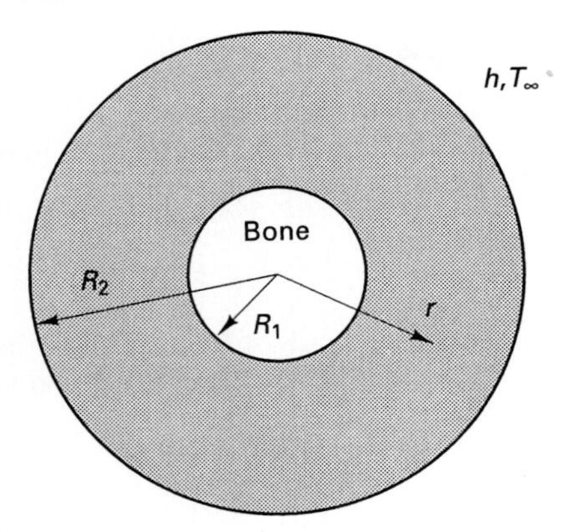


**Figure P2.7** Cross section of a cylinder modeling the human leg.

**2.8.** If the human torso is approximated as the long slab shown in Fig. P2.8, construct (do not solve) a mathematical model for the study of heat conduction in the human torso. Allow for the dependence of the thermal conductivity on location [$k = k(x)$].

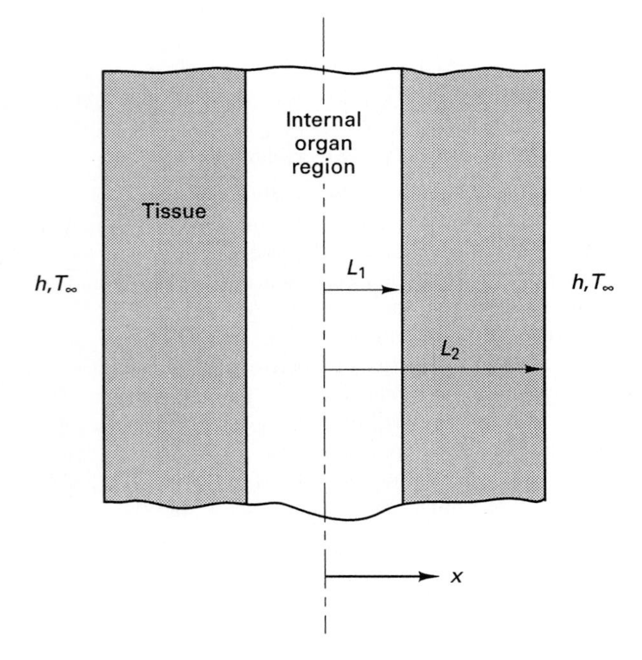


**Figure P2.8** Cross section of a slab modeling the human torso.

**2.9.** Obtain an expression for a boundary condition that is a combination of boundary conditions of the second and third kind. More specifically, the boundary under consideration is in contact with a moving fluid ($h$, $T_\infty$) and is subjected to a heat flux $\dot{q}''$. Write this boundary condition in the three major coordinate systems (Cartesian, cylindrical, and spherical).

**2.10.** Obtain a condition for the heat flux at the interface between two different solids in perfect thermal contact. Express this boundary condition in terms of the temperature of the solids. If one of these solids slides onto the other, frictional heating, $\dot{Q}''_{fr}$, takes place at the interface. Modify the heat flux boundary condition above to account for the frictional heating. Write your final expression in the three major coordinate systems.

## REFERENCES


1. *Temperature Measurement Handbook,* Omega Engineering Inc., Stamford, CT, 1990.
2. C. VEST, *Holographic Interferometry,* Wiley, New York, 1979.
3. A. BEJAN, *Convection Heat Transfer,* Wiley, New York, 1984.
4. A. BEJAN, *Advanced Engineering Thermodynamics,* Wiley, New York, 1988.
5. H. B. CALLEN, *Thermodynamics,* Wiley, New York, 1960.
6. J. C. CHATO, Fundamentals of bioheat transfer, in *Thermal Dosimetry and Treatment Planning,* M. Gautherie, ed., Springer-Verlag, New York, pp. 1–56, 1990.
7. H. H. PENNES, Analysis of tissue and arterial blood temperatures in the resting human forearm, *J. Appl. Physiol.,* Vol. 1, pp. 93–122, 1948.
8. W.-J. YANG, *Biothermal Fluid Sciences,* Hemisphere, New York, 1989.
9. M. M. CHEN and K. R. HOLMES, Microvascular contributions in tissue heat transfer, *Ann. N.Y. Acad. Sci.,* Vol. 335, pp. 137–150, 1980.
10. M. GAUTHERIE, Etude par thermométrie infrarouge des properties thermiques de tissues humains "in vivo." Influence de la température et de la vascularisation, *Rev. Fr. Etud. Clin. Biol.,* Vol. 14, pp. 585–901, 1969.
11. E. HURLBURT and J. C. CHATO, Computer aided modelling of tissue temperatures using finite difference methods, in *B. T. Chao Symposium Volume,* Department of Mechanical and Industrial Engineering, University of Illinois, Urbana, IL, 1987.
12. K. H. KELLER and L. SEILER, An analysis of peripheral heat transfer in man, *J. Appl. Physiol.,* Vol. 30, pp. 779–786, 1971.
13. F. P. INCROPERA and D. P. DEWITT, *Fundamentals of Heat and Mass Transfer,* 3rd ed. Wiley, New York, 1990.
14. A. J. CHAPMAN, *Fundamentals of Heat Transfer,* Macmillan, New York, 1987.

CHAPTER 3

# REVIEW OF REPRESENTATIVE BASIC PROBLEMS IN HEAT CONDUCTION

In this chapter a number of representative basic heat conduction problems are reviewed. These problems are commonly used by the practicing engineer to obtain *estimates* of heat losses or gains through configurations such as simple or laminated walls, insulated or bare pipes, and finned surfaces. The problems are divided into three general categories: steady-state one-dimensional conduction, spatially isothermal transient conduction, and one-dimensional conduction in extended surfaces (fins). The material reviewed in this section is typically discussed at great length in undergraduate heat transfer textbooks [1–4].

## 3.1 STEADY-STATE ONE-DIMENSIONAL HEAT CONDUCTION

Three applications are considered here, each dealing with one of the three major coordinate systems used in the modeling of heat conduction processes: the plane wall problem associated with one-dimensional steady conduction in Cartesian coordinates, the cylindrical pipe problem associated with one-dimensional steady conduction in cylindrical coordinates, and the spherical shell problem associated with one-dimensional steady conduction in spherical coordinates.

### The Plane Wall Problem

This problem is shown schematically in Fig. 3.1. It represents a plane wall each side of which is in contact with a moving fluid. The conductivity of the wall, $k$, is as-

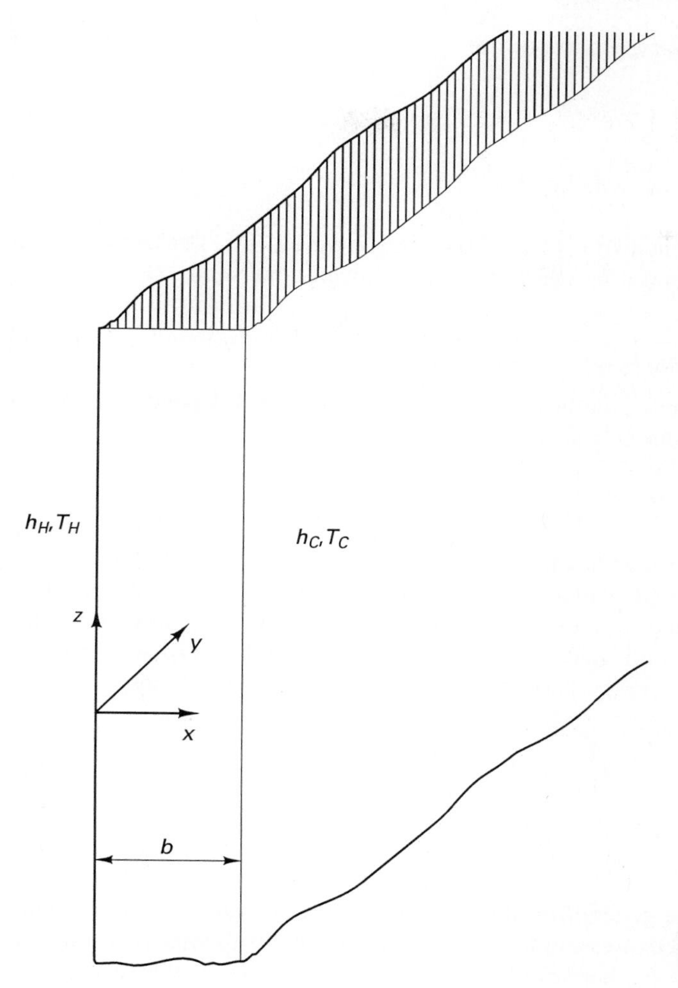


**Figure 3.1** Flat wall with covering fluid at both its sides.

sumed to be constant. With no loss of generality, we state that the temperature of the fluid on the left side of the wall and far away from it, $T_H$, is higher than the temperature of the fluid on the right side of the wall and far away from it, $T_C$. This statement assures the constancy of temperature outside the thermal boundary layers on each side of the wall. The average heat transfer coefficient on the hot side is denoted by $h_H$ and on the cold side by $h_C$. The wall extends to infinity in both the $y$ and $z$ directions and no heat is generated within it. With the aforementioned assumptions, the energy equation (2.33) reduces to

$$\frac{d^2 T}{dx^2} = 0 \tag{3.1}$$

The corresponding boundary conditions are

$$\text{at } x = 0: \qquad h_H(T - T_H) = k\frac{dT}{dx} \tag{3.2}$$

$$\text{at } x = b: \qquad h_C(T - T_C) = -k\frac{dT}{dx} \tag{3.3}$$

Solving the set of equations (3.1)–(3.3) [i.e., integrating eq. (3.1) twice and using eqs. (3.2) and (3.3) to obtain the two constants of integration] yields

$$T = \frac{-(T_H - T_C)}{(h_H/h_C) + (h_H/k)b + 1}\left(1 + \frac{h_H}{k}x\right) + T_H \tag{3.4}$$

Since eq. (3.4) is linear in $x$, the heat flux through the wall is independent of $x$ and can be calculated from Fourier's law:

$$\dot{Q}'' = -k\frac{\partial T}{\partial x} = \frac{h_H(T_H - T_C)}{(h_H/h_C) + (h_H/k)b + 1} \tag{3.5}$$

Note that the heat transfer coefficient on either side of the wall is a measure of the imperfection of the thermal contact between the wall and the fluid. As the heat transfer coefficients $h_H$ and $h_C$ increase and approach infinity, the thermal contact becomes perfect and the wall surface attains the fluid temperature ($T_H$ at $x = 0$ and $T_C$ at $x = b$, respectively). In the perfect thermal contact limit [1–4], eqs. (3.4) and (3.5) reduce to

$$T = -\frac{T_H - T_C}{b}x + T_H \tag{3.6}$$

$$\dot{Q}'' = k\frac{T_H - T_C}{b} \tag{3.7}$$

The analysis above can be generalized for the case of a wall of variable thermal conductivity. After a general expression for the temperature in a wall with thermal conductivity that depends on the horizontal coordinate, $x$, is obtained, the special case of a multilayered wall (insulated or laminated) will be considered.

The energy equation for the case where the thermal conductivity of the material depends on $x$ takes the form

$$\frac{d}{dx}\left[k(x)\frac{dT}{dx}\right] = 0 \tag{3.8}$$

The boundary conditions remain identical to eqs. (3.2) and (3.3). Integrating eq. (3.8) twice and applying boundary conditions (3.2) and (3.3) to obtain the two constants of integration (the simple algebraic details are left as an exercise to the reader) yields

$$T = T_H - \frac{(T_H - T_C)[h_H I(x) - h_H I(0) + 1]}{h_H[I(b) - I(0)] + (h_H/h_C) + 1} \tag{3.9}$$

where the function $I(x)$ is an integral defined as

$$I(x) = \int_0^x \frac{d\xi}{k(\xi)} \tag{3.10}$$

In Eq. (3.10) $\xi$ is a dummy variable. Note also that $I(0) = 0$. The heat flux at any location is obtained invoking Fourier's law and eq. (3.9):

$$\dot{Q}'' = -k(x)\frac{dT}{dx} = \frac{h_H(T_H - T_C)}{h_H[I(b) - I(0)] + (h_H/h_C) + 1} \tag{3.11}$$

In the special case of a multilayered wall (a common configuration in heat transfer applications) consisting of $n$ layers each of different but constant thermal conductivity as shown in Fig. 3.2, the thermal conductivity is given by the following expression:

$$k(x) = \begin{cases} k_1 & 0 \le x \le x_1 \\ k_2 & x_1 < x < x_2 \\ \vdots & \\ k_i & x_{i-1} < x < x_i \\ \vdots & \\ k_{n-1} & x_{n-2} < x < x_{n-1} \\ k_n & x_{n-1} < x < x_n \end{cases} \tag{3.12}$$

Assume that the temperature distribution in the $i$th layer of the composite wall (i.e., $x_{i-1} < x < x_i$) needs to be determined. Combining eqs. (3.10) and (3.12) to evaluate the $I(0)$, $I(b)$, and $I(x)$ yields

$$I(x) = \sum_{m=1}^{i-1} \frac{b_m}{k_m} + \frac{x - x_{i-1}}{k_i} \qquad x_{i-1} < x < x_i \tag{3.13}$$

$$I(0) = 0 \tag{3.14}$$

$$I(b) = \sum_{m=1}^{n} \frac{b_m}{k_m} \tag{3.15}$$

Next, eqs. (3.9), (3.13), and (3.15) are combined and the final expression for the temperature distribution in the $i$th layer is obtained:

$$T_i = T_H - \frac{(T_H - T_C)\left[h_H\left(\dfrac{x - x_{i-1}}{k_i} + \displaystyle\sum_{m=1}^{i-1} \frac{b_m}{k_m}\right) + 1\right]}{h_H \displaystyle\sum_{m=1}^{n} \frac{b_m}{k_m} + \dfrac{h_H}{h_C} + 1} \tag{3.16}$$

The heat flux through the wall is calculated in a straightforward manner after substituting eqs. (3.14) and (3.15) into eq. (3.11):

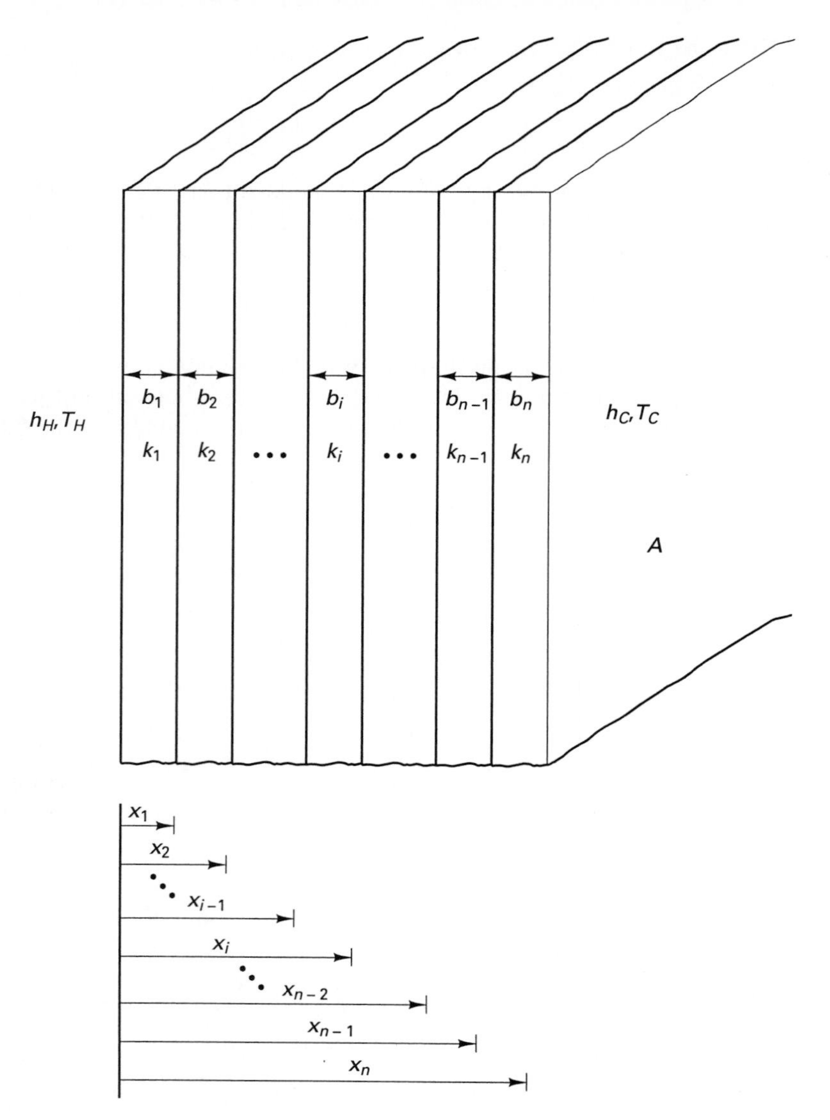


**Figure 3.2** Multilayered flat wall.

$$\dot{Q}'' = \frac{T_H - T_C}{\sum_{m=1}^{n} \frac{b_m}{k_m} + \frac{1}{h_C} + \frac{1}{h_H}} \tag{3.17}$$

Note that in the plane wall example outlined above, since the temperature is a linear function of position, $x$, the heat flux through the wall is constant.

Before closing this section, the concept of thermal resistance will be introduced. This concept originates from drawing the analogy between the expression for one-dimensional heat transfer rate through a wall and Ohm's law of electricity for one-dimensional current flow, which states that

$$I = \frac{\Delta V}{R_e} \tag{3.18}$$

As shown in Fig. 3.3, $\Delta V$(volts) is the applied voltage difference across an electrical resistance $R_e$ (ohms), and $I$ (amperes) is the intensity of the current passing through its resistance.

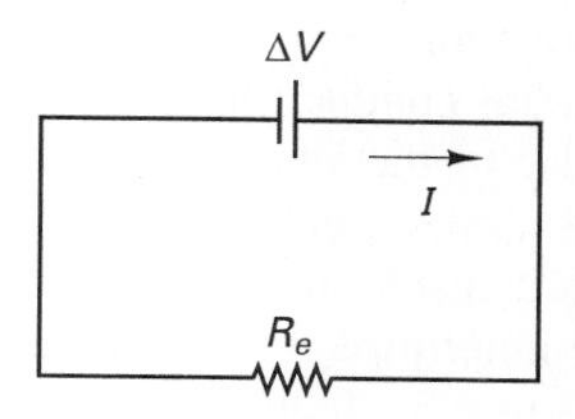


**Figure 3.3** Schematic of the electrical circuit based on which the analogy between Ohm's law and Fourier's law is drawn.

If the heat transfer rate through the composite wall is denoted by $\dot{Q} = A\dot{Q}''$, where $A$ is the area of the wall perpendicular to which heat is flowing (Fig. 3.2), the following conclusions can be drawn by comparing, by inspection, eqs. (3.17) and (3.18). First, the temperature difference $\Delta T = T_H - T_C$ is analogous to the voltage difference. Second, the heat transfer rate $\dot{Q} = A\dot{Q}''$ flowing through the wall is analogous to the electric current flowing through the resistance. Therefore, eq. (3.17) can be written as

$$\dot{Q} = \frac{T_H - T_C}{\sum_{m=1}^{n} \frac{b_m}{Ak_m} + \frac{1}{Ah_C} + \frac{1}{Ah_H}} \tag{3.19}$$

Comparing eq. (3.18) to eq. (3.19), one easily realizes that the denominator of eq. (3.19) plays a role analogous to that of the electrical resistance of eq. (3.18). It is because of this reason that the concept of the thermal resistance to the flow of the heat in the configuration of Fig. 3.2 is introduced and is defined as follows:

$$R_{\text{th}} = \sum_{m=1}^{n} \frac{b_m}{Ak_m} + \frac{1}{Ah_C} + \frac{1}{Ah_H} \tag{3.20}$$

Three contributors to the thermal resistance of eq. (3.20) can be identified. First, thermal resistance offered by the material of the various layers in the wall (represented by the various terms in the series expression). Second, thermal resistance offered by the presence of imperfect thermal contact between the right (cold) side of the wall and the convecting fluid. Third, thermal resistance offered by the presence of imperfect thermal contact between the left (hot) side of the wall and the convecting fluid. Note that in the limit of perfect thermal contact between the fluid and the wall ($h_H$, $h_C \to \infty$) the corresponding contributions to the overall thermal resistance vanish.

A more general expression for the thermal resistance can be obtained if eq. (3.11) is used in place of eq. (3.17) for the heat flux. Since $I(0) = 0$ from eq. (3.10), the reader can easily show that, in this case,

$$R_{\text{th}} = \frac{I(b)}{A} + \frac{1}{Ah_C} + \frac{1}{Ah_H} \tag{3.21}$$

### The Cylindrical Pipe Problem

The analysis presented above for the plane wall problem can be repeated to study the problem of steady-state radial conduction from a cylindrical pipe. This geometry is of great importance because of the vast use of pipes in heat transfer engineering applications. A schematic of a cross section of a pipe is shown in Fig. 3.4. Assuming that there is fluid flow both inside and outside the pipe, the heat transfer from the fluid inside the pipe to the inner pipe wall is characterized by an average heat transfer coefficient $h_H$ and temperature $T_H$. Outside the pipe the corresponding notation is $h_C$ and $T_C$. Of interest is the special case of radial conduction (i.e., heat transfer around the periphery or along the length of the pipe are neglected). Based on the foregoing assumptions, the heat conduction equation for constant thermal conductivity of the pipe wall material reduces to

$$\frac{d}{dr}\left(r\frac{dT}{dr}\right) = 0 \tag{3.22}$$

The corresponding boundary conditions read

$$\text{at } r = R_1: \qquad h_H(T - T_H) = k\frac{dT}{dr} \tag{3.23}$$

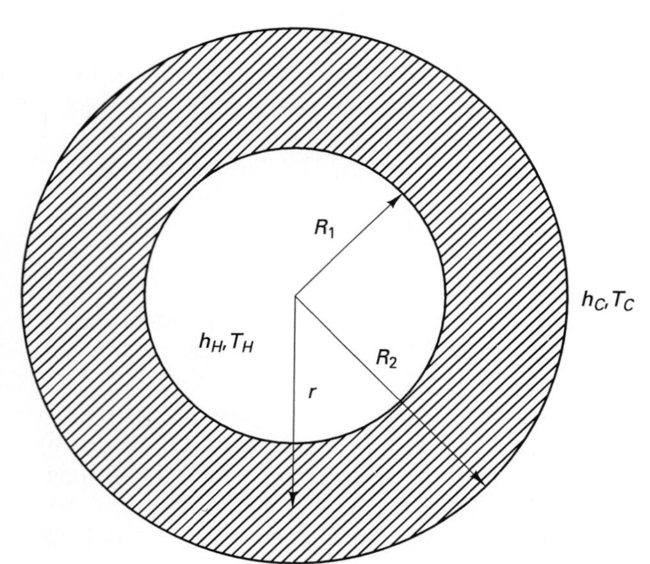


**Figure 3.4** Cross section of a cylindrical pipe.

$$\text{at } r = R_2: \qquad h_C(T - T_C) = -k\frac{dT}{dr} \tag{3.24}$$

Solving (3.22)–(3.24) is straightforward and yields the temperature distribution of the pipe wall:

$$T = \frac{-(T_H - T_C)}{\ln\dfrac{R_2}{R_1} + \dfrac{k}{h_C R_2} + \dfrac{k}{h_H R_1}}\left(\ln\frac{r}{R_1} + \frac{k}{h_H R_1}\right) + T_H \tag{3.25}$$

The heat transfer rate through the wall is calculated by applying Fourier's law:

$$\dot{Q} = -k(2\pi r L)\frac{dT}{dr} = \frac{2\pi L(T_H - T_C)}{\dfrac{1}{R_1 h_H} + \dfrac{\ln(R_2/R_1)}{k} + \dfrac{1}{R_2 h_C}} \tag{3.26}$$

where $L$ is the length of the pipe (perpendicular to the plane of Fig. 3.4). Note that the heat transfer rate through the pipe wall is constant, as it should be for a steady-state process to exist [the temperature gradient in eq. (3.26) is inversely proportional to the radial distance $r$, while the lateral area of the pipe is proportional to $r$.]

In the limit of perfect thermal contact or very large heat transfer coefficients $h_H$ and $h_C$ the temperature of the inner wall approaches $T_H$, and of the outer wall $T_C$. In this limit [1–4], eqs. (3.25) and (3.26) reduce to

$$T = T_H - \frac{(T_H - T_C)\ln(r/R_1)}{\ln(R_2/R_1)} \tag{3.27}$$

$$\dot{Q} = \frac{2\pi L(T_H - T_C)}{\ln(R_2/R_1)/k} \tag{3.28}$$

Much like in the case of a plane wall, the analysis above for the pipe can be generalized for the case of a cylindrical shell with variable thermal conductivity. A common heat transfer example of this type is an insulated pipe or a pipe in which scale has been deposited on its inner surface. The heat conduction equation for the case of a pipe with variable thermal conductivity takes the form

$$\frac{d}{dr}\left[k(r)r\frac{dT}{dr}\right] = 0 \tag{3.29}$$

The boundary conditions remain identical to eqs. (3.23) and (3.24). Proceeding in a manner identical to what was outlined in conjunction with the plane wall problem, we obtain

$$T = T_H - \frac{(T_H - T_C)\left\{h_H[G(r) - G(R_1)] + \dfrac{1}{R_1}\right\}}{h_H[G(R_2) - G(R_1)] + \dfrac{h_H}{R_2 h_C} + \dfrac{1}{R_1}} \tag{3.30}$$

$$\dot{Q} = \frac{2\pi L h_H (T_H - T_C)}{h_H[G(R_2) - G(R_1)] + \dfrac{h_H}{R_2 h_C} + \dfrac{1}{R_1}} \tag{3.31}$$

where the function $G(r)$ is an integral defined as

$$G(r) = \int_{R_1}^{r} \frac{d\xi}{\xi k(\xi)} \tag{3.32}$$

In eq. (3.32), $\xi$ is a dummy variable. Note that $G(R_1) = 0$. Consider now the special case of a pipe surrounded by a multilayered insulation, each layer having a different but constant thermal conductivity (Fig. 3.5). The thermal conductivity of the composite wall consisting of the pipe wall and the several insulation layers is given by the following expression:

$$k(r) = \begin{cases} k_1 & R_1 < r < R_2 \\ k_2 & R_2 < r < R_3 \\ \vdots & \\ k_i & R_i < r < R_{i+1} \\ \vdots & \\ k_{n-2} & R_{n-2} < r < R_{n-1} \\ k_{n-1} & R_{n-1} < r < R_n \end{cases} \tag{3.33}$$

If of interest is the temperature distribution in the $i$th layer, combining eqs. (3.30), (3.32), and (3.33) [note that $R_2$ in eq. (3.30) needs to be replaced by $R_n$] yields

$$T = T_H - \frac{(T_H - T_C)\left(\displaystyle\sum_{m=1}^{i-1} \frac{1}{k_m} \ln \frac{R_{m+1}}{R_m} + \frac{1}{k_i} \ln \frac{r}{R_i} + \frac{1}{h_H R_1}\right)}{\displaystyle\sum_{m=1}^{n-1} \frac{1}{k_m} \ln \frac{R_{m+1}}{R_m} + \frac{1}{R_n h_C} + \frac{1}{R_1 h_H}} \qquad R_i < r < R_{i+1} \tag{3.34}$$

Similarly, the heat flux through the wall is obtained by combining eqs. (3.31)–(3.33) and by replacing $R_2$ in eq. (3.31) by $R_n$:

$$\dot{Q} = \frac{2\pi L (T_H - T_C)}{\displaystyle\sum_{m=1}^{n-1} \frac{1}{k_m} \ln \frac{R_{m+1}}{R_m} + \frac{1}{R_n h_C} + \frac{1}{R_1 h_H}} \tag{3.35}$$

In the limit of a single layer, eqs. (3.34) and (3.35) reduce to eqs. (3.25) and (3.26).

Following the discussion of the preceding section for the thermal resistance, invoking the analogy between eq. (3.35) and Ohm's law, eq. (3.18), an expression can be obtained for the thermal resistance in a pipe wall surrounded by a multilayered insulation:

$$R_{\text{th}} = \sum_{m=1}^{n-1} \frac{1}{2\pi L k_m} \ln \frac{R_{m+1}}{R_m} + \frac{1}{2\pi L R_n h_C} + \frac{1}{2\pi L R_1 h_H} \tag{3.36}$$

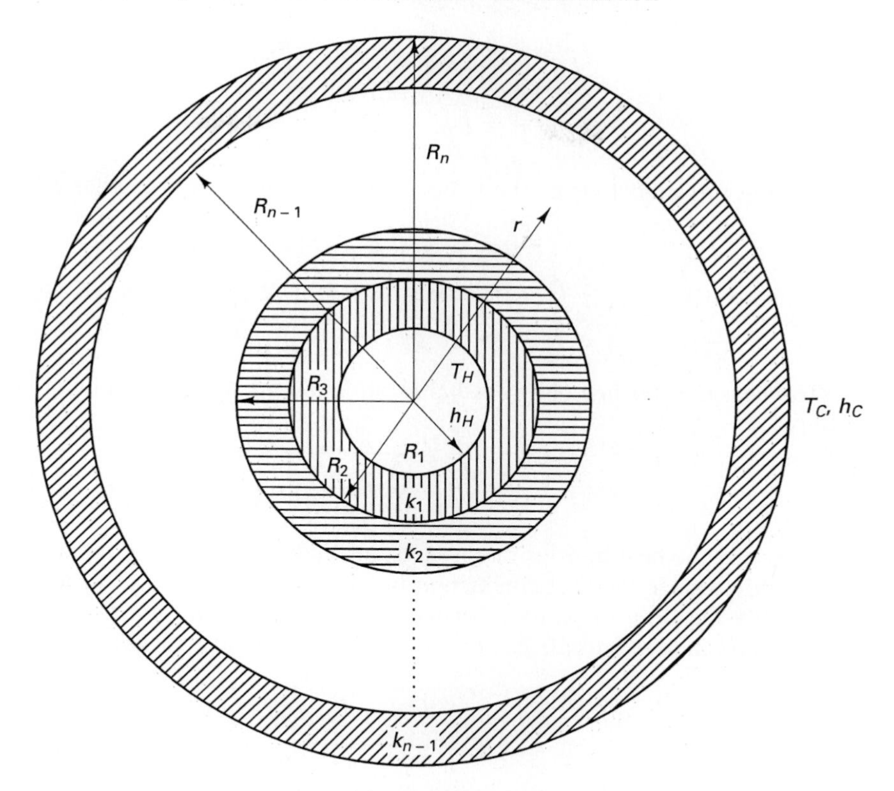


**Figure 3.5** Cross section of a cylindrical pipe surrounded by multilayered insulation.

A more general equation for the thermal resistance results from comparing eq. (3.31) to Ohm's law:

$$R_{\text{th}} = \frac{G(R_2)}{2\pi L} + \frac{1}{2\pi L R_2 h_C} + \frac{1}{2\pi L R_1 h_H} \tag{3.37}$$

### The Spherical Shell Problem

An analysis similar to that above for plane walls and cylindrical pipes can be performed for unidirectional steady heat conduction through a spherical shell. Figure 3.4 can serve the purpose of illustrating the present problem as well, if one interprets this figure as showing the cross section of a spherical shell instead of that of a cylindrical pipe. With the aforementioned background in mind and assuming that the thermal conductivity of the spherical shell is constant, the unidirectional steady conduction through the shell is described by the following model:

$$\frac{d}{dr}\left(r^2 \frac{dT}{dr}\right) = 0 \tag{3.38}$$

$$\text{at } r = R_1: \qquad h_H(T - T_H) = k\frac{dT}{dr} \tag{3.39}$$

$$\text{at } r = R_2: \qquad h_C(T - T_C) = -k\frac{dT}{dr} \tag{3.40}$$

Solving this model yields the temperature distribution in the spherical shell.

$$T = \frac{(T_H - T_C)\left(\dfrac{1}{r} - \dfrac{1}{R_1} - \dfrac{k}{R_1^2 h_H}\right)}{\dfrac{1}{R_1} - \dfrac{1}{R_2} + \dfrac{k}{h_H R_1^2} + \dfrac{k}{h_C R_2^2}} + T_H \tag{3.41}$$

The heat transfer rate at any radial location is calculated by applying Fourier's law:

$$\dot{Q} = -k(4\pi r^2)\frac{dT}{dr} = \frac{T_H - T_C}{\dfrac{1}{4\pi}\left(\dfrac{1}{h_H R_1^2} + \dfrac{1}{h_C R_2^2} + \dfrac{1}{kR_1} - \dfrac{1}{kR_2}\right)} \tag{3.42}$$

Note that the heat transfer rate is constant (independent of the radial position $r$) as it should be since the heat conduction in the spherical shell is at steady state.

In the limit of perfect thermal contact between the fluids at both sides of the shell and the shell itself ($h_H$, $h_C \to \infty$), eqs. (3.41) and (3.42) reduce to

$$T = \frac{(T_H - T_C)\left(\dfrac{1}{r} - \dfrac{1}{R_1}\right)}{\dfrac{1}{R_1} - \dfrac{1}{R_2}} \tag{3.43}$$

$$\dot{Q} = \frac{T_H - T_C}{\dfrac{1}{4\pi k}\left(\dfrac{1}{R_1} - \dfrac{1}{R_2}\right)} \tag{3.44}$$

A more general analysis of the spherical shell problem can be performed if the constant thermal conductivity assumption is relaxed. To this end, we allow the thermal conductivity to vary with the radial coordinate. The conduction equation in this case reads

$$\frac{d}{dr}\left[k(r)\, r^2 \frac{dT}{dr}\right] = 0 \tag{3.45}$$

The boundary conditions (3.39) and (3.40) remain the same. Solving this model in a manner identical to that for plane walls and cylindrical pipes yields

$$T = \frac{-(T_H - T_C)\left[F(r) - F(R_1) + \dfrac{1}{h_H R_1^2}\right]}{\dfrac{1}{h_H R_1^2} + \dfrac{1}{h_C R_2^2} - F(R_1) + F(R_2)} + T_H \tag{3.46}$$

$$\dot{Q} = \frac{T_H - T_C}{\dfrac{1}{4\pi}\left[\dfrac{1}{h_H R_1^2} + \dfrac{1}{h_C R_2^2} - F(R_1) + F(R_2)\right]} \tag{3.47}$$

The function $F(r)$ is an integral defined as

$$F(r) = \int_{R_1}^{r} \frac{d\xi}{\xi^2 k(\xi)} \tag{3.48}$$

where $\xi$ is a dummy variable. Note that $F(R_1) = 0$.

A special case of the variable thermal conductivity model is that of a multilayered spherical shell, each layer having a different but constant thermal conductivity. This situation is depicted in Fig. 3.5, which should now be interpreted as the schematic of the cross section of a multilayered spherical shell instead of a cylindrical pipe. The expression for the thermal conductivity of the multilayered spherical shell is identical to that of the multilayered cylindrical pipe and is given by eq. (3.33). Substituting eq. (3.33) into eqs. (3.46) and (3.47), after replacing $R_2$ in these equations by $R_n$, and omitting the algebraic detail, for brevity we obtain

$$T = \frac{-(T_H - T_C)\left[\displaystyle\sum_{m=1}^{i-1} \frac{1}{k_m}\left(\frac{1}{R_m} - \frac{1}{R_{m+1}}\right) + \frac{1}{k_i}\left(\frac{1}{R_i} - \frac{1}{r}\right) + \frac{1}{h_H R_1^2}\right]}{\dfrac{1}{h_H R_1^2} + \dfrac{1}{h_C R_n^2} + \displaystyle\sum_{m=1}^{n-1} \frac{1}{k_m}\left(\frac{1}{R_m} - \frac{1}{R_{m+1}}\right)} + T_H$$

$$R_i < r < R_{i+1} \tag{3.49}$$

$$\dot{Q} = \frac{T_H - T_C}{\dfrac{1}{4\pi}\left[\dfrac{1}{h_H R_1^2} + \dfrac{1}{h_C R_n^2} + \displaystyle\sum_{m=1}^{n-1} \frac{1}{k_m}\left(\frac{1}{R_m} - \frac{1}{R_{m+1}}\right)\right]} \tag{3.50}$$

Finally, invoking the analogy between unidirectional electricity and heat conduction as we did for the plane wall and cylindrical pipe problems, we derive, by inspection, the following expression for the thermal resistance of a multilayered shell structure:

$$R_{\text{th}} = \frac{1}{4\pi}\left[\frac{1}{h_H R_1^2} + \frac{1}{h_C R_n^2} + \sum_{m=1}^{n-1} \frac{1}{k_m}\left(\frac{1}{R_m} - \frac{1}{R_{m+1}}\right)\right] \tag{3.51}$$

A more general expression for the thermal resistance accounting for arbitrary variation of $k$ with $r$ results from comparing eq. (3.47) to Ohm's law:

$$R_{\text{th}} = \frac{1}{4\pi}\left[\frac{1}{h_H R_1^2} + \frac{1}{h_C R_2^2} - F(R_1) + F(R_2)\right] \tag{3.52}$$

## 3.2 TRANSIENT HEAT TRANSFER FROM A SPATIALLY ISOTHERMAL BODY

In this case, the interest is focused on obtaining the temperature history of a body exchanging heat with its surroundings, in the limit where no temperature gradients exist in the body. To obtain a criterion as to when such a limit is realistic, one needs to compare the order of magnitude of conduction in the body to the order of magnitude of the heat transfer to the surroundings. To this end, assume that a body of conductivity $k$ and initial temperature $T_i$ is suddenly cooled convectively ($h$) by a fluid of temperature $T_\infty$. The characteristic dimension of the body is denoted by $D$. Let $T_0$ denote the surface temperature of the body. The magnitude of conduction within the body is $O(k[(T_i - T_0)/D]$. The magnitude of convection from the body to the fluid is $O(h[T_0 - T_\infty])$. Equating the conduction heat transfer estimate to the convection heat transfer estimate yields

$$k\frac{T_i - T_0}{D} = h(T_0 - T_\infty) \tag{3.53}$$

Solving for the characteristic temperature difference in the body, we find that

$$T_i - T_0 = \mathrm{Bi}(T_0 - T_\infty) \tag{3.54}$$

where the Biot number is defined as

$$\mathrm{Bi} = \frac{hD}{k} \tag{3.55}$$

Clearly, the criterion for the temperature differences within the body to be negligible is that the Biot number is very small:

$$\mathrm{Bi} << O(1) \quad \text{or} \quad \mathrm{Bi} \to 0 \tag{3.56}$$

This criterion is often satisfied by bodies of very large thermal conductivity.

The process of transient conduction in a spatially isothermal body is illustrated in the following example.

**Example 3.1**

The schematic of the problem of interest is shown in Fig. 3.6. A body of mass $M$, specific heat $c$, surface area $A$, and initial temperature $T_0$ is immersed in a fluid of temperature $T_f(t)$ that varies with time as follows:

$$T_f(t) = \begin{cases} T_0 & \text{for} \quad t \le 0 \\ T_0 + Bt & \text{for} \quad t \ge 0 \end{cases} \tag{3.57}$$

where $B$ is a known constant. The heat transfer coefficient $h$ at the body–fluid interface is assumed constant and known. It is of interest to obtain the temperature history of the body if $\mathrm{Bi} << O(1)$ (the heat transfer in the body is spatially isothermal).

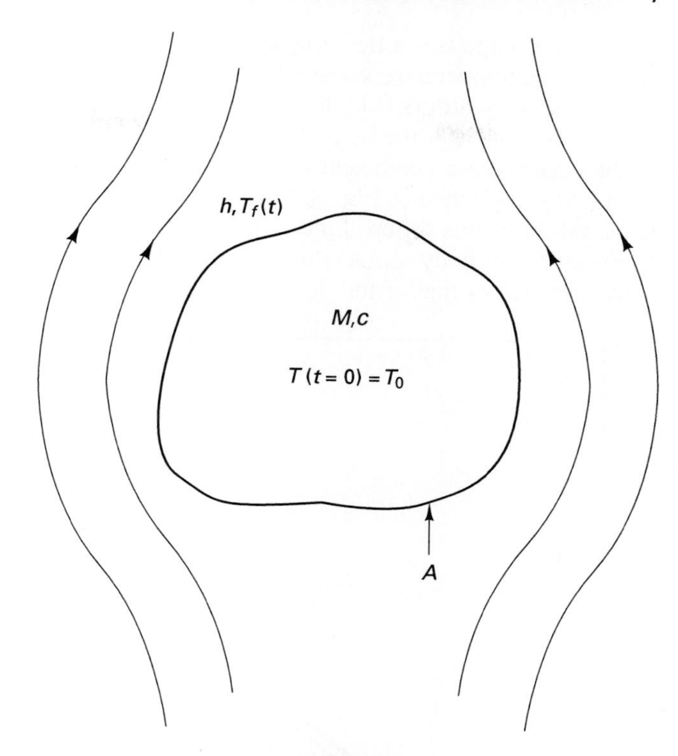


**Figure 3.6**

**Solution**

If the boundary of the body is viewed as a fixed control volume, one can apply the first law of thermodynamics (energy balance) to that control volume to obtain

$$Mc\frac{dT}{dt} = hA(T_f - T) \tag{3.58}$$

Combining eqs. (3.57) and (3.58) yields

$$\frac{dT}{dt} + \frac{hA}{Mc}T = \frac{hA}{Mc}(T_0 + Bt) \tag{3.59}$$

Initially, the temperature of the body equals $T_0$:

$$t = 0: \qquad T = T_0 \tag{3.60}$$

The temperature history of the body results from solving the first-order linear differential equation (3.59) subject to initial condition (3.60):

$$T = T_0 + Bt + \frac{B}{hA/Mc}\left(e^{-(hA/Mc)t} - 1\right) \tag{3.61}$$

Note that at large times the temperature variation with time in the body is linear, like the temperature variation of the fluid. Actually, the temperature variation in the body differs from that of the fluid by a constant that accounts for the thermal resistance at the body–fluid interface, reflected by the finite magnitude of the heat transfer coefficient $h$. The temperature history given analytically in eq. (3.61) is graphed in Fig. 3.7. The effect of the group $hA/Mc$ diminishes at large values of this group. Figure 3.7 also indicates that for large heat transfer coefficients the body temperature varies with time in a linear manner, practically identical to that of the fluid temperature.

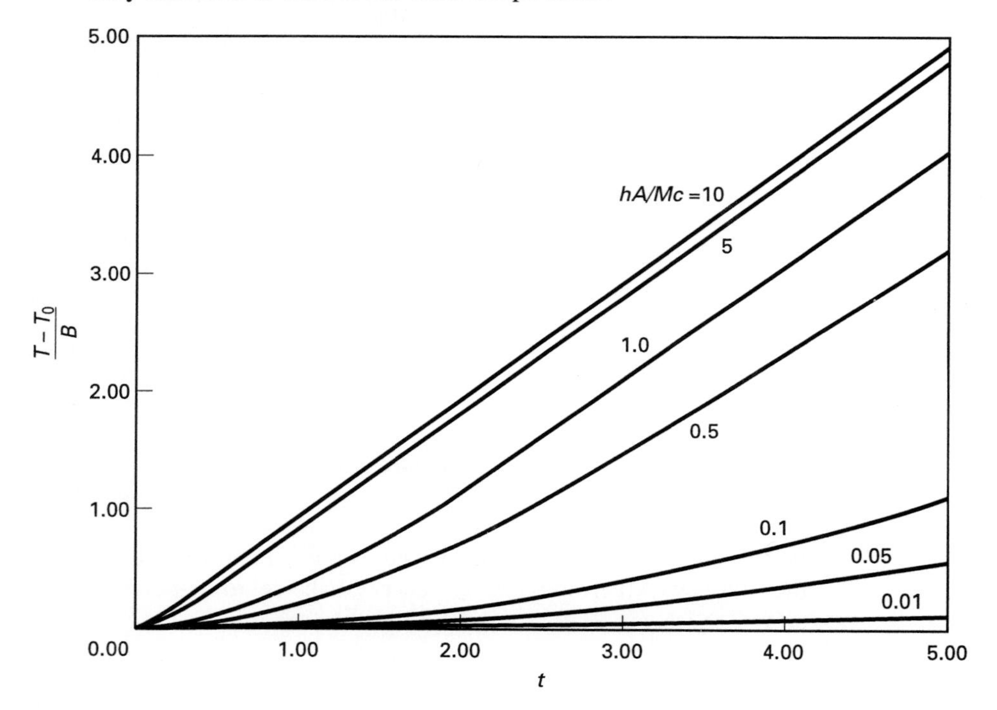


**Figure 3.7** Temperature history of the body.

## 3.3 ONE-DIMENSIONAL HEAT TRANSFER IN FINS

Fins are commonly used devices that serve to extend a heat transfer surface. A heat transfer surface with fins attached to it is called an *extended surface*. Naturally, a considerably larger amount of heat can be transferred from or to an extended surface at a given time period than from or to a bare surface (without the fins attached to it). In what follows, an analysis is presented for the process of one-dimensional conduction heat transfer in a single fin [1–5]. The range of validity of the one-dimensional conduction approximation is defined at the end of this section.

Consider the schematic of a general fin assembly presented in Fig. 3.8a. This fin has variable cross section $S(x)$, where $x$ measures the distance from the heat transfer surface on which the fin is attached or the distance from the base of the fin. In addition, it is assumed that the fin is immersed in a moving fluid of temperature $T_\infty$ away from the fin surface. The average heat transfer coefficient at the fin surface is denoted by $h$. It is also assumed that the fin material generates heat of volumetric rate $\dot{Q}'''_{\text{gen}}$ and that it exchanges heat with the environment by radiation at a rate $\dot{Q}''_{\text{rad}}$.

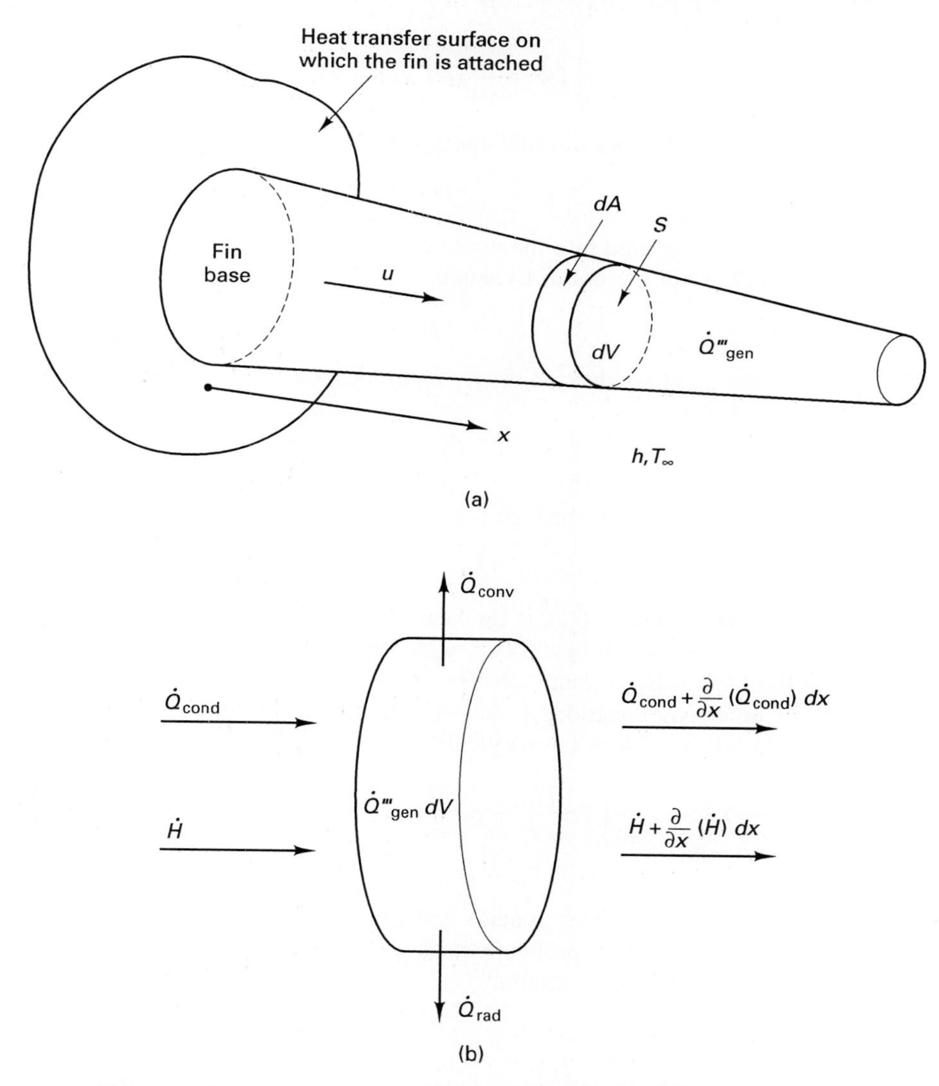


**Figure 3.8** (a) Long fin of arbitrary geometry. (b) Differential control volume.

Finally, for the sake of generality, let us also assume that the fin moves with respect to the surface on which it is attached with constant velocity $u$. This last assumption is useful when extrusion processes (such as of a wire or metal sheet from a die) need to be studied.

If an energy balance is applied to the fixed control volume shown in Fig. 3.8a and enlarged in Fig. 3.8b, we obtain

$$\frac{\partial}{\partial t}(U) = \dot{Q}_{\text{cond}} + \dot{H} + \dot{Q}'''_{\text{gen}}\, dV - \left[\dot{Q}_{\text{cond}} + \frac{\partial}{\partial x}(\dot{Q}_{\text{cond}})\, dx\right]$$
$$- \left[\dot{H} + \frac{\partial}{\partial x}(\dot{H})\, dx\right] - \dot{Q}_{\text{conv}} - \dot{Q}''_{\text{rad}}\, dA \qquad (3.62)$$

In eq. (3.62), $U$ is the internal energy of the control volume:

$$U = \rho\, dV\, cT \qquad (3.63)$$

$H$ is the enthalpy rate into the control volume because of the "flow" of the fin material in and out of the control volume:

$$\dot{H} = \rho u S c T \qquad (3.64)$$

$\dot{Q}_{\text{cond}}$ is the axial conduction at any axial location $x$. Using Fourier's law, we have

$$\dot{Q}_{\text{cond}} = -kS\,\frac{\partial T}{\partial x} \qquad (3.65)$$

$\dot{Q}_{\text{conv}}$ is the convection through the lateral surface of the control volume:

$$\dot{Q}_{\text{conv}} = h\, dA(T - T_\infty) \qquad (3.66)$$

As mentioned earlier, $\dot{Q}'''_{\text{gen}}$ is the heat generation per unit volume and $\dot{Q}''_{\text{rad}}$ is the net radiative heat flux (for blackbody radiation, for example, $\dot{Q}''_{\text{rad}} = \sigma(T^4 - T_\infty^4)$, where $\sigma$ is the Stefan–Boltzmann constant [1–4]).

Combining equations (3.62)–(3.66) and assuming constant properties for the fin material ($\rho$, $c$, $k$ = const) yields

$$\rho c\,\frac{\partial}{\partial t}\left(\frac{dV}{dx}T\right) = k\,\frac{\partial}{\partial x}\left(S\,\frac{\partial T}{\partial x}\right) - \rho c\,\frac{\partial}{\partial x}(SuT) - \frac{dA}{dx}h(T - T_\infty) + \dot{Q}'''_{\text{gen}}\frac{dV}{dx} - \dot{Q}''_{\text{rad}}\frac{dA}{dx} \qquad (3.67)$$

Equation (3.67) is rather general and can be applied to solve a host of one-dimensional heat conduction problems in fins. A group of representative problems is considered in the following example.

**Example 3.2**

Obtain the temperature distribution and the heat transfer rate through the base of the four fins shown schematically in Fig. 3.9.

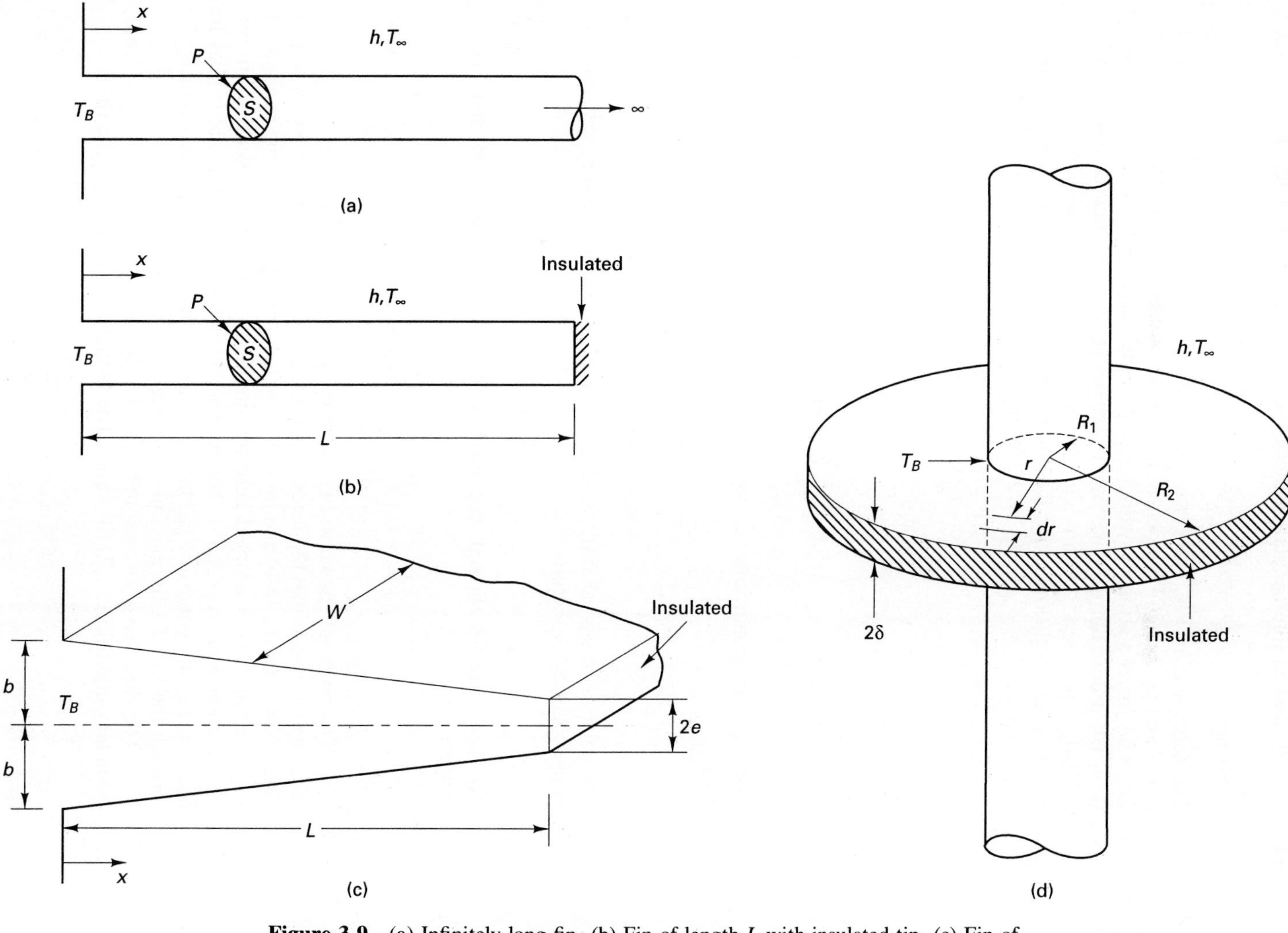


**Figure 3.9** (a) Infinitely long fin. (b) Fin of length $L$ with insulated tip. (c) Fin of trapezoidal cross section. (d) Circular fin.

**Solution**

The schematic in Fig. 3.9a shows an infinitely long fin of constant cross section $S$ and perimeter $P$ immersed in a fluid of temperature $T_\infty$ far away from the fin. The fin is stationary with respect to the surface on which it is attached ($u = 0$), it does not generate heat internally ($\dot{Q}'''_{gen} = 0$), and the radiation heat exchange between the fin and the environment is negligible ($\dot{Q}''_{rad} = 0$). Steady-state conditions exist. Taking the above into account and noticing that in this case $dV = S\,dx$, $dA = P\,dx$, the general fin equation (3.67) yields

$$\frac{d^2\Theta}{dx^2} - m^2\Theta = 0 \tag{3.68}$$

where

$$\Theta = T - T_\infty \tag{3.69}$$

$$m^2 = \frac{hP}{kS} \tag{3.70}$$

The boundary conditions necessary to complete the formulation of the problem are

$$x = 0: \qquad \Theta = T_B - T_\infty = \Theta_B \tag{3.71}$$

$$x \to \infty: \qquad \Theta \to 0 \tag{3.72}$$

The solution to eqs. (3.68), (3.71), and (3.72) reads (for details, see Appendix A and references therein)

$$\Theta = \Theta_B e^{-mx} \tag{3.73}$$

The heat transfer through the fin base can be calculated with the help of Fourier's law,

$$\dot{q}_B = -kS\left(\frac{d\Theta}{dx}\right)_{x=0} = (kSPh)^{1/2}\,\Theta_B \tag{3.74}$$

The problem shown in Fig. 3.9b is identical to the problem of Fig. 3.9a with the exception that the fin is now finite, of length $L$, and that its tip is insulated. Note that the insulated tip condition is an approximation that may not always be valid. It is adopted here for simplicity. An improvement of this approximation was proposed by Harper and Brown [6]. Experimental studies by Sparrow [7] in cylindrical pin fins have shown that neglecting the heat transfer through the tip of a fin may introduce significant errors.

The heat transfer through the fin is still governed by eq. (3.68). Boundary condition (3.71) is valid. The boundary condition at the tip is

$$x = L: \qquad \frac{d\Theta}{dx} = 0 \tag{3.75}$$

The final expressions for the temperature distribution and heat flux through the base of the fin read

$$\Theta = \Theta_B \frac{\cosh[m(L - x)]}{\cosh(mL)} \tag{3.76}$$

$$\dot{q}_B = \Theta_B (PhkS)^{1/2} \tanh(mL) \tag{3.77}$$

Figure 3.9c shows a fin of trapezoidal cross section. The width of the fin is denoted by $W$. The major assumptions made above in connection with the geometries shown in Fig. 3.9a and b hold here as well. The main difference is that the cross section of the fin is variable ($S = 2W\{e + [(L - x)(b - e)]/L\}$, $P \approx 2W$ if $W >> b$). The boundary conditions are the same as those in the configuration shown in Fig. 3.9b. Based on the above, the general equation (3.67) when applied to the trapezoidal fin of Fig. 3.9c reduces to

$$\frac{d^2\Theta}{dx_*^2} + \frac{1}{x_*}\frac{d\Theta}{dx_*} - B^2 \frac{\Theta}{x_*} = 0 \tag{3.78}$$

where

$$x_* = e + \frac{b - e}{L}(L - x) \tag{3.79}$$

$$B^2 = \frac{h/k}{[(b - e)/L]^2} \tag{3.80}$$

The solution of eq. (3.80) can be obtained with the help of Example A.1 in Appendix A. The final results for the temperature distribution and the heat flux through the base of the fin are

$$\frac{\Theta}{\Theta_B} = \frac{I_0(2Bx_*^{1/2})K_1(2Be^{1/2}) + K_0(2Bx_*^{1/2})I_1(2Be^{1/2})}{K_1(2Be^{1/2})I_0(2Bb^{1/2}) + K_0(2Bb^{1/2})I_1(2Be^{1/2})} \tag{3.81}$$

$$\dot{q} = -k(2bW)\left(\frac{d\Theta}{dx}\right)_{x=0}$$

$$= 2Bkb^{1/2}(b - e)\frac{W}{L}\Theta_B \frac{I_1(2Bb^{1/2})K_1(2Be^{1/2}) - K_1(2Bb^{1/2})I_1(2Be^{1/2})}{K_1(2Be^{1/2})I_0(2Bb^{1/2}) + K_0(2Bb^{1/2})I_1(2Be^{1/2})} \tag{3.82}$$

where $I_0$ and $I_1$ are the modified Bessel functions of the first kind and zeroth order and of the first kind and first order, respectively. Similarly $K_0$ and $K_1$ are the modified Bessel functions of the second kind and zeroth order and of the second kind and first order, respectively.

Figure 9d shows a circular fin of rectangular profile of thickness $2\delta$ attached on a pipe. In this case, the cross-sectional area is $S = 4\pi r\delta$ and the elementary area in contact with the fluid accounting for both the top and bottom surfaces of the fin $dA = 4\pi r\, dr$. In addition, steady state is assumed, the fin is motionless, and there is no heat generation or radiation effects. Based on the above, eq. (3.67) becomes

$$\frac{d^2\Theta}{dr^2} + \frac{1}{r}\frac{d\Theta}{dr} - n^2\Theta = 0 \tag{3.83}$$

where $r$ replaced $x$ to denote the fact that the conduction heat transfer occurs in the radial direction and where

$$n = \left(\frac{h}{k\delta}\right)^{1/2} \tag{3.84}$$

The corresponding boundary conditions are

$$r = R_1: \qquad \Theta = T_B - T_\infty = \Theta_B \tag{3.85}$$

$$r = R_2: \qquad \frac{d\Theta}{dr} = 0 \tag{3.86}$$

The solution for the temperature is again obtained with the help of Appendix A in terms of the modified Bessel functions:

$$\frac{\Theta}{\Theta_B} = \frac{I_0(nr)K_1(nR_2) + I_1(nR_2)K_0(nr)}{I_0(nR_1)K_1(nR_2) + I_1(nR_2)K_0(nR_1)} \tag{3.87}$$

The heat transfer rate at the fin base is obtained by invoking Fourier's law:

$$\begin{aligned} \dot{q} &= -4\pi k R_1 \delta\left(\frac{d\Theta}{dr}\right)_{r=R_1} \\ &= -4\pi R_1 (k\,\delta h)^{1/2} \frac{I_1(nR_1)K_1(nR_2) - I_1(nR_2)K_1(nR_1)}{I_0(nR_1)K_1(nR_2) + I_1(nR_2)K_0(nR_1)} \end{aligned} \tag{3.88}$$

## Validity of the Unidirectional Conduction Approximation in Fins

The assumption of unidirectional heat conduction in a fin is not always valid. To define the domain of validity of this assumption, consider a fin of length $L$ (in the $x$-direction). Assume that heat is removed from the fin convectively ($h$, $T_\infty$) and that the fin base temperature is denoted by $T_B$. The heat flux by conduction in the lengthwise direction of the fin is approximated by

$$\dot{q}_x'' = O\left(k\frac{T_B - T_\infty}{L}\right) \tag{3.89}$$

where $k$ is the conductivity of the fin material. The heat flux by conduction in the $y$-direction (transversal direction) is set equal to the convective heat flux removal through the surface of the fin and is approximated by

$$\dot{q}_y'' = O(h(T_B - T_\infty)) \tag{3.90}$$

For the unidirectional heat conduction assumption to be valid $\dot{q}_x'' >> \dot{q}_y''$, or taking into account eqs. (3.89) and (3.90), we obtain

$$\text{Bi} << O(1) \tag{3.91}$$

where Bi is the Biot number defined as

$$\text{Bi} = \frac{hL}{k} \tag{3.92}$$

Before applying the unidirectional conduction approximation to study heat transfer in fins, the validity of the criterion (3.91) needs to be verified.

## PROBLEMS

**3.1.** In a laboratory experiment an electric current is passed through a slab of thickness $2\delta$. As a result, heat generation ($\dot{q}'''$) takes place in the slab. Modeling the slab as a plane wall and assuming that the heat conduction process in the slab is unidirectional, show that if the thermal conductivity of the slab is constant, the steady-state energy equation describing the conduction phenomenon in the slab is

$$\frac{d^2 T}{dx^2} + \frac{\dot{q}'''}{k} = 0$$

Next, obtain the temperature distribution in the slab if it is cooled on both sides by a fluid of temperature $T_c$. The heat transfer coefficient is denoted by $h_c$. What is the heat flux at each side of the slab?

**3.2.** Re-solve Problem 3.1 allowing for the thermal conductivity of the slab to be a function of $x$.

**3.3.** A layer of electrical insulation of thickness $t$ is applied on both sides of the slab considered in Problem 3.1 (Fig. P3.3). The purpose of this layer is to satisfy safety consider-

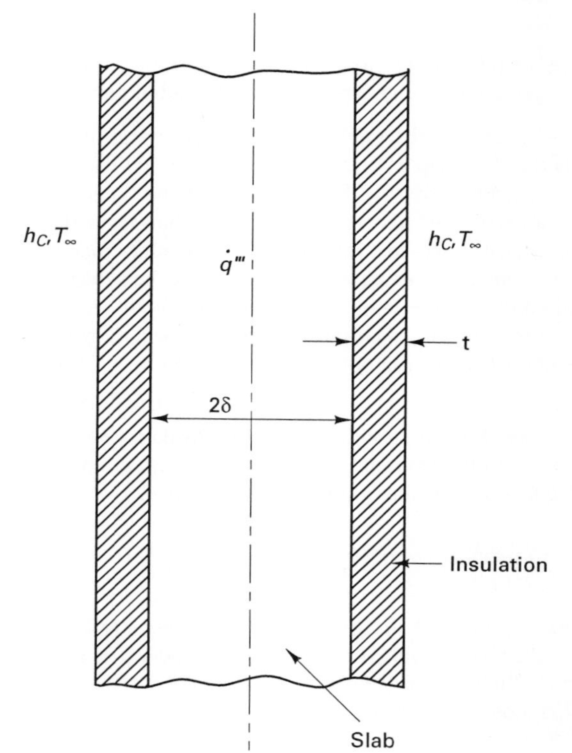


**Figure P3.3** Cross section of an electrically insulated slab.

ations related to the danger stemming from the fact that an electric current is passing through the slab. It is still desired that the slab is cooled by the fluid described in Problem 3.1, as shown in Fig. P3.3. Therefore, the insulating material is chosen such that it conducts heat well but does not conduct electricity. The thermal conductivity of the electrical insulation is denoted by $k_e$. Obtain the temperature distribution in the slab and in the insulation layer as well as the heat flux through the insulation surface.

**3.4.** In a heat exchange apparatus, the wall of a pipe of inner radius $R_1$ and outer radius $R_2$ is heated electrically (by passing an electric current through it). As a result, heat ($\dot{q}'''$) is generated within the pipe. This heat is used to warm up a fluid flowing through the pipe. The bulk temperature of this fluid is denoted by $T_1$ and the corresponding heat transfer coefficient by $h_1$. At the same time, the air (of temperature $T_2$) surrounding the pipe is set into motion by natural convection. The natural convection heat transfer coefficient on the outside of the pipe is denoted by $h_2$. Prove that for the problem described above, if the thermal conductivity of the pipe material ($k$) is constant and for steady-state conditions, the general heat conduction equation reads

$$\frac{1}{r}\frac{d}{dr}\left(r\frac{dT}{dr}\right) + \frac{\dot{q}'''}{k} = 0$$

After obtaining the corresponding convection boundary conditions (at $r = R_1$ and $r = R_2$), solve the conduction model for the temperature field in the pipe wall. What is the heat transfer rate at the inner surface of the pipe? What is the temperature distribution in the pipe and the heat transfer rate at $r = R_1$ in the limit of perfect thermal contact at $r = R_1$ and $r = R_2$ ($h_1, h_2 \rightarrow \infty$)?

**3.5.** Re-solve Problem 3.4 after relaxing the assumption of constant thermal conductivity. Allow for the thermal conductivity to be dependent on the radial coordinate, $k = k(r)$.

**3.6.** For safety considerations, a layer of thermally conducting electrical insulation is placed around the pipe described in Problem 3.4, as shown in Fig. P3.6. The thermal conductivity of this insulation is denoted by $k_e$. The remainder of the "scenario" outlined in Problem 3.4 stays the same. Obtain the temperature distribution in the pipe wall as well as in the insulation. What is the heat transfer rate at the inner pipe wall ($r = R_1$)? What are the above-mentioned temperature distribution and heat transfer rate in the limit of perfect thermal contact at the inner pipe surface ($h_1 \rightarrow \infty$) as well as the outer insulation surface ($h_2 \rightarrow \infty$)? If the maximum allowable insulation temperature (to avoid burnout) is denoted by $T_m$, can you obtain an upper limit for $\dot{q}'''$ such that burnout is avoided?

**3.7.** The schematic shown in Fig. P3.7 represents a blood-perfused tissue layer with negligible metabolic heat production. Assuming steady-state unidirectional conduction in this tissue, the bioheat equation explained in Chapter 2 yields

$$k_t\frac{d^2T}{dx^2} - \dot{m}_b'''c_{pb}(T - T_A) = 0$$

The temperatures at the two sides of the tissue ($T_0$ and $T_L$) are known. Note that the conduction equation for the tissue resembles in form the conduction equation for a fin discussed in this chapter. Obtain the temperature distribution in the tissue and the heat transfer rates through the sides of the tissue.

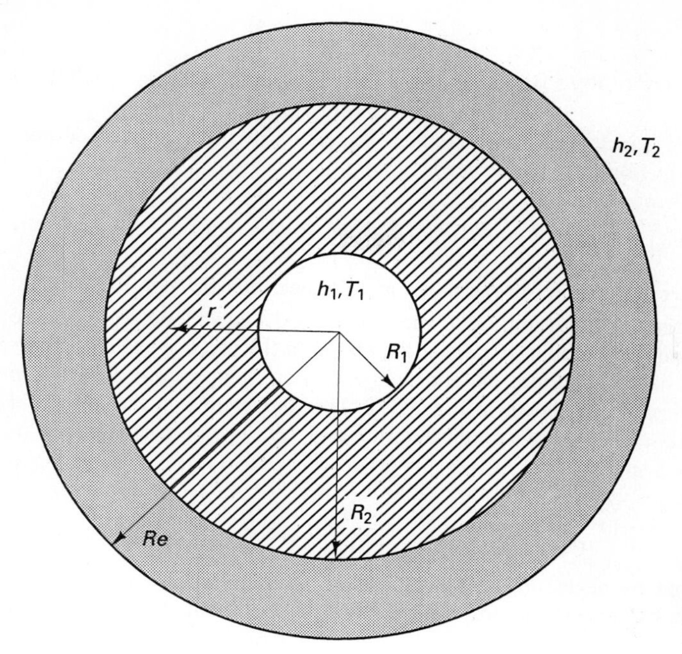


**Figure P3.6** Cross section of a pipe with an electrical insulation layer.

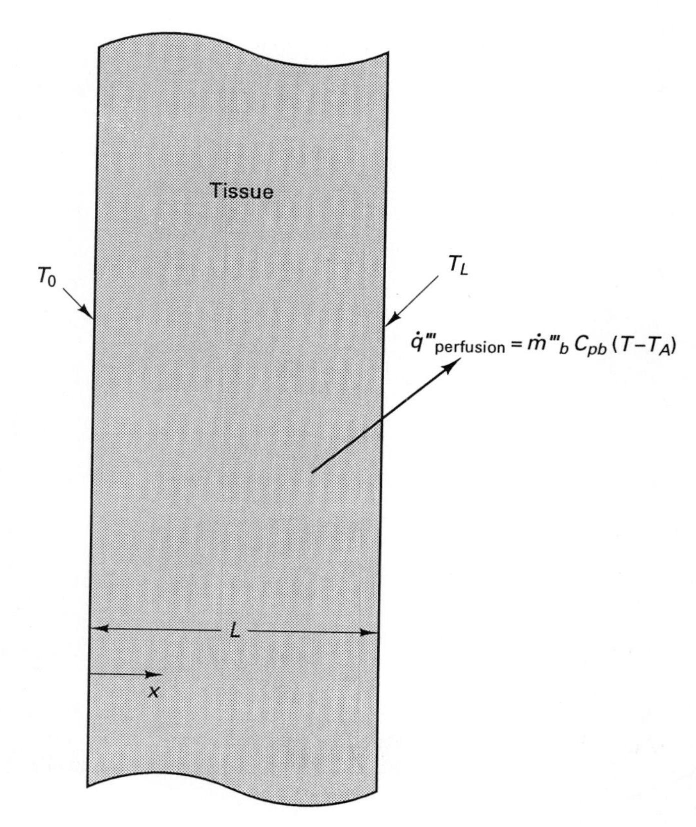


**Figure P3.7** Schematic of a tissue layer.

**3.8.** Reconsider Problem 3.7, including metabolic heat generation in addition to blood perfusion. The bioheat equation in this case reads

$$k_t \frac{d^2 T}{dx^2} - \dot{m}_b''' c_{pb}(T - T_A) + \dot{q}''' = 0$$

Obtain the temperature distribution in the tissue and the heat flux through its sides.

**3.9.** Heat transfer in the tissue of extremities of the human body (arms and legs) can be treated approximately as conduction in a cylindrical body in the presence of blood perfusion and metabolic heat generation. Obtain the steady-state temperature distribution in the hollow cylindrical tissue shown in Fig. P3.9. The temperature at the inner and outer surfaces of the tissue is known as indicated in Fig. P3.9. The bioheat equation in this case reduces to

$$\frac{k_t}{r} \frac{d}{dr}\left( r \frac{dT}{dr} \right) - \dot{m}_b''' c_{pb}(T - T_A) + \dot{q}''' = 0$$

Solve the problem first by neglecting the metabolic heat generation ($\dot{q}''' = 0$). Next, improve the solution by accounting for the presence of metabolic heat generation ($\dot{q}''' \neq 0$).

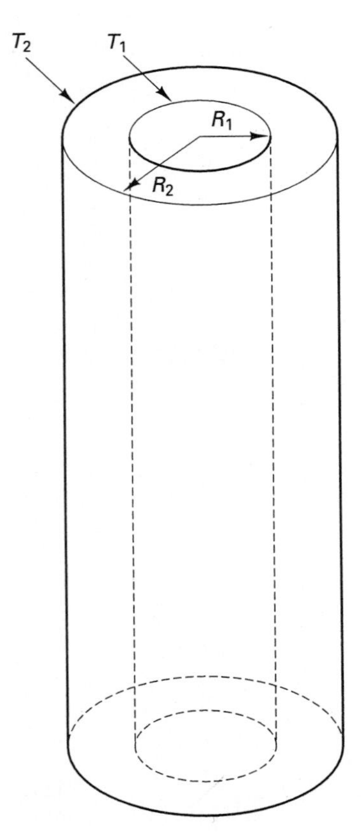


**Figure P3.9** Schematic of a cylindrical body modeling a human arm or leg.

**3.10.** Figure P3.10 shows the cross section of a system of two tissues modeled as concentrical hollow cylinders. Neglecting the metabolic heat generation effect, obtain the steady-state temperature distribution in the tissues. The bioheat equation for this case is

$$k_i \frac{1}{r}\frac{d}{dr}\left(r\frac{dT_i}{dr}\right) - \dot{m}'''_{bi}c_{pbi}(T - T_A) = 0$$

where $i = 1$ denotes the inner tissue and $i = 2$ denotes the outer tissue. What is the heat flux at $r = R$?

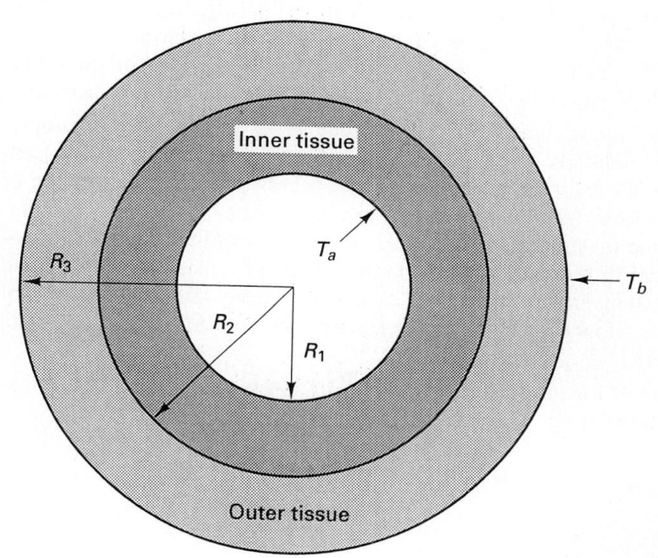


**Figure P3.10**

**3.11.** In the extrusion process shown in Fig. P3.11, the wire that is produced can be modeled as a very long fin moving with velocity $U$. The thermophysical properties of the wire material are known. Cooling of the wire is attained by forced convection ($h$, $T_\infty$). The

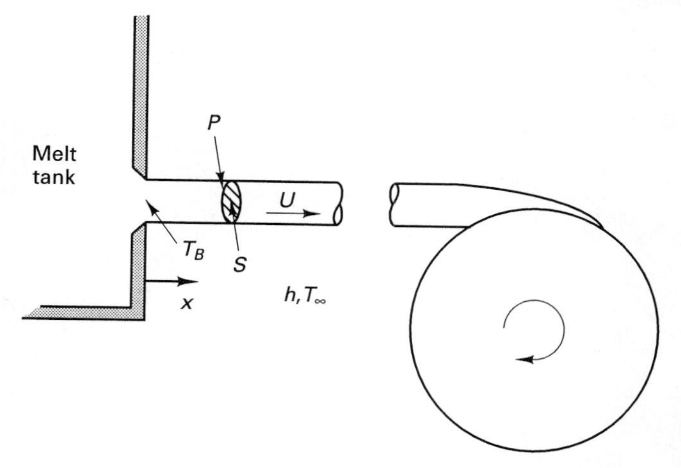


**Figure P3.11** Schematic of the extrusion process.

temperature of the wire at the exit from the melt tank can be estimated and is denoted by $T_B$. What is the temperature distribution in the wire, and what is the overall heat transfer rate through the lateral surface of the wire?

**3.12.** Re-solve Problem 3.11 accounting for the effect of blackbody radiation, $\dot{Q}''_{\text{rad}} = \sigma(T^4 - T_\infty^4)$, in addition to convection in the cooling process of the wire.

**3.13.** In a quenching process a red-hot metal billet of diameter $D$ and density $\rho$ is immersed into a large bath of coolant of constant temperature $T_c$. The initial temperature of the billet is denoted by $T_i$. It is observed that boiling takes place around the billet for a certain period of time $0 < t < t_B$. The heat transfer coefficient during the boiling process is denoted by $h_B$. After that, the cooling occurs mainly by natural convection characterized by a heat transfer coefficient $h_c$. Obtain the temperature history of the billet. The billet is assumed to be spatially isothermal, due to its high thermal conductivity.

**3.14.** During the startup of an electric heater, electric current passes through a large number of identical thermally conductive wires generating internal heat at a rate $\dot{q}'''W/m^3$ in each wire. The cooling of each wire, at the same time, occurs by natural convection of the surrounding air, characterized by a heat transfer coefficient $h$. The initial temperature of each wire is $T_i$ and the air temperature is $T_\infty$. Each wire can be modeled as a spatially isothermal cylinder of diameter $D$ and length $L$ made of a material with known properties. Obtain the temperature history of each wire during the transient startup process described above. How does the time it takes for each wire to reach steady state depend on $\dot{q}'''$ and on $h$?

**3.15.** Consider the conical fin of known material of Fig. P3.15. Obtain its temperature distribution and the heat transfer rate through its base.

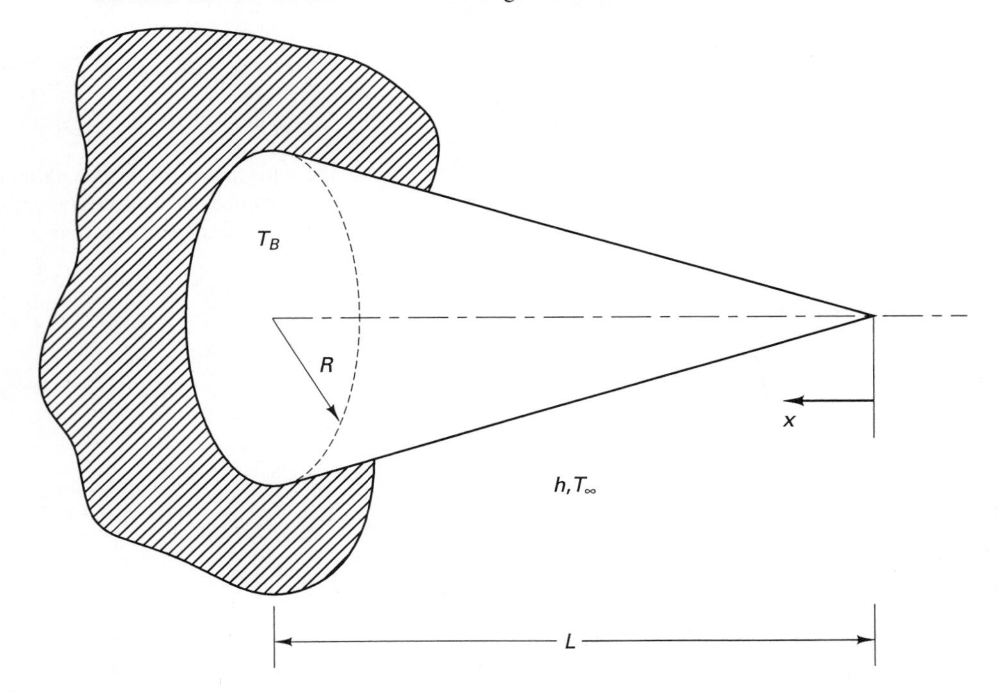


**Figure P3.15** Conical fin.

**3.16.** Consider the rectangular fin of triangular profile of known material shown in Fig. P3.16. Obtain the temperature distribution in the fin and the heat transfer rate through its base.

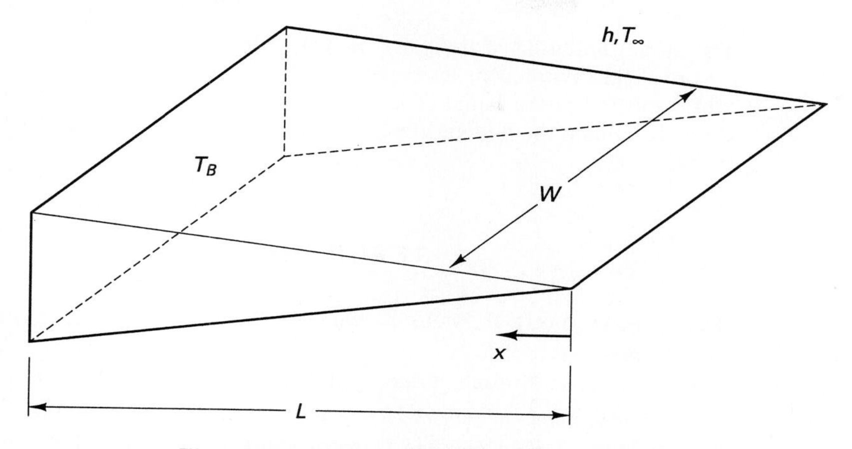


**Figure P3.16** Fin of triangular cross section.

**3.17.** A circular fin (Fig. P3.17) with thickness and heat transfer coefficient varying radially is to be considered. The half-thickness of the fin, $y$, is given by

$$y = \delta\left(\frac{r}{r_i}\right)^{-n}$$

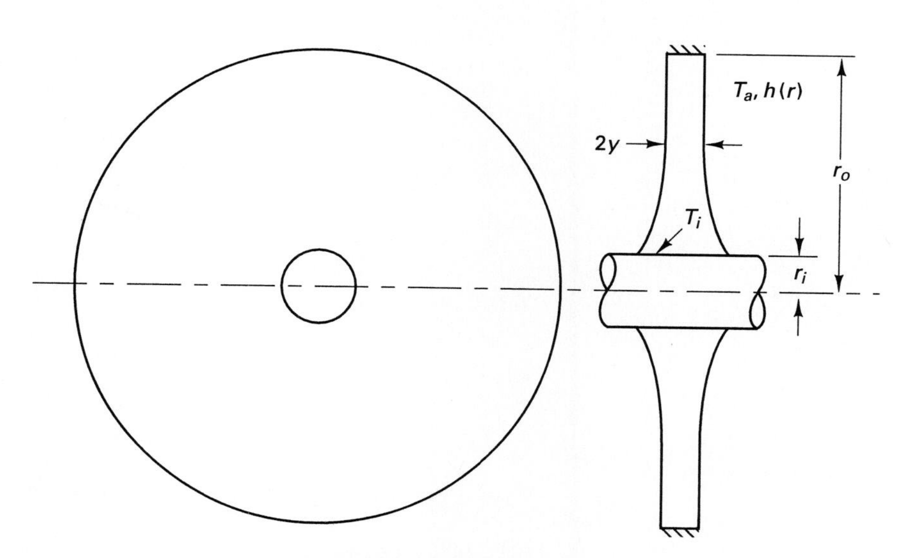


**Figure P3.17** Circular fin.

and the local heat transfer coefficient is given by

$$h = h_0\left(\frac{r}{r_i}\right)^P$$

The fin has an insulated tip at $r = r_0$ and its base temperature is maintained at $T_i$ (at $r = r_i$). The environment temperature is $T_a$. It is assumed that the fin is sufficiently thin compared to the radial coordinate, $r$, and all the thermophysical properties are constant. Obtain the fin temperature distribution and the heat transfer rate through the fin base.

## REFERENCES


1. F. P. INCROPERA AND D. P. DEWITT, *Fundamentals of Heat and Mass Transfer*, 3rd ed., Wiley, New York, 1990.
2. J. P. HOLMAN, *Heat Transfer*, 6th ed., McGraw-Hill, New York, 1986.
3. A. J. CHAPMAN, *Fundamentals of Heat Transfer*, Macmillan, New York, 1987.
4. J. H. LIENHARD, *A Heat Transfer Textbook*, 2nd ed., Prentice Hall, Englewood Cliffs, NJ, 1987.
5. D. O. KERN AND A. D. KRAUS, *Extended Surface Heat Transfer*, McGraw-Hill, New York, 1972.
6. W. B. HARPER AND D. R. BROWN, *Mathematical Equations for Heat Conduction in Fins of Air-Cooled Engines*, NACA Report 158, 1922.
7. E. M. SPARROW, Heat transfer in fluid flows which do not follow the contour of boundary walls, *J. Heat Transfer*, Vol. 110, pp. 1145–1153, 1988.

CHAPTER 4

# MULTIDIMENSIONAL STEADY CONDUCTION: THE METHOD OF SEPARATION OF VARIABLES

In this chapter the phenomenon of heat conduction occurring simultaneously in more than one direction within a body is considered. In particular, two-dimensional steady conduction is studied in detail in three basic coordinate systems commonly used in thermal engineering: Cartesian, cylindrical, and spherical. Three-dimensional steady conduction is also discussed, but in less detail, since the fundamentals of the solution process for three-dimensional conduction follow as a direct extension of the solution process for two-dimensional conduction. This chapter features the method of separation of variables [1–3], a powerful and popular tool in solving multidirectional heat conduction problems.

## 4.1 TWO-DIMENSIONAL CONDUCTION IN CARTESIAN COORDINATES

The fundamental principles needed for the solution of problems of this type are highlighted next, in Example 4.1.

### Example 4.1

Consider the arrangement shown schematically in Fig. 4.1a. A long (in the $z$-direction) conductive bar of rectangular cross section ($H \times b$) is attached to a hot surface with one of its sides. A cold fluid of temperature $T_\infty$ flows around the three remaining sides of the bar. Based on the above, the bar plays the role of a *fin,* augmenting the heat transfer from the hot surface to the fluid. The heat

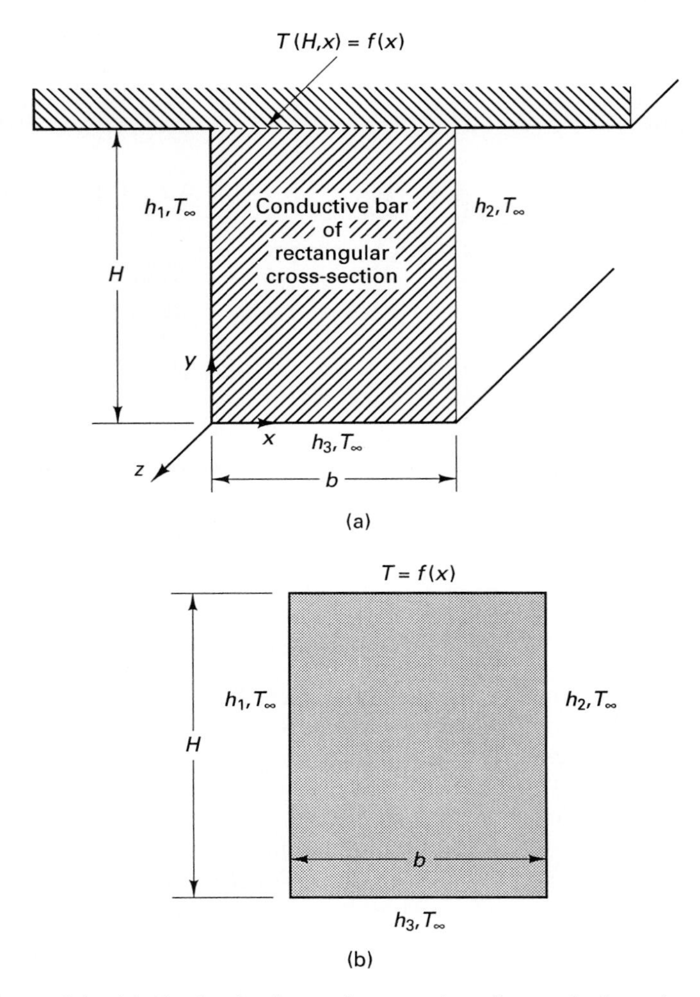


**Figure 4.1** (a) Conductive bar acting as a short fin attached to a hot surface. (b) Cross section of the bar.

transfer process in the fin is assumed to be two-dimensional (in the $x$ and $y$ directions); that is, it is assumed that the temperature changes in the $z$-direction are negligible. This assumption is justified from the physics of the problem and from the fact that the length of the fin in the $z$-direction is very large compared to both $H$ and $b$, making *end effects* unimportant. The heat transfer coefficients along the three sides of the fin are constant and are denoted by $h_1$, $h_2$, and $h_3$, as shown in Fig. 4.1a. It is also assumed that the temperature at $y = H$ is known and it is denoted in general by $f(x)$. In reality, $f(x)$ can be measured by, say, installing thermocouples at the base of the fin, where the rectangular conductive bar joins the hot surface. Of interest is the heat flux through the base of the fin as well as the temperature distribution within the fin. To model the problem mathematically, it is enough to consider the cross section in the $x$–$y$ plane of the rectangular conductive bar, shown in Fig. 4.1b. The conduction equation (2.23) for the two-dimensional steady-state problem outlined above reduces to

$$\frac{\partial^2 T}{\partial x^2} + \frac{\partial^2 T}{\partial y^2} = 0 \tag{4.1}$$

The corresponding boundary conditions are

$$\text{at } x = 0: \quad k\frac{\partial T}{\partial x} = h_1(T - T_\infty) \tag{4.2}$$

$$\text{at } x = b: \quad -k\frac{\partial T}{\partial x} = h_2(T - T_\infty) \tag{4.3}$$

$$\text{at } y = 0: \quad k\frac{\partial T}{\partial y} = h_3(T - T_\infty) \tag{4.4}$$

$$\text{at } y = H: \quad T = f(x) \tag{4.5}$$

The system of equations (4.1)–(4.5) completely describes the conduction phenomenon of interest.

Instead of proceeding with the solution of this system of equations with the method of separation of variables, it is pedagogically worthwhile to define a number of limiting cases of the problem and to study these cases. This approach reduces the algebraic complexities and allows for a clear presentation of the methodology. The three limiting cases are shown in Fig. 4.2 and will be treated sequentially.

### Case (a)

This case represents the limit of $H \to \infty$ (a very "tall" rectangular geometry) and $h_1 \to \infty$, $h_2 \to \infty$, $h_3 \to \infty$ (the fact that the heat transfer coefficients are very large implies "perfect" thermal contact and therefore allows us to set the surface temperature of the body equal to the fluid temperature in contact with body, $T_\infty$).

The governing equation of the conduction problem is still eq. (4.1). The corresponding boundary conditions according to the Cartesian coordinate system of Fig. 4.2 are:

$$\text{at } x = 0: \quad T = T_\infty \tag{4.6}$$

$$\text{at } x = b: \quad T = T_\infty \tag{4.7}$$

$$\text{at } y = 0: \quad T = f(x) \tag{4.8}$$

$$\text{at } y \to \infty: \quad T = T_\infty \tag{4.9}$$

Note that when one deals with conduction in a rectangle, one has a choice of four "corners" in which to place the Cartesian coordinate system (in the

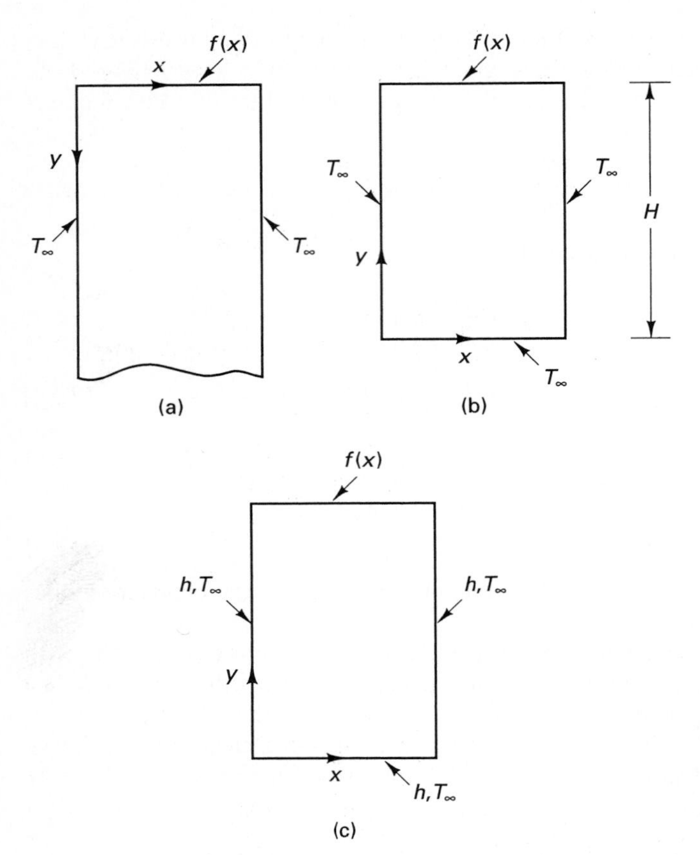


**Figure 4.2** (a) Bar of infinite length with isothermal sides. (b) Bar of finite length with three isothermal surfaces. (c) Bar of finite length in contact with a convecting fluid.

present case we have a choice of two corners because the body is semi-infinite in the $y$- direction). Appropriate placement of the coordinate system may result in simplifications of the algebra required to obtain the solution. At the end of this example we discuss placement of the coordinate system.

The set of equations (4.1) and (4.6)–(4.9) completely defines the problem of Fig. 4.2a. For the method of separation of variables to work in two-dimensional conduction, it is required that the governing equation as well as no less than *three* out of the four boundary conditions are *homogeneous* (recall the definition of the homogeneous boundary condition in Chapter 2). The governing equation (4.1) is homogeneous. However, none of the four conditions (4.6)–(4.9) is homogeneous. Note, for example, that $T_\infty$ is a known constant, in general different than zero and of *zeroth* degree in the unknown function (temperature) and its derivatives. The definition of a homogeneous boundary condition requires that all terms of a boundary condition other than zero itself are of the first degree in the temperature ($T$) and its derivatives. Fortunately, in the present problem a variable transformation is possible that renders three boundary conditions [(4.6), (4.7), and (4.9)] homogeneous. This simple transformation is common in heat conduction problems and is recommended as a first attempt if one is interested in transforming nonhomogeneous boundary conditions to homogeneous boundary conditions. The transformation consists of defining a new variable

$$\Theta = T - T_\infty \tag{4.10}$$

and expressing the model of the problem (governing equation and boundary conditions) in terms of the new variable. Performing this task yields

$$\frac{\partial^2 \Theta}{\partial x^2} + \frac{\partial^2 \Theta}{\partial y^2} = 0 \tag{4.11}$$

$$x = 0: \quad \Theta = 0 \tag{4.12}$$

$$x = b: \quad \Theta = 0 \tag{4.13}$$

$$y = 0: \quad \Theta = f(x) - T_\infty = \phi(x) \tag{4.14}$$

$$y \to \infty: \quad \Theta = 0 \tag{4.15}$$

Indeed, after the transformation the governing equation remained homogeneous and three boundary conditions [(4.12) (4.13), and (4.15)] became homogeneous. The method of separation of variables can now be applied directly to the system of eqs. (4.11)–(4.15).

In the first step of the application of the method it is assumed that the unknown temperature ($\Theta$) can be written as the product of two functions, each of which depends on only one of the independent variables $x$ and $y$, that is,

$$\Theta = X(x)Y(y) \tag{4.16}$$

Substituting eq. (4.16) into eq. (4.11) and dividing both sides by the product $X(x)Y(y)$ yields

$$\underbrace{\frac{X''}{X}}_{-\alpha^2} + \underbrace{\frac{Y''}{Y}}_{\alpha^2} = 0 \tag{4.17}$$

The primes in eq. (4.17) denote differentiation with respect to the pertinent independent variable ($x$ or $y$). Observing that the first term in eq. (4.17) is a function of $x$ only and the second term is a function of $y$ only we conclude that for eq. (4.17) to hold in general, each of these terms must equal a constant. Setting the first term equal to $-\alpha^2$ and the second term equal to $\alpha^2$, where $\alpha^2$ is a nonzero unknown real constant, yields two ordinary differential equations for $X$ and $Y$:

$$X'' + \alpha^2 X = 0 \tag{4.18}$$

$$Y'' - \alpha^2 Y = 0 \tag{4.19}$$

The decision to set the first term (which depends on $x$ only) in eq. (4.17) equal to $-\alpha^2$ and the second term (which depends on $y$ only) equal to $\alpha^2$ is *crucial* and it is not made arbitrarily. The problem is not solvable if the opposite choice is made. The reason behind this choice will be explained fully later in this example. At this point it will suffice to say that the solution of the ordinary differential equation corresponding to the direction in which both boundary conditions are homogeneous (usually termed the "homogeneous direction") needs to be obtained in terms of *orthogonal* functions for the method of separation of variables to work. In our example, $x$ is the homogeneous direction; therefore, the solution of eq. (4.18) needs to be obtained in terms of orthogonal functions in $x$. A general definition of the concept of orthogonality and orthogonal functions is given in Appendix C.

The general solutions to eqs. (4.18) and (4.19) are

$$X = C_1 \sin(\alpha x) + C_2 \cos(\alpha x) \tag{4.20}$$

$$Y = D_1 e^{\alpha y} + D_2 e^{-\alpha y} \tag{4.21}$$

A summary of solutions of second-order differential equations with constant coefficients such as eqs. (4.18) and (4.19) is presented in Appendix B.

Combining eqs. (4.16), (4.20), and (4.21) yields

$$\Theta = [C_1 \sin(\alpha x) + C_2 \cos(\alpha x)][D_1 e^{\alpha y} + D_2 e^{-\alpha y}] \tag{4.22}$$

The object now is to obtain the unknown constants $C_1$, $C_2$, $D_1$, $D_2$, and $\alpha$ in eq. (4.22). Applying boundary conditions (4.12) and (4.15) yields $C_2 = 0$ and $D_1 = 0$, respectively. Combining the product of the remaining constants into one constant ($C_1 D_2 = C$) yields

$$\Theta = C \sin(\alpha x) e^{-\alpha y} \tag{4.23}$$

Next, we apply boundary condition (4.13) to obtain

$$\sin(\alpha b) = 0 \tag{4.24}$$

or

$$\alpha_n = \frac{n\pi}{b} \qquad n = 1, 2, \ldots \tag{4.25}$$

All values obtained from eq. (4.25) satisfy eq. (4.24). These values are commonly termed *eigenvalues*. To account for this fact, in the semantics of the presentation $\alpha$ was replaced by $\alpha_n$, where each value of the subscript corresponds to a solution of eq. (4.24). In addition, we note that each value of $\alpha_n$ from eq. (4.25) yields a linearly independent solution for the temperature in eq. (4.23). Therefore, the general solution for the temperature is the linear superposition of all these partial linearly independent solutions and reads

$$\Theta = \sum_{n=1}^{\infty} C_n \sin\left(\frac{n\pi}{b} x\right) e^{-(n\pi/b)y} \tag{4.26}$$

Note that since $\sin(\alpha_n x)$ is an odd function of its argument, negative values of $n$ were not included in eq. (4.25). These values do not yield additional linearly independent solutions for the temperature expression. The last unknown constant to be obtained is $C_n$ in eq. (4.26). There is only *one* boundary condition left, eq. (4.14), while $C_n$ takes on an *infinite* number of values, one for each value of the index $n$. What makes the solution possible from this point on is the *orthogonality* property of the sine function (Appendix C). According to this property, the following is true:

$$\int_0^b \sin\left(\frac{n\pi x}{b}\right)\sin\left(\frac{m\pi x}{b}\right) dx = \begin{cases} 0 & n \neq m \\ \dfrac{b}{2} & n = m \end{cases} \tag{4.27}$$

Equation (4.27) can be verified by the reader simply by evaluating the integral. The orthogonality property will be used together with the remaining boundary condition, (4.14).

Applying this boundary condition [i.e., substituting eq. (4.26) into eq. (4.14)] yields

$$\phi(x) = \sum_{n=1}^{\infty} C_n \sin\left(\frac{n\pi x}{b}\right) \tag{4.28}$$

Equation (4.28) can be recognized as the Fourier sine series expansion of the function $\phi(x)$, where $C_n$ are the Fourier coefficients [1, 2, 4, 5]. To evaluate these coefficients, we multiply both sides of eq. (4.28) by sin $(m\pi x/b)$ and integrate both sides of the resulting equation over the range of the variable $x$ (from $x = 0$ to $x = b$):

$$\int_0^b \phi(x) \sin\left(\frac{m\pi}{b}x\right) dx = \sum_{n=1}^{\infty} C_n \int_0^b \sin\left(\frac{n\pi}{b}x\right)\sin\left(\frac{m\pi}{b}x\right) dx \tag{4.29}$$

Note that in eq. (4.29) the integration was inserted inside the summation sign. Such an operation is possible if the series converges uniformly and the terms of the series are continuous functions of the interval $(0, b)$ [1, 2, 4, 5].

Writing out the summation (mentally) on the right-hand side of eq. (4.29) and making use of eq. (4.27), we realize that all terms in the summation are *zero,* except for the $m$th term (i.e., the term obtained by setting $n = m$.) According to eq. (4.27), the integral in this term equals $b/2$. Based on the above, eq. (4.29) yields

$$\int_0^b \phi(x) \sin\left(\frac{m\pi}{b}x\right) dx = C_m \frac{b}{2} \tag{4.30}$$

Renaming $m$ to $n$ (after all, both $m$ and $n$ denote an arbitrary integer and $n = m$) and solving for $C_n$, we obtain

$$C_n = \frac{2\displaystyle\int_0^b \phi(x) \sin(n\pi x/b)\, dx}{b} \tag{4.31}$$

The general solution for the temperature is a direct result of substituting eq. (4.31) into eq. (4.26):

$$\Theta = \frac{2}{b}\sum_{n=1}^{\infty}\left[\int_0^b \phi(x)\sin\left(\frac{n\pi}{b}x\right)dx\right]\sin\left(\frac{n\pi}{b}x\right)e^{-(n\pi/b)y} \tag{4.32}$$

Recalling the definition of $\Theta$, eq. (4.10), and $\phi(x)$, eq. (4.14), yields the temperature distribution $T(x, y)$ in the configuration of Fig. 4.2a.

$$T(x, y) = T_\infty + \frac{2}{b}\sum_{n=1}^{\infty}\left[\int_0^b (f(x) - T_\infty)\sin\left(\frac{n\pi}{b}x\right)dx\right]\sin\left(\frac{n\pi}{b}x\right)e^{-(n\pi/b)y} \tag{4.33}$$

The local heat flux through the base of the fin is obtained by using Fourier's law together with eq. (4.33). Before proceeding with the evaluation of the local heat flux it is worth recalling an important theorem from calculus pertinent to the differentiation of an infinite series [4, 5]. According to this theorem, *if each term of an infinite series has a derivative and the series of derivatives converges uniformly, the series can be differentiated term by term.* Assuming that the expression (4.33) satisfies this theorem, we have

$$\dot{q}''(x) = -k\left(\frac{\partial T}{\partial y}\right)_{y=0} = \frac{2\pi}{b^2}k\sum_{n=1}^{\infty}\left[\int_0^b (f(x) - T_\infty)\sin\left(\frac{n\pi}{b}x\right)dx\right]n\sin\left(\frac{n\pi}{b}x\right) \tag{4.34}$$

The total heat transfer through the base of the fin cross section is given by

$$\dot{q}' = \int_0^b \dot{q}''(x)\,dx \tag{4.35}$$

Combining eqs. (4.34) and (4.35) yields

$$\dot{q}' = \frac{2\pi}{b^2}k\int_0^b\left\{\sum_{n=1}^{\infty}\left[\int_0^b (f(x) - T_\infty)\sin\left(\frac{n\pi}{b}x\right)dx\right]n\sin\left(\frac{n\pi}{b}x\right)\right\}dx \tag{4.36}$$

In the simple case where $f(x) = T_0$ (constant fin base temperature), eqs. (4.33) and (4.34), respectively, reduce to

$$T(x, y) = T_\infty + \frac{4}{\pi}(T_0 - T_\infty)\sum_{n=0}^{\infty}\frac{1}{2n+1}\sin\left[\frac{(2n+1)\pi x}{b}\right]e^{-[(2n+1)\pi y]/b} \tag{4.37}$$

$$\dot{q}''(x) = \frac{4k}{b}(T_0 - T_\infty)\sum_{n=0}^{\infty}\sin\left[\frac{(2n+1)\pi x}{b}\right] \tag{4.38}$$

Note that if the expression of the local heat flux [eq. (4.38)] is evaluated at $x = 0$ or $x = b$, it yields the value of zero. This is a physically unrealistic result: Since at the corners of the fin base (near $y = 0$) the temperature at $x = 0$ and $x = b$ changes from $T_\infty$ to $T_0$ over a very small distance $\Delta y$ ($\Delta y = 0$ in the present idealized problem), one would expect very large (infinite) values of the temperature gradient $\partial T/\partial y$ in the neighborhood of $y = 0$ for both sides, at $x = 0$ and $x = b$. The reason for the physically unrealistic result mentioned earlier is that the boundary conditions (4.12) and (4.13) were enforced first in the solution process [prior to condition (4.14)]. Hence the expression (4.37) yields a constant-temperature value for the entire sides of the fin at $x = 0$ and $x = b$, including the neighborhood of $y = 0$. This results in zero values of the temperature gradient, $\partial T/\partial y$ (and the heat flux) at $y = 0$ and $x = 0$ or $b$.

Before closing the discussion of the problem in Fig. 4.2a, it is necessary to examine the case of $\alpha^2 = 0$. Note that this case was excluded from the solution so far since it was assumed earlier that $\alpha^2 \neq 0$. If $\alpha^2 = 0$, the solutions to equations (4.18) and (4.19) are

$$X = A_1 x + B_1 \tag{4.39}$$

$$Y = C_1 y + D_1 \tag{4.40}$$

Therefore,

$$\Theta = (A_1 x + B_1)(C_1 y + D_1) \tag{4.41}$$

Boundary conditions (4.12), (4.13), and (4.15) require that $B_1 = 0$, $A_1 = 0$, and $C_1 = 0$, respectively; therefore,

$$\Theta = 0 \qquad \text{for} \quad \alpha^2 = 0 \tag{4.42}$$

which is a trivial solution and has no effect on the general solution for the temperature field. Note that such is *not* always the case. The value $\alpha^2 = 0$ should always be considered carefully and its effect on the solution determined.

**Case (b)**

The configuration of Fig. 4.2b differs from the configuration of Fig. 4.2a only in that the fin height in the $y$-direction ($H$) is finite. Therefore, in modeling the problem, eqs. (4.11)–(4.13) remain valid for the configuration of Fig. 4.2b as well. The boundary condition at $y = 0$ (the tip of the fin) is given by

$$y = 0, \qquad \Theta = 0 \tag{4.43}$$

Boundary condition (4.14) also remains essentially unchanged. However, since the coordinate system is now attached to the tip of the fin (Fig. 4.2b), boundary condition (4.14) has to be evaluated at $y = H$, the base of the fin.

In summary, the conduction problem shown schematically in Fig. 4.2b is described completely by the set of equations (4.11)–(4.14) and (4.43). Clearly, three of the boundary conditions are homogeneous and we can proceed with the solution using the method of separation of variables. Repeating the steps outlined earlier in connection with the problem of case (a), we obtain the following expression for the temperature:

$$\Theta = [C_1 \sin(\alpha x) + C_2 \cos(\alpha x)][D_1 \sinh(\alpha y) + D_2 \cosh(\alpha y)] \tag{4.44}$$

Note that the $y$ part of the solution is expressed in terms of the hyperbolic sine and cosine functions instead of the exponential function of eq. (4.22). Both choices are mathematically equivalent and correct; however, in the case of a semi-infinite direction [e.g., the $y$-direction in case (a)] the use of exponentials simplifies the algebra involved and it is generally recommended. To continue, we apply boundary conditions (4.12), (4.13), and (4.43) and proceed in the manner discussed in the solution of case (a). Omitting the details for brevity, the expression for the temperature after applying the above-mentioned three boundary conditions reads

$$\Theta = \sum_{n=1}^{\infty} C_n \sin\left(\frac{n\pi x}{b}\right) \sinh\left(\frac{n\pi y}{b}\right) \tag{4.45}$$

The remaining unknown constant $C_n$ is obtained with the help of the last boundary condition (4.14) and by making use of the orthogonality property of the sine function. Since the details are identical to those of case (a), we show only the final result for $C_n$:

$$C_n = \frac{2\int_0^b \phi(x) \sin(n\pi x/b)\, dx}{b \sinh(n\pi H/b)} \tag{4.46}$$

The expression for the temperature, therefore, reads:

$$T = T_\infty + \frac{2}{b}\sum_{n=1}^{\infty} \frac{\int_0^b (f(x) - T_\infty) \sin(n\pi x/b) dx}{\sinh(n\pi H/b)} \sin\left(\frac{n\pi x}{b}\right) \sinh\left(\frac{n\pi y}{b}\right) \tag{4.47}$$

Using Fourier's law, the local heat flux through the base of the fin is obtained:

$$\begin{aligned} \dot{q}''(x) &= -k\left(\frac{\partial T}{\partial y}\right)_{y=H} \\ &= -k\frac{2\pi}{b^2}\sum_{n=1}^{\infty} \frac{\int_0^b (f(x) - T_\infty) \sin(n\pi x/b)\, dx}{\sinh(n\pi H/b)} n \sin\left(\frac{n\pi x}{b}\right) \cosh\left(\frac{n\pi H}{b}\right) \end{aligned} \tag{4.48}$$

The total heat flux through the base of the fin results from combining eqs. (4.35) and (4.48):

$$\dot{q}' = -k\frac{2\pi}{b^2}\int_0^b \left[\sum_{n=1}^{\infty} \frac{\int_0^b (f(x) - T_\infty) \sin(n\pi x/b)\, dx}{\sinh\,(n\pi H/b)} n \sin\left(\frac{n\pi x}{b}\right) \cosh\left(\frac{n\pi H}{b}\right)\right] dx \tag{4.49}$$

If the base temperature of the fin is constant, say $f(x) = T_0$, eqs. (4.47)–(4.49) reduce to

$$T(x, y) = T_\infty + \frac{4(T_0 - T_\infty)}{\pi} \sum_{n=0}^{\infty} \frac{1}{(2n + 1)\sinh\left[\frac{(2n+1)\pi H}{b}\right]} \sin\left[\frac{(2n+1)\pi x}{b}\right] \cdot \sinh\left[\frac{(2n+1)\pi y}{b}\right] \tag{4.50}$$

$$\dot{q}''(x) = -\frac{4k(T_0 - T_\infty)}{b} \sum_{n=0}^{\infty} \coth\left[\frac{(2n+1)\pi H}{b}\right] \sin\left[\frac{(2n+1)\pi x}{b}\right] \tag{4.51}$$

The earlier discussion following eq. (4.38) regarding the value of $\dot{q}''(0)$ and $\dot{q}''(b)$ holds in this case as well.

All the results for the temperature and the heat flux in this case in the limit of $H \to \infty$ match those of case (a). The discussion pertaining to the possibility of an eigenvalue being equal to zero ($\alpha^2 = 0$) at the end of the problem considered in case (a) holds here as well, and the result is that $\alpha^2 = 0$ yields the trivial solution $\Theta = 0$. In the remaining examples, if the zero-eigenvalue case exists, it will be discussed only if it yields a nontrivial solution. If no mention is made of the zero eigenvalue, it is implied that it yields a trivial ($\Theta = 0$) contribution to the general solution.

**Case (c)**

The last configuration shown in Fig. 4.2c represents a situation in which the heat transfer coefficient between the fin and the surrounding fluid is finite. For the sake of simplicity, we assume that the heat transfer coefficient on all three sides of the fin in contact with the fluid is constant and is denoted by $h$. In other words, the configuration in Fig. 4.2c represents the physical situation of Fig. 4.1 in the limit of $h_1 = h_2 = h_3 = h$.

The mathematical model differs from that of Fig. 4.2a and b only in the expression for the boundary conditions at the three sides of the fin in contact with the fluid. According to the coordinate system shown in Fig. 4.2c, these boundary conditions are

$$x = 0: \qquad k\frac{\partial T}{\partial x} = h(T - T_\infty) \tag{4.52}$$

$$x = b: \qquad -k\frac{\partial T}{\partial x} = h(T - T_\infty) \tag{4.53}$$

$$y = 0: \qquad k\frac{\partial T}{\partial y} = h(T - T_\infty) \tag{4.54}$$

The boundary condition at the base is, as before:

$$y = H: \qquad T = f(x) \tag{4.55}$$

Equation (4.1) remains the governing equation of the problem. To proceed with the problem solution we first note that none of the boundary conditions

(4.52)–(4.55) is homogeneous. However, three of the boundary conditions can be transformed to homogeneous with the help of the variable transformation defined in eq. (4.10). Applying this coordinate transformation to the boundary conditions (4.52)–(4.55) yields

$$x = 0: \qquad k\frac{\partial \Theta}{\partial x} = h\Theta \tag{4.56}$$

$$x = b: \qquad -k\frac{\partial \Theta}{\partial x} = h\Theta \tag{4.57}$$

$$y = 0: \qquad k\frac{\partial \Theta}{\partial y} = h\Theta \tag{4.58}$$

$$y = H: \qquad \Theta = f(x) - T_\infty = \phi(x) \tag{4.59}$$

The governing equation in terms of the variable $\Theta$ was reported earlier in expression (4.11). At this point, the governing equation (4.11) and the three boundary conditions (4.56)–(4.58) are homogeneous and the method of separation of variables can be applied. Proceeding as shown earlier in connection with the solution to the problem of configuration shown in Fig. 4.1a, we obtain the following expression for the temperature:

$$\Theta = [A \sin(\alpha x) + B \cos(\alpha x)][C \sinh(\alpha y) + D \cosh(\alpha y)] \tag{4.60}$$

Applying boundary conditions (4.56) and (4.58) yields

$$A = \frac{h}{k\alpha}B \tag{4.61}$$

$$C = \frac{h}{k\alpha}D \tag{4.62}$$

Substituting into eq. (4.60) and combining constants, we obtain

$$\Theta = K\left[\frac{h}{\alpha k}\sin(\alpha x) + \cos(\alpha x)\right]\left[\frac{h}{\alpha k}\sinh(\alpha y) + \cosh(\alpha y)\right] \tag{4.63}$$

where $K$ is an unknown constant. Next, boundary condition (4.57) is applied to yield, after simple algebraic manipulations,

$$\cot(\alpha b) = \frac{\alpha k}{2h} - \frac{h}{2k\alpha} = \frac{\alpha b}{2\text{Bi}} - \frac{\text{Bi}}{2(\alpha b)} \tag{4.64}$$

where $\text{Bi} = hb/k$ is the Biot number.

For any prescribed numerical values of $b$, $h$, and $k$, equation (4.64), which is a simple nonlinear algebraic equation in $\alpha$, can be solved numerically or even graphically to yield the eigenvalues. A graphical solution of eq. (4.64) is shown in Fig. 4.3 for three values of the Biot number: $\text{Bi} = 0.1$, 1.0, and 10. The values of $\alpha b$ that satisfy eq. (4.64) are the intersections between the

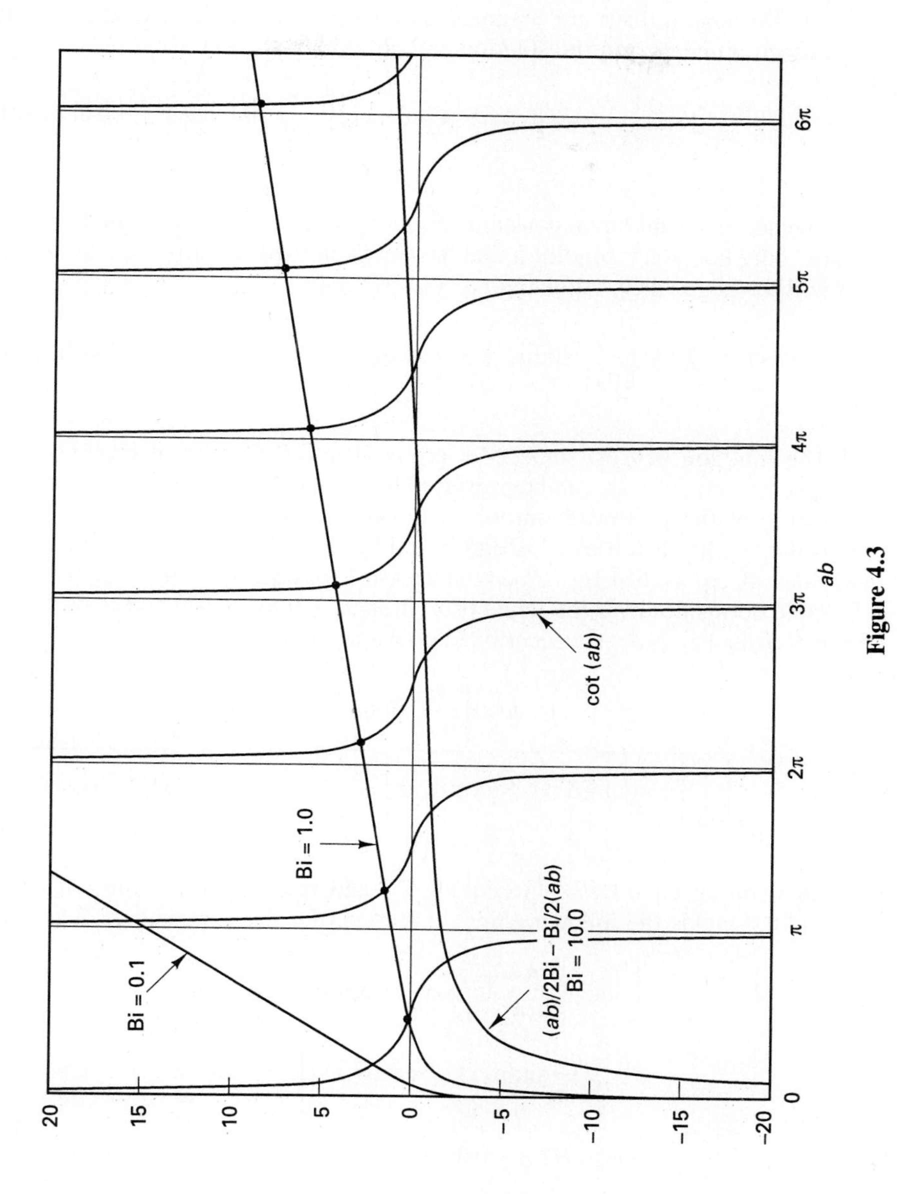

20
15
10
5
0
−5
−10
−15
−20
0
π
2π
3π
4π
5π
6π
ab
Bi = 0.1
Bi = 1.0
(ab)/2Bi − Bi/2(ab)
Bi = 10.0
cot (ab)


**Figure 4.3**

curves corresponding to the left- and the right-hand side of eq. (4.64). These points are indicated by dark circles for the case of Bi = 1.0. From this point on, the eigenvalues are assumed to be known; that is, $\alpha_1$, $\alpha_2$, $\alpha_3$, are all known numbers and the solution can be written as

$$\Theta = \sum_{n=1}^{\infty} K_n \left[\frac{h}{\alpha_n k} \sin(\alpha_n x) + \cos(\alpha_n x)\right]\left[\frac{h}{\alpha_n k} \sinh(\alpha_n y) + \cosh(\alpha_n y)\right] \tag{4.65}$$

The series of unknown constants $K_n$, $n = 1, 2, \ldots$, is obtained by applying the only boundary condition that has not been used so far [eq. (4.59)], together with orthogonality. Applying eq. (4.59) yields

$$\phi(x) = \sum_{n=1}^{\infty} K_n \left[\frac{h}{\alpha_n k} \sin(\alpha_n x) + \cos(\alpha_n x)\right]\left[\frac{h}{\alpha_n k} \sinh(\alpha_n H) + \cosh(\alpha_n H)\right] \tag{4.66}$$

The function $(h/\alpha_n k) \sin(\alpha_n x) + \cos(\alpha_n x)$ in the first set of brackets, with $\alpha_n$ given by eq. (4.64), can be proved to be orthogonal by a straightforward application of the general definition of orthogonality (Appendix C). The proof of orthogonality is left as an exercise to the reader. To proceed, we multiply both sides of eq. (4.66) by $(h/\alpha_m k) \sin(\alpha_m x) + \cos(\alpha_m x)$, $m = 1, 2, 3, \ldots$ integrate both sides over the range of $x$ (from $x = 0$ to $x = b$) and make use of the orthogonality property mentioned above to obtain

$$K_n = \frac{\displaystyle\int_0^b \phi(x)\left[\frac{h}{\alpha_n k} \sin(\alpha_n x) + \cos(\alpha_n x)\right] dx}{\left[\displaystyle\int_0^b \left[\frac{h}{\alpha_n k} \sin(\alpha_n x) + \cos(\alpha_n x)\right]^2 dx\right]\left[\frac{h}{\alpha_n k} \sinh(\alpha_n H) + \cosh(\alpha_n H)\right]} \tag{4.67}$$

Substituting eq. (4.67) into eq. (4.65) and recalling the definition of $\Theta$ [eq. (4.10)] yields the following final expression for the temperature distribution:

$$T = T_\infty + \sum_{n=1}^{\infty} \left\{\int_0^b \phi(x)\left[\frac{h}{\alpha_n k} \sin(\alpha_n x) + \cos(\alpha_n x)\right] dx\right\} \cdot \left[\frac{h}{\alpha_n k} \sin(\alpha_n x) + \cos(\alpha_n x)\right]\left[\frac{h}{\alpha_n k} \sinh(\alpha_n y) + \cosh(\alpha_n y)\right] \Big/ \left(\left[\frac{h}{\alpha_n k} \sinh(\alpha_n H) + \cosh(\alpha_n H)\right]\left\{\int_0^b \left[\frac{h}{\alpha_n k} \sin(\alpha_n x) + \cos(\alpha_n x)\right]^2 dx\right\}\right) \tag{4.68}$$

The local heat flux through the base of the fin is obtained once again by applying Fourier's law:

$$\dot{q}''(x) = -k\left(\frac{\partial T}{\partial y}\right)_{y=H}$$

$$= -k \sum_{n=1}^{\infty} \left\{\int_0^b \phi(x)\left[\frac{h}{\alpha_n k}\sin(\alpha_n x) + \cos(\alpha_n x)\right] dx\right\}$$

$$\cdot \left[\frac{h}{\alpha_n k}\sin(\alpha_n x) + \cos(\alpha_n x)\right]\left[\frac{h}{k}\cosh(\alpha_n H) + \alpha_n \sinh(\alpha_n H)\right] \Big/$$

$$\left(\left[\frac{h}{\alpha_n k}\sinh(\alpha_n H) + \cosh(\alpha_n H)\right]\left\{\int_0^b \left[\frac{h}{\alpha_n k}\sin(\alpha_n x) + \cos(\alpha_n x)\right]^2 dx\right\}\right) \tag{4.69}$$

The expression for the total heat transfer through the base of the fin can be obtained easily if eq. (4.69) is substituted into expression (4.35).

In the special case, where the base temperature is constant, $f(x) = T_0$ or $\phi(x) = T_0 - T_\infty$, the expressions for the temperature, the local heat flux through the base, and the total heat flux through base, respectively, become

$$T(x, y) = T_\infty + 2(T_0 - T_\infty) \sum_{n=1}^{\infty} \left\{\frac{h}{\alpha_n k}[1 - \cos(\alpha_n b)] + \sin(\alpha_n b)\right\}$$

$$\cdot \left[\frac{h}{\alpha_n k}\sin(\alpha_n x) + \cos(\alpha_n x\right]$$

$$\cdot \left[\frac{h}{\alpha_n k}\sinh(\alpha_n y) + \cosh(\alpha_n y)\right] \Big/ \left\{\alpha_n b\left[1 + \frac{h^2}{\alpha_n^2 k^2} + \left(1 - \frac{h^2}{\alpha_n^2 k^2}\right)\right.\right.$$

$$\left.\left.\cdot \frac{\sin(2\alpha_n b)}{2\alpha_n b} + \frac{2h}{\alpha_n k}\frac{\sin^2(\alpha_n b)}{\alpha_n b}\right]\left[\frac{h}{\alpha_n k}\sinh(\alpha_n H) + \cosh(\alpha_n H)\right]\right\} \tag{4.70}$$

$$\dot{q}''(x) = -\frac{2k(T_0 - T_\infty)}{b} \sum_{n=1}^{\infty} \left\{\frac{h}{\alpha_n k}[1 - \cos(\alpha_n b)] + \sin(\alpha_n b)\right\}\left[\frac{h}{\alpha_n k} + \tanh(\alpha_n H)\right]$$

$$\cdot \left[\frac{h}{\alpha_n k}\sin(\alpha_n x) + \cos(\alpha_n x)\right] \Big/ \left\{\left[\left(1 + \frac{h^2}{\alpha_n^2 k^2}\right) + \left(1 - \frac{h^2}{\alpha_n^2 k^2}\right)\right.\right.$$

$$\left.\left.\cdot \frac{\sin(2\alpha_n b)}{2\alpha_n b} + \frac{2h}{\alpha_n k}\frac{\sin^2(\alpha_n b)}{\alpha_n b}\right]\left[\frac{h}{\alpha_n k}\tanh(\alpha_n H) + 1\right]\right\} \tag{4.71}$$

$$\dot{q}' = -2k(T_0 - T_\infty) \sum_{n=1}^{\infty} \frac{1}{\alpha_n b}\left\{\frac{h}{\alpha_n k}[1 - \cos(\alpha_n b)] + \sin(\alpha_n b)\right\}^2\left[\frac{h}{\alpha_n k} + \tanh(\alpha_n H)\right] \Big/$$

$$\left\{\left[\left(1 + \frac{h^2}{\alpha_n^2 k^2}\right) + \left(1 - \frac{h^2}{\alpha_n^2 k^2}\right)\frac{\sin(2\alpha_n b)}{2\alpha_n b} + \frac{2h}{\alpha_n k}\frac{\sin^2(\alpha_n b)}{\alpha_n b}\right]\right.$$

$$\left.\cdot \left[\frac{h}{\alpha_n k}\tanh(\alpha_n H) + 1\right]\right\} \tag{4.72}$$

It is worth noting that $\alpha_n = 0$ is not an eigenvalue for the present problem. Therefore, it is not necessary to go through the analysis outlined after case (a). Figure 4.4a shows graphically the temperature distribution obtained from eq. (4.70) at the midheight of the fin ($y = H/2$) for a fin of square cross section ($H/L = 1$). Increasing the value of the Biot number, which is proportional to

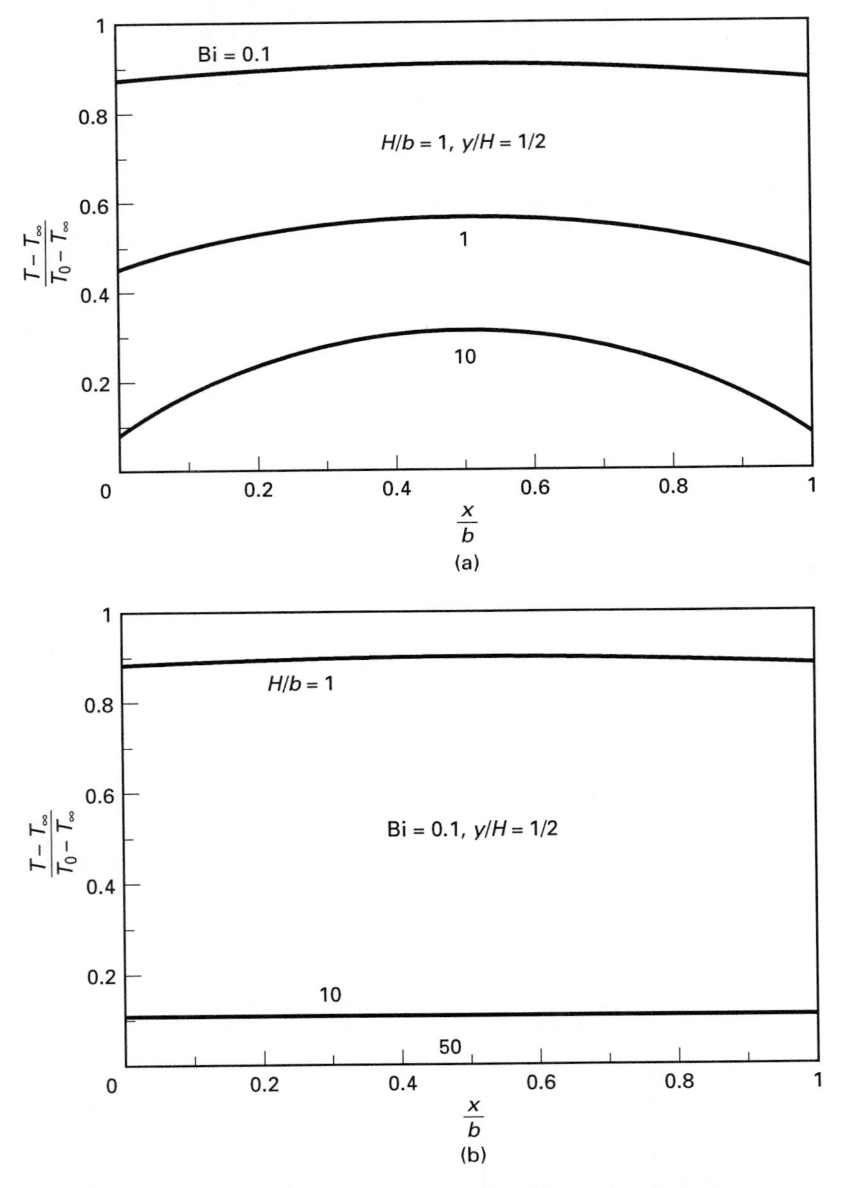


**Figure 4.4** (a) Effect of Bi on temperature distribution. (b) Effect of geometric aspect ratio on temperature distribution.

the heat transfer coefficient, enhances the heat removal through the sides of the fin and decreases the temperature of the fin. As the Biot number decreases, the dependence of the temperature on $x$ diminishes and the temperature field becomes one-dimensional. Figure 4.4b illustrates the effect of the geometric aspect ratio of the fin cross section on the temperature distribution at midheight for $Bi = 0.1$. As the fin becomes more slender, the temperature distribution at midheight becomes less affected by the heating at the base of the fin. For $H/b = 50$ the fin temperature at midheight is constant and equal to the fluid temperature $T_\infty$.

Before closing, a general comment on the appropriate placement of the coordinate system is useful. In rectangular geometries such as those considered in this example, one has a choice of four corners in which to place the origin of the coordinate system. It is usually recommended that the origin of the coordinate system is placed in the corner with the simplest homogeneous boundary conditions. Such choice simplifies the algebra associated with the problem solution. The zero temperature ($\Theta = 0$) and the zero temperature gradient ($\partial\Theta/\partial x = 0$ or $\partial\Theta/\partial y = 0$) are two of the simplest homogeneous boundary conditions. A corner having a combination of these two types of homogeneous boundary conditions, for example, is generally preferable to a corner having one or two convection boundary conditions of the type $[h\Theta = \pm k(\partial\Theta/\partial x)$ or $h\Theta = \pm k(\partial\Theta/\partial y)]$. The spirit of the discussion above on the placement of the origin of the coordinate system is valid not only for Cartesian coordinates but also for cylindrical coordinates. In spherical coordinates the center of the sphere of interest is usually chosen as the origin of the coordinate system.

## 4.2 AXISYMMETRIC CONDUCTION IN CYLINDRICAL COORDINATES

The solution of conduction problems in cylindrical coordinates will be illustrated with the help of two representative examples that find application in heat transfer engineering.

**Example 4.2**

Consider the *short* cylindrical fin of length $L$ and radius $R$ shown in Fig. 4.5. It is in contact with a flowing fluid of temperature $T_\infty$ far away from the fin surface. The temperature of the fin base is a known function of the radial coordinate and is denoted by $f(r)$. With no loss of generality, we can assume that $f(r)$ is higher than $T_\infty$ and that the purpose of the fin is to aid the cooling of the surface on which it is attached. The average heat transfer coefficient between the fin outer surface and the fluid is denoted by $h$. The coordinate system $(r, \theta, z)$ is attached at the center of the tip of the fin. It is of interest to obtain the temperature distribution inside the fin as well as the total heat flux through the base of the fin. Two separate cases are considered. In the first case, it is assumed that the heat transfer coefficient $h$ is very large ($h \to \infty$). Therefore, the

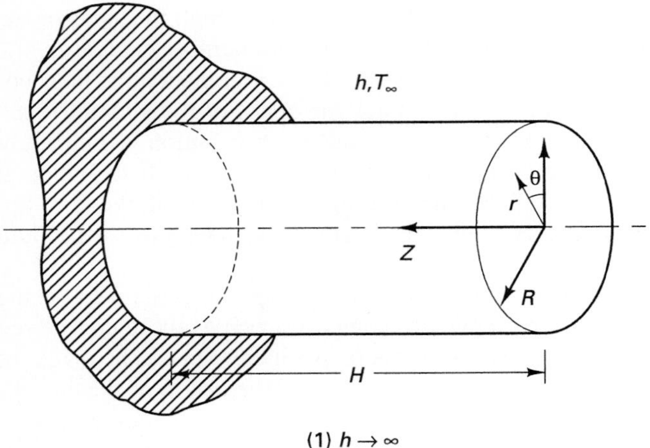


**Figure 4.5** Cylindrical "pin" fin.

surface temperature of the fin is set equal to the fluid temperature, $T_\infty$. In the second case, a finite value is assigned to $h$.

**Case (a) The heat transfer coefficient is very large (h → ∞)**

Noting that the temperature field for the problem described above does not depend on the angular coordinate, $\theta$, and introducing the variable $\Theta = T - T_\infty$ to transform the boundary conditions at the fin surface to homogeneous boundary conditions yields the following heat conduction model:

$$\frac{\partial^2\Theta}{\partial r^2} + \frac{1}{r}\frac{\partial\Theta}{\partial r} + \frac{\partial^2\Theta}{\partial z^2} = 0 \tag{4.73}$$

$$r = 0: \quad \frac{\partial\Theta}{\partial r} = 0 \tag{4.74}$$

$$r = R: \quad \Theta = 0 \tag{4.75}$$

$$z = 0: \quad \Theta = 0 \tag{4.76}$$

$$z = L: \quad \Theta = f(r) - T_\infty = \phi(r) \tag{4.77}$$

Note that the equation (4.73) is a special case of eq. (2.34) in the limit of no time dependence, no angular dependence, and no heat generation. Equation (4.74) results from the symmetry of the problem about the $z$-axis and the fact that the temperature gradient is continuous at any point inside the cylinder. The other three boundary conditions are self-explanatory. Clearly, conditions (4.74)–(4.76) as well as the governing equation (4.73) are homogeneous and the method of separation of variables can be applied. As it is customary we assume that

$$\Theta = G(r)H(z) \tag{4.78}$$

where $G$ and $H$ are unknown functions of $r$ and $z$, respectively. Substituting eq. (4.78) into eq. (4.73) and dividing through by the product $GH$ yields

$$\overbrace{\frac{G''}{G} + \frac{G'}{rG}}^{-\alpha^2} + \overbrace{\frac{H''}{H}}^{\alpha^2} = 0 \tag{4.79}$$

While the sum of the first two terms on the left-hand side of eq. (4.79) depends on the radial coordinate, $r$, only, the third term depends on the axial coordinate, $z$, only. Therefore, for eq. (4.79) to hold in general, both the sum of terms that depend on $r$ and the term that depends on $z$ should equal a constant, independent of both $r$ and $z$. With the foregoing argument in mind, we set the group of two terms that depend on $r$ equal to $-\alpha^2$ and term that depends on $z$ equal to $\alpha^2$ [for eq. (4.79) to be satisfied] and obtain the following pair of equations:

$$H'' - \alpha^2 H = 0 \tag{4.80}$$

$$G'' + \frac{1}{r}G' + \alpha^2 G = 0 \tag{4.81}$$

Clearly, adding eqs. (4.80) and (4.81) side by side, we recover eq. (4.79). Much as in Example 4.1, it is crucial that eqs. (4.80) and (4.81) are constructed such that orthogonal functions are obtained in the direction with two homogeneous boundary conditions (the radial direction in our problem). As discussed next, the solution of eq. (4.81) is, indeed, in terms of orthogonal functions. The reasoning above led us to set the $r$-dependent group of terms in eq. (4.79) equal to $-\alpha^2$ and the $z$-dependent term equal to $\alpha^2$, not the other way around. Both eqs. (4.80) and (4.81) are ordinary differential equations. Their solutions are (Appendixes A and B)

$$H = A \sinh(\alpha z) + B \cosh(\alpha z) \tag{4.82}$$

$$G = CJ_0(\alpha r) + DY_0(\alpha r) \tag{4.83}$$

where $J_0(\alpha r)$ and $Y_0(\alpha r)$ are the Bessel functions of order zero of the first and second kind, respectively. Hence, the solution for the temperature reads [eq. (4.78)]

$$\Theta = [CJ_0(\alpha r) + DY_0(\alpha r)][A \sinh(\alpha z) + B \cosh(\alpha z)] \tag{4.84}$$

Applying boundary conditions (4.74) and (4.76) yields

$$D = B = 0 \tag{4.85}$$

Combining the remaining constants and renaming the product $AC$ to $K$ in eq. (4.84), we obtain

$$\Theta = KJ_0(\alpha r) \sinh(\alpha z) \tag{4.86}$$

Substituting eq. (4.86) into boundary condition (4.75) results in

$$0 = KJ_0(\alpha R) \sinh(\alpha z) \tag{4.87}$$

For a nontrivial solution to the temperature field to exist, (4.87) implies that

$$J_0(\alpha_n R) = 0 \tag{4.88}$$

Equation (4.88) needs to be solved numerically to yield the values of $\alpha_n$ satisfying it (eigenvalues). Alternatively, existing tables of zeros of Bessel functions [4] can be consulted. Indeed, an infinite number of values of $\alpha_n$ exist ($\alpha_1$, $\alpha_2$, $\alpha_3$, . . .) that satisfy eq. (4.88). To proceed with the presentation, we assume that $\alpha_1$, $\alpha_2$, . . . are known. A solution for the temperature exists for each value of $\alpha_n$. All these solutions are linearly independent. Hence the general solution is given by the following superposition:

$$\Theta = \sum_{n=1}^{\infty} K_n J_0(\alpha_n r) \sinh(\alpha_n z) \tag{4.89}$$

To obtain $K_n$ we make use of the fact that $J_0(\alpha_n r)$ is an orthogonal function with respect to a weighting factor $r$. A brief discussion as well as references for additional reading on Bessel functions are included in Appendix A. First, we apply boundary condition (4.77):

$$\phi(r) = \sum_{n=1}^{\infty} K_n J_0(\alpha_n r) \sinh(\alpha_n L) \tag{4.90}$$

Next, we multiply both sides of eq. (4.90) by $rJ_0(\alpha_m r)$, integrate both sides from $r = 0$ to $r = R$, and invoke the orthogonality property of $J_0(\alpha_m r)$ to obtain

$$K_n = \frac{\displaystyle\int_0^R \phi(r) J_0(\alpha_n r) r\, dr}{\sinh(\alpha_n L) \displaystyle\int_0^R J_0^2(\alpha_n r) r\, dr} \tag{4.91}$$

The integral in the denominator of eq. (4.91) is

$$\int_0^R J_0^2(\alpha_n r) r\, dr = \frac{R^2}{2}\left[J_1^2(\alpha_n R) + J_0^2(\alpha_n R)\right] \tag{4.92}$$

The second term on the right-hand side of eq. (4.92) is equal to zero since the values of $\alpha_n$, $n = 1, 2, \ldots$, are solutions of eq. (4.88). The expression for $K_n$ then reduces to

$$K_n = \frac{2\displaystyle\int_0^R \phi(r) J_0(\alpha_n r) r\, dr}{\sinh(\alpha_n L) R^2 J_1^2(\alpha_n R)} \tag{4.93}$$

The final expression for the temperature field is obtained by substituting eq. (4.93) into eq. (4.89) and by recalling that $\Theta = T - T_\infty$.

$$T = T_\infty + 2\sum_{n=1}^{\infty} \frac{\int_0^R \phi(r) J_0(\alpha_n r) r\, dr}{\sinh(\alpha_n L) R^2 J_1^2(\alpha_n R)} J_0(\alpha_n r) \sinh(\alpha_n z) \tag{4.94}$$

If numerical values are assigned to the problem parameters, expression (4.94) can be used to provide the temperature at any point in the fin. The overall heat flux through the base of the fin ($z = L$) is determined by integrating the local heat flux evaluated with Fourier's law, over the fin base

$$\dot{q} = \int_0^R -k\left(\frac{\partial T}{\partial z}\right)_{z=L} 2\pi r\, dr \tag{4.95}$$

Combining eqs. (4.94) and (4.95), we obtain

$$\dot{q} = -4\pi k \int_0^R \left[\sum_{n=1}^{\infty} \frac{\int_0^R \phi(r) J_0(\alpha_n r) r\, dr}{\sinh(\alpha_n L) R^2 J_1^2(\alpha_n R)} J_0(\alpha_n r)\alpha_n \cosh(\alpha_n L)\right] r\, dr \tag{4.96}$$

In the special case where the base temperature is constant $T_0(\phi = T_0 - T_\infty)$, equations (4.94) and (4.96) reduce to

$$T = T_\infty + 2(T_0 - T_\infty)\sum_{n=1}^{\infty} \frac{J_0(\alpha_n r)\sinh(\alpha_n z)}{(\alpha_n R) J_1(\alpha_n R)\sinh(\alpha_n L)} \tag{4.97}$$

$$\dot{q} = -4\pi k L (T_0 - T_\infty)\sum_{n=1}^{\infty} [(\alpha_n L)\tanh(\alpha_n L)]^{-1} \tag{4.98}$$

This result for $\dot{q}$ for the present idealized problem does not account for the large heat flux at the periphery of the fin base encountered in a real application with similar boundary conditions, as discussed earlier, after eq. (4.38).

**Case (b) The heat transfer coefficient is finite**

If the heat transfer coefficient is finite, the boundary conditions at the lateral surface (eq. (4.75) and at the tip (eq. 4.76) of the fin need to be modified as follows:

$$r = R: \qquad -k\frac{\partial \Theta}{\partial r} = h\Theta \tag{4.99}$$

$$z = 0: \qquad k\frac{\partial \Theta}{\partial z} = h\Theta \tag{4.100}$$

The governing equation (4.73) as well as boundary conditions (4.74) and (4.77) remain unchanged and together with the new boundary conditions (4.99) and (4.100) constitute the conduction model of the problem.

Since the governing equation (4.73) and the boundary conditions (4.74), (4.99), and (4.100) are homogeneous in terms of the variable $\Theta$, the method of separation of variables can be applied directly. The sequence of steps followed earlier in case (a) are repeated, but details are not given here for brevity. After application of boundary conditions (4.74) and (4.100) the expression for the temperature reads

$$\Theta = K\left[\frac{h}{k\alpha}\sinh(\alpha z) + \cosh(\alpha z)\right]J_0(\alpha r) \tag{4.101}$$

The values of $\alpha$ that satisfy boundary condition (4.99) are obtained after substituting expression (4.101) into the above-mentioned boundary condition. The resulting expression reads

$$J_1(\alpha_n R) = \frac{h}{k\alpha_n}J_0(\alpha_n R) \tag{4.102}$$

Equation (4.102) can be solved numerically if the exact numerical values of $h$, $k$, and $R$ are prescribed to yield the eigenvalues $\alpha_n$, $n = 1, 2, 3, \ldots$ . From this point on, we assume that $\alpha_1$, $\alpha_2$, $\alpha_3$, are known. The general solution for $\Theta$ is the linear superposition of the solutions corresponding to $\alpha_1$, $\alpha_2$, $\alpha_3$, . . . ; that is,

$$\Theta = \sum_{n=1}^{\infty} K_n\left[\frac{h}{k\alpha_n}\sinh(\alpha_n z) + \cosh(\alpha_n z)\right]J_0(\alpha_n r) \tag{4.103}$$

To proceed, we apply the last boundary condition (4.77), together with the orthogonality property of $J_0(\alpha_n r)$. Omitting the details since they are identical to what was reported earlier in case (a), we show only the final result for $K_n$:

$$K_n = \frac{\displaystyle\int_0^R \phi(r)J_0(\alpha_n r)r\,dr}{\displaystyle\left[\frac{h}{k\alpha_n}\sinh(\alpha_n L) + \cosh(\alpha_n L)\right]\int_0^R J_0^2(a_n r)r\,dr} \tag{4.104}$$

Making use of result (4.92) for the integral of the denominator in eq. (4.104), as well as of eq. (4.102) yields

$$K_n = \frac{\displaystyle 2\int_0^R \phi(r)J_0(\alpha_n r)r\,dr}{\displaystyle\left[\frac{h}{k\alpha_n}\sinh(\alpha_n L) + \cosh(\alpha_n L)\right]J_0^2(\alpha_n R)\left[1 + \left(\frac{h}{k\alpha_n}\right)^2\right]R^2} \tag{4.105}$$

Combining eqs. (4.103) and (4.105), we obtain the final expression for the temperature:

$$T = T_\infty + 2\sum_{n=1}^{\infty} \frac{\left[\int_0^R \phi(r)J_0(\alpha_n r)r\,dr\right]\left[\frac{h}{k\alpha_n}\sinh(\alpha_n z) + \cosh(\alpha_n z)\right]J_0(\alpha_n r)}{R^2\left[1 + \left(\frac{h}{k\alpha_n}\right)^2\right]\left[\frac{h}{k\alpha_n}\sinh(\alpha_n L) + \cosh(\alpha_n L)\right]J_0^2(\alpha_n R)} \tag{4.106}$$

The heat flux through the fin base is calculated by using Fourier's law, that is, by substituting eq. (4.106) into eq. (4.95):

$$\dot{q} = -4\pi k \int_0^R \sum_{n=1}^{\infty} \frac{\left[\int_0^R \phi(r)J_0(\alpha_n r)r\,dr\right]\left[\frac{h}{\alpha_n k}\cosh(\alpha_n L) + \sinh(\alpha_n L)\right]\alpha_n J_0(\alpha_n r)}{R^2\left[1 + \left(\frac{h}{\alpha_n k}\right)^2\right]\left[\frac{h}{\alpha_n k}\sinh(\alpha_n L) + \cosh(\alpha_n L)\right]J_0^2(\alpha_n R)}\, r\,dr$$

$$= -4\pi k \sum_{n=1}^{\infty} \frac{\left[\int_0^R \phi(r)J_0(\alpha_n r)r\,dr\right]\left[\frac{h}{\alpha_n k}\cosh(\alpha_n L) + \sinh(\alpha_n L)\right]J_1(\alpha_n r)}{R^2\left[1 + \left(\frac{h}{\alpha_n k}\right)^2\right]\left[\frac{h}{\alpha_n k}\sinh(\alpha_n L) + \cosh(\alpha_n L)\right]J_0^2(\alpha_n R)} \tag{4.107}$$

It can easily be shown that eqs. (4.106) and (4.107) reduce to eqs. (4.94) and (4.96), respectively, in the limit $h \to \infty$.

If the base temperature is constant [$f(x) = T_0$ or $\phi(r) = T_0 - T_\infty$], the expressions for the temperature in the fin and the heat flux through the fin base become

$$T = T_\infty + 2(T_0 - T_\infty)\sum_{n=1}^{\infty} \frac{J_1(\alpha_n r)\left[\frac{h}{\alpha_n k}\cosh(\alpha_n z) + \sinh(\alpha_n z)\right]J_0(\alpha_n r)}{\alpha_n R J_0^2(\alpha_n r)\left[1 + \left(\frac{h}{\alpha_n k}\right)^2\right]\left[\frac{h}{\alpha_n k}\sinh(\alpha_n L) + \cosh(\alpha_n L)\right]} \tag{4.108}$$

$$\dot{q} = -4\pi k(T_0 - T_\infty)\sum_{n=1}^{\infty} \frac{J_1^2(\alpha_n r)\left[\frac{h}{\alpha_n k}\cosh(\alpha_n z) + \sinh(\alpha_n z)\right]}{J_0^2(\alpha_n r)\left[1 + \left(\frac{h}{\alpha_n k}\right)^2\right]\left[\frac{h}{\alpha_n k}\sinh(\alpha_n L) + \cosh(\alpha_n L)\right]} \tag{4.109}$$

**Example 4.3**

The lateral surface of the cylinder shown in Fig. 4.6 is exposed to a radiation flux $\dot{q}''(z)$. Such a flux can be imposed with the help of a ring heater surrounding the cylinder. The two ends of the cylinder are cooled by being in contact

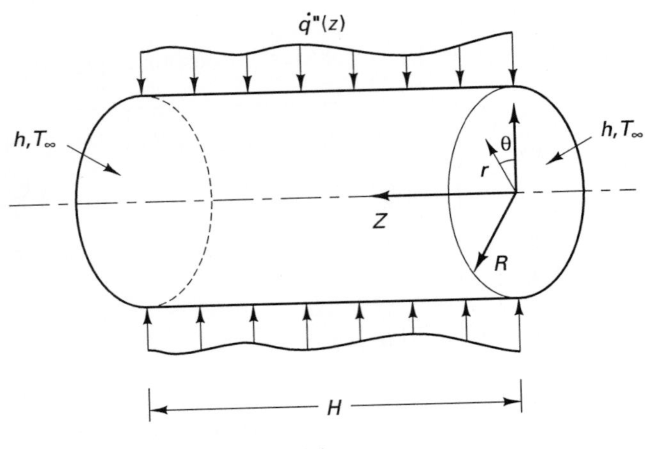


**Figure 4.6** Cylinder exposed to a radiation heat flux.

with a fluid of temperature $T_\infty$. The corresponding heat transfer coefficient is denoted by $h$ and is the same for both ends of the cylinder. It is of interest to obtain the temperature distribution in the cylinder and the total heat removal at both ends, $z = 0$ and $z = L$. Two separate cases will be investigated. In case (a) the heat transfer coefficient $h$ will be assumed to be very large ($h \to \infty$). In case (b) the heat transfer coefficient will be of finite value.

**Case (a) The heat transfer coefficient is infinite (h $\to \infty$)**

The mathematical model in this case (governing equations and boundary conditions) reads as follows:

$$\frac{\partial^2 \Theta}{\partial r^2} + \frac{1}{r}\frac{\partial \Theta}{\partial r} + \frac{\partial^2 \Theta}{\partial z^2} = 0 \tag{4.110}$$

$$r = 0: \quad \frac{\partial \Theta}{\partial r} = 0 \tag{4.111}$$

$$r = R: \quad k\frac{\partial \Theta}{\partial r} = \dot{q}'' \tag{4.112}$$

$$z = 0: \quad \Theta = 0 \tag{4.113}$$

$$z = L: \quad \Theta = 0 \tag{4.114}$$

where once again we introduced, $\Theta = T - T_\infty$ to render three of the boundary conditions homogeneous. Proceeding as shown in Example 4.2, case (a), we assume that $\Theta = G(r)\ H(z)$, substitute this expression into eq. (4.110) and eventually obtain

$$\underbrace{\frac{G''}{G} + \frac{G'}{rG}}_{\alpha^2} + \underbrace{\frac{H''}{H}}_{-\alpha^2} = 0 \tag{4.115}$$

Equation (4.115) yields two ordinary differential equations, by setting the sum of the $r$-dependent terms equal to $\alpha^2$ and the $z$-dependent term equal to $-\alpha^2$. Note that this choice is dictated by the fact that in this problem the axial direction ($z$) is homogeneous.

$$H'' + \alpha^2 H = 0 \tag{4.116}$$

$$G'' + \frac{1}{r}G' - \alpha^2 G = 0 \tag{4.117}$$

The solutions of eqs. (4.116) and (4.117) are (Appendixes A and B)

$$H = A \sin(\alpha z) + B \cos(\alpha z) \tag{4.118}$$

$$G = CI_0(\alpha r) + DK_0(\alpha r) \tag{4.119}$$

where $I_0(\alpha r)$ and $K_0(\alpha r)$ are the modified Bessel functions of order zero and of the first and second kind, respectively.

It is worth mentioning that the part of the solution for the temperature corresponding to the homogeneous direction, $z$, is in terms of the orthogonal functions [$\sin(\alpha z)$ and $\cos(\alpha z)$]. On the other hand, the part of solution corresponding to the nonhomogeneous direction, $r$, is in terms of nonorthogonal functions [$I_0(\alpha z)$ and $K_0(\alpha z)$] since the modified Bessel functions, unlike the Bessel functions, are not orthogonal (Appendix C). The solution for the temperature at this point reads

$$\Theta = [A \sin(\alpha z) + B \cos(\alpha z)][CI_0(\alpha r) + DK_0(\alpha r)] \tag{4.120}$$

Applying boundary conditions (4.111) and (4.113) and combining constants, we obtain

$$\Theta = K \sin(\alpha z) I_0(\alpha r) \tag{4.121}$$

Next, condition (4.114) yields

$$\sin(\alpha L) = 0 \tag{4.122}$$

which is satisfied by

$$\alpha_n = \frac{n\pi}{L} \qquad n = 1, 2, \ldots \tag{4.123}$$

Therefore, the general solution for the temperature field becomes

$$\Theta = \sum_{n=1}^{\infty} K_n \sin\left(\frac{n\pi}{L} z\right) I_0\left(\frac{n\pi}{L} r\right) \tag{4.124}$$

Note that for $n = 0$ or $\alpha_0 = 0$ the trivial solution $\Theta = 0$ is obtained and it is not included in expression (4.124); that is, the summation starts from $n = 1$. The constant $K_n$ is obtained by applying the boundary condition (4.112) and by making use of the orthogonality property of the sine function. Since the procedure was explained in detail earlier [Example 4.1, cases (a) and (b)], we show only the final result:

$$K_n = \frac{2\int_0^L \dot{q}''(z)\sin\left(\frac{n\pi}{L}z\right)dz}{n\pi k I_1\left(\frac{n\pi}{L}R\right)} \tag{4.125}$$

Therefore, the expression for the temperature after substituting eq. (4.125) into eq. (4.124) and using the definition of $\Theta$ reads

$$T = T_\infty + \frac{2}{\pi}\sum_{n=1}^{\infty}\frac{\int_0^L \dot{q}''(z)\sin\left(\frac{n\pi}{L}z\right)dz}{nkI_1\left(\frac{n\pi}{L}R\right)}\sin\left(\frac{n\pi}{L}z\right)I_0\left(\frac{n\pi}{L}r\right) \tag{4.126}$$

To obtain the heat removed at $z = 0$, we apply Fourier's law:

$$\dot{q}_{z=0} = \int_{r=0}^{R} k\left(\frac{\partial T}{\partial z}\right)_{z=0} 2\pi r\,dr \tag{4.127}$$

Combining eqs. (4.126), and (4.127) and performing the algebra yields

$$\dot{q}_{z=0} = 4R\sum_{n=1}^{\infty}\frac{1}{n}\left[\int_0^L \dot{q}''(z)\sin\left(\frac{n\pi z}{L}\right)dz\right] \tag{4.128}$$

Similarly, the heat removal at $z = L$ is

$$\begin{aligned}\dot{q}_{z=L} &= \int_0^R -k\left(\frac{\partial T}{\partial z}\right)_{z=L} 2\pi r\,dr \\ &= -4R\sum_{n=1}^{\infty}\frac{(-1)^n}{n}\left[\int_0^L \dot{q}''(z)\sin\left(\frac{n\pi z}{L}\right)dz\right]\end{aligned} \tag{4.129}$$

In the special case where the heat flux through the lateral surface of the cylinder is constant ($\dot{q}'' =$ const), the reader can easily show that eqs. (4.126), (4.128), and (4.129), respectively, become

$$T = T_\infty + \frac{4}{\pi}\frac{\dot{q}''}{k}\sum_{n=0}^{\infty}\frac{L}{\pi(2n+1)^2 I_1\left[\frac{(2n+1)\pi}{L}R\right]}\sin\left[\frac{(2n+1)\pi z}{L}\right]I_0\left[\frac{(2n+1)\pi r}{L}\right] \tag{4.130}$$

$$\dot{q}_{z=0} = 8R\dot{q}'' \sum_{n=0}^{\infty} \frac{L}{\pi(2n+1)^2} \tag{4.131}$$

$$\dot{q}_{z=L} = -8R\dot{q}'' \sum_{n=0}^{\infty} (-1)^{2n+1} \frac{L}{\pi(2n+1)^2} \tag{4.132}$$

Note that the total heat transfer rate inputed to the cylinder equals $2\pi RL\dot{q}''$. Based on the symmetry of the problem, half of this heat transfer rate should be removed at $z = 0$ and the other half at $z = L$. Hence the result of the series expression (4.131) is $\dot{q}_{z=0} = \pi RL\dot{q}''$ and the result of the series expression (4.132), $\dot{q}_{z=L} = \pi RL\dot{q}''$.

**Case (b) The heat transfer coefficient is finite**

If the heat transfer coefficient is finite, the boundary conditions at the two ends of the cylinder become

$$z = 0: \qquad k\frac{\partial\Theta}{\partial z} = h\Theta \tag{4.133}$$

$$z = L: \qquad -k\frac{\partial\Theta}{\partial z} = h\Theta \tag{4.134}$$

The remaining two boundary conditions (4.111) and (4.112), as well as the governing equation (4.110), remain unchanged and complete the problem formulation. The solution procedure is the same as that discussed in case (a). We show only the key steps here.

After applying boundary conditions (4.111) and (4.133), the solution reads

$$\Theta = KI_0(\alpha r)\left[\frac{h}{\alpha k}\sin(\alpha z) + \cos(\alpha z)\right] \tag{4.135}$$

The eigenvalues are obtained by applying boundary condition (4.134), which results in

$$\cot(\alpha_n L) = -\frac{1}{2}\left[\frac{h}{\alpha_n k} - \frac{\alpha_n k}{h}\right] \qquad n = 1, 2, \ldots \tag{4.136}$$

After numerical values are assigned to $L$, $h$, and $k$, eq. (4.136) can be solved numerically to yield the eigenvalues $\alpha_n$, $n = 1, 2, 3, \ldots$ . To proceed, we assume that from this point on $\alpha_n$, $n = 1, 2, 3, \ldots$ , are known. Therefore,

$$\Theta = \sum_{n=1}^{\infty} K_n I_0(\alpha_n r)\left[\frac{h}{\alpha_n k}\sin(\alpha_n z) + \cos(\alpha_n z)\right] \tag{4.137}$$

The constant $K_n$ is obtained by applying boundary condition (4.112) together with the orthogonality property of the function in brackets in [eq. (4.137)].

The fact that this function possesses the orthogonality property can be proved with the help of Appendix C. This function was also shown to possess the orthogonality property earlier in Example 4.1, case(c). The final expression for $K_n$ reads

$$K_n = \frac{\displaystyle\int_0^L \left[\frac{h}{\alpha_n k}\sin(\alpha_n z) + \cos(\alpha_n z)\right]\dot{q}''(z)\,dz}{kI_1(\alpha_n R)\displaystyle\int_0^L \left[\frac{h}{\alpha_n k}\sin(\alpha_n z) + \cos(\alpha_n z)\right]^2 dz} \tag{4.138}$$

Substituting eq. (4.138) into eq. (4.137) and performing the algebra yields the final expression for the temperature field:

$$T = T_\infty + \sum_{n=1}^{\infty} \frac{\left\{\displaystyle\int_0^L \left[\frac{h}{\alpha_n k}\sin(\alpha_n z) + \cos(\alpha_n z)\right]\dot{q}''(z)\,dz\right\} I_0(\alpha_n r)\left[\frac{h}{\alpha_n k}\sin(\alpha_n z) + \cos(\alpha_n z)\right]}{\alpha_n k I_1(\alpha_n R)\left[\left(\frac{h^2}{\alpha_n^2 k^2} + 1\right)\frac{L}{2} + \frac{\sin(2\alpha_n L)}{4\alpha_n}\left(1 - \frac{h^2}{\alpha_n^2 k^2}\right) + \frac{h}{\alpha_n^2 k}\sin^2(\alpha_n L)\right]} \tag{4.139}$$

Proceeding as in case (a), we are now in a position to evaluate the heat fluxes at each end of the cylinder by applying Fourier's law. The final result of this operation is

$$\dot{q}_{z=0} = 2\pi \sum_{n=1}^{\infty} \frac{h}{\alpha_n k}\frac{R}{\alpha_n} \frac{\displaystyle\int_0^L \left[\frac{h}{\alpha_n k}\sin(\alpha_n z) + \cos(\alpha_n z)\right]\dot{q}''(z)\,dz}{\left(\frac{h^2}{\alpha_n^2 k^2} + 1\right)\frac{L}{2} + \frac{\sin(2\alpha_n L)}{4\alpha_n}\left(1 - \frac{h^2}{\alpha_n^2 k^2}\right) + \frac{h}{\alpha_n^2 k}\sin^2(\alpha_n L)} \tag{4.140}$$

$$\dot{q}_{z=L} = -2\pi \sum_{n=1}^{\infty} \frac{R}{\alpha_n} \frac{\left\{\displaystyle\int_0^L \left[\frac{h}{\alpha_n k}\sin(\alpha_n z) + \cos(\alpha_n z)\right]\dot{q}''(z)\,dz\right\}\left[\frac{h}{\alpha_n k}\cos(\alpha_n L) - \sin(\alpha_n L)\right]}{\left(\frac{h^2}{\alpha_n^2 k^2} + 1\right)\frac{L}{2} + \frac{\sin(2\alpha_n L)}{4\alpha_n}\left(1 - \frac{h^2}{\alpha_n^2 k^2}\right) + \frac{h}{\alpha_n^2 k}\sin^2(\alpha_n L)} \tag{4.141}$$

When the heat flux incident to the lateral surface is constant ($\dot{q}''$ = const), eqs. (4.139) to (4.141) reduce to

$$T = T_\infty + \dot{q}'' \sum_{n=1}^{\infty} \frac{\left\{\frac{h}{\alpha_n^2 k}[1 - \cos(\alpha_n z)] + \frac{\sin(\alpha_n L)}{\alpha_n}\right\}\left[\frac{h}{\alpha_n k}\sin(\alpha_n z) + \cos(\alpha_n z)\right] I_0(\alpha_n r)}{\alpha_n k\left[\left(\frac{h^2}{\alpha_n^2 k^2} + 1\right)\frac{L}{2} + \frac{\sin(2\alpha_n L)}{4\alpha_n}\left(1 - \frac{h^2}{\alpha_n^2 k^2}\right) + \frac{h}{\alpha_n^2 k}\sin^2(\alpha_n L)\right] I_1(\alpha_n R)} \tag{4.142}$$

$$\dot{q}_{z=0} = 2\pi\dot{q}'' \sum_{n=1}^{\infty} \frac{\dfrac{h}{\alpha_n k}\dfrac{R}{\alpha_n}\left[\dfrac{h}{\alpha_n^2 k}(1 - \cos(\alpha_n z)) + \dfrac{\sin(\alpha_n L)}{\alpha_n}\right]}{\left(\dfrac{h^2}{\alpha_n^2 k^2} + 1\right)\dfrac{L}{2} + \dfrac{\sin(2\alpha_n L)}{4\alpha_n}\left(1 - \dfrac{h^2}{\alpha_n^2 k^2}\right) + \dfrac{h}{\alpha_n^2 k}\sin^2(\alpha_n L)} \tag{4.143}$$

$$\dot{q}_{z=L} = -2\pi\dot{q}'' \sum_{n=1}^{\infty} \frac{R}{\alpha_n} \frac{\left\{\dfrac{h}{\alpha_n^2 k}[1 - \cos(\alpha_n L)] + \dfrac{\sin(\alpha_n L)}{\alpha_n}\right\}\left[\dfrac{h}{\alpha_n k}\cos(\alpha_n L) - \sin(\alpha_n L)\right]}{\left(\dfrac{h^2}{\alpha_n^2 k^2} + 1\right)\dfrac{L}{2} + \dfrac{\sin(2\alpha_n L)}{4\alpha_n}\left(1 - \dfrac{h^2}{\alpha_n^2 k^2}\right) + \dfrac{h}{\alpha_n^2 k}\sin^2(\alpha_n L)} \tag{4.144}$$

## 4.3 TWO-DIMENSIONAL CONDUCTION IN SPHERICAL COORDINATES

The method of separation of variables can also be applied to conduction problems in spherical coordinates. The essential characteristics of the method discussed in Sections 4.1 and 4.2 pertaining to Cartesian and cylindrical coordinates, respectively, remain unchanged. In the following example, we stress the features that apply particularly to conduction in spherical coordinates.

**Example 4.4**

A condensing droplet on a horizontal flat surface can be modeled as the hemisphere shown in Fig. 4.7. The bottom surface temperature of the droplet, $T_c$, is assumed to be lower than the temperature of the convex surface. It is further

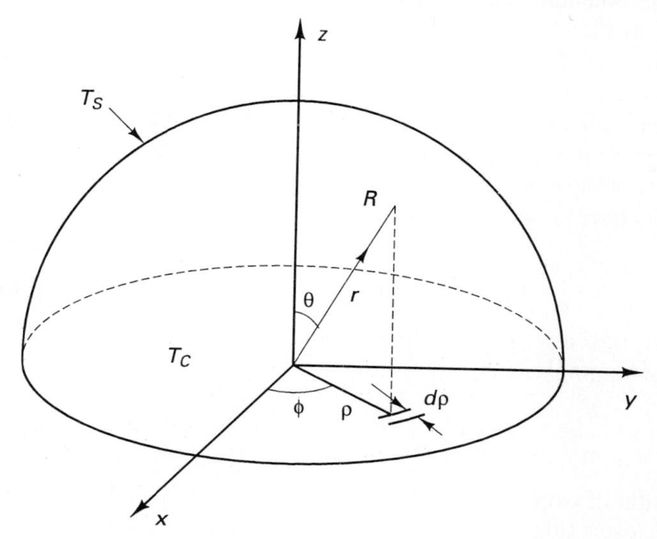


**Figure 4.7** Hemispherical droplet condensing on an isothermal surface.

assumed that the convex surface is at the saturation temperature of the liquid, $T_s$. We need to determine the temperature distribution in the droplet as well as the cooling rate through the bottom surface at the droplet that is in contact with the horizontal surface. All convection effects are negligible and the mass of the droplet changes very slowly with time.

**Solution** For the problem described above, the temperature is independent of coordinate $\varphi$. Therefore, the conduction equation in spherical coordinates reduces to

$$\frac{\partial}{\partial r}\left(r^2 \frac{\partial \Theta}{\partial r}\right) + \frac{1}{\sin\theta}\frac{\partial}{\partial\theta}\left(\sin\theta \frac{\partial \Theta}{\partial \theta}\right) = 0 \tag{4.145}$$

where $\Theta = T - T_c$.

Two boundary conditions, one in the radial and one in the angular direction are routine to express in mathematical form:

$$r = R: \quad \Theta = T_s - T_c = \Theta_c \tag{4.146}$$

$$\theta = \pi/2: \quad \Theta = 0 \tag{4.147}$$

The remaining two boundary conditions are not clear a priori and deserve a brief commentary. First, since the boundary condition at $r = 0$ falls on the $\theta = \pi/2$ plane, it is reasonable to expect that for $r = 0$, $\Theta = 0$ from condition (4.147). Next, we observe that the temperature distribution at $\theta = 0$ is unknown. None of the "common" types of boundary conditions discussed so far can be applied there in a plausible manner. In this case we make use of the fact that from the physics of the problem, the temperature must be *bounded* at $\theta = 0$. It is worth noting in advance that the $\theta$-direction with one bounded and one homogeneous boundary condition plays the role of the homogeneous direction (the direction that featured two homogeneous boundary conditions) encountered earlier in this chapter in connection with problems in Cartesian and cylindrical coordinate systems.

This is true because instead of the bounded boundary condition, one may utilize the symmetry of the problem and state that $(\partial\Theta/\partial\theta)_{\theta=0} = 0$. This boundary condition is homogeneous and shows that the $\theta$-direction is indeed the homogeneous direction for this problem.

Letting

$$\Theta = G(r)H(\theta) \tag{4.148}$$

substituting the above into eq. (4.145), dividing the resulting equation by the product $GH$, and rearranging yields

$$\frac{1}{G}\frac{d}{dr}\left(r^2 \frac{dG}{dr}\right) = -\frac{1}{H\sin\theta}\frac{d}{d\theta}\left(\sin\theta \frac{dH}{d\theta}\right) = \alpha^2 \tag{4.149}$$

Since the left-hand side of expression (4.119) depends on $r$ only and the right-hand side on $\theta$ only, each side was set equal to a constant, $\alpha^2$. Hence we recover two ordinary differential equations:

$$r^2 \frac{d^2 G}{dr^2} + 2r \frac{dG}{dr} - \alpha^2 G = 0 \tag{4.150}$$

$$\frac{d}{d\theta}\left(\sin\theta \frac{dH}{d\theta}\right) + \alpha^2(\sin\theta)H = 0 \tag{4.151}$$

The solution of these equations is discussed in Appendixes B and D. Here it will suffice to say that eq. (4.150) is an *Euler equation* and eq. (4.151) is a *Legendre equation* [1,5,6]. The solution to eq. (4.150) is

$$G(r) = Ar^n + \frac{B}{r^{n+1}} \tag{4.152}$$

where

$$\alpha^2 = n(n + 1) \tag{4.153}$$

This solution is determined by letting $G(r) = r^p$, substituting this expression into eq. (4.150) and solving the resulting quadratic algebraic equation for $p$. In our problem, two distinct values of $p$ were obtained. The details of the solution process are included in Appendix B (Example B.2).

To proceed, the following variable transformation is customarily applied:

$$\eta = \cos\theta \tag{4.154}$$

Then

$$\frac{dH}{d\theta} = \frac{dH}{d\eta}\frac{d\eta}{d\theta} = -\sin\theta \frac{dH}{d\eta} \tag{4.155}$$

Multiplying both sides of eq. (4.155) by $\sin\theta$ and noting that

$$\sin^2\theta = 1 - \cos^2\theta = 1 - \eta^2 \tag{4.156}$$

it follows that

$$\frac{d}{d\theta}\left(\sin\theta \frac{dH}{d\theta}\right) = \frac{d}{d\theta}\left[(\eta^2 - 1)\frac{dH}{d\eta}\right]$$

$$= \frac{d}{d\eta}\left[(\eta^2 - 1)\frac{dH}{d\eta}\right]\frac{d\eta}{d\theta} = \frac{d}{d\eta}\left[(1 - \eta^2)\frac{dH}{d\eta}\right]\sin\theta \tag{4.157}$$

Substituting eq. (4.157) into eq. (4.151), canceling $\sin\theta$, and making use of eq. (4.153), we obtain

$$\frac{d}{d\eta}\left[(1 - \eta^2)\frac{dH}{d\eta}\right] + n(n + 1)H = 0 \tag{4.158}$$

Equation (4.158) can be rewritten as

$$(1 - \eta^2)H'' - 2\eta H' + n(n + 1)H = 0 \tag{4.159}$$

where the primes denote differentiation with respect to $\eta$.

As mentioned earlier, eq. (4.159) is Legendre's equation, and its solution is expressed in terms of the Legendre polynomials and the Legendre functions of the second kind $P_n(\eta)$ and $Q_n(\eta)$, respectively (Appendix D). Hence

$$H(\eta) = CP_n(\eta) + DQ_n(\eta) \tag{4.160}$$

Combining eqs. (4.148), (4.152), and (4.160) yields the expression for the temperature distribution:

$$\Theta = \left(Ar^n + \frac{B}{r^{n+1}}\right)[CP_n(\eta) + DQ_n(\eta)] \tag{4.161}$$

To obtain the unknown constants, we make use of the boundary conditions. First, the fact that the temperature is bounded (zero) as $r \to 0$ yields that

$$B = 0 \tag{4.162}$$

Also, since the Legendre functions of the second kind $Q_n(\eta)$ are unbounded for $\eta \pm 1$ or $\theta = 0, \pi, 2\pi, \ldots$, we require that

$$D = 0 \tag{4.163}$$

to satisfy the condition that the temperature is bounded at $\theta = 0$.

Combining eqs. (4.161)–(4.163) yields

$$\Theta = Kr^n P_n(\eta) \tag{4.164}$$

By superposition over all values of $n$ ($n = 1, 2, \ldots$),

$$\Theta = \sum_{n=1}^{\infty} K_n r^n P_n(\eta) \tag{4.165}$$

The unknown constant $K_n$ is obtained by applying boundary condition (4.147) together with the orthogonality property of the Legendre polynomials of the first kind (Appendix D). Equation (4.147) yields

$$\Theta_s = \sum_{n=1}^{\infty} K_n R^n P_n(\eta) \tag{4.166}$$

Multiplying both sides of eq. (4.166) by $P_m(\eta)$ (the weighting factor is unity), integrating from $\eta = 0$ ($\theta = \pi/2$) to $\eta = 1$($\theta = 0$), and making use of the orthogonality property of $P_m(\eta)$ (Appendix D) yields

$$K_n = \frac{\Theta_s \displaystyle\int_0^1 P_n(\eta)\, d\eta}{R^n \displaystyle\int_0^1 [P_n(\eta)]^2\, d\eta}$$

$$= \frac{\Theta_s}{R^n}[P_{n+1}(1) - P_{n-1}(1) - P_{n+1}(0) + P_{n-1}(0)] \tag{4.167}$$

The final expression for the temperature is obtained by combining equations (4.165) and (4.167):

$$T = T_c + (T_s - T_c) \sum_{n=1}^{\infty} [P_{n+1}(1) - P_{n-1}(1) - P_{n+1}(0) + P_{n-1}(0)]\left(\frac{r}{R}\right)^n P_n(\cos\theta) \tag{4.168}$$

The heat flux through the bottom surface of the sphere is obtained with the help of Fourier's law. Based on the Cartesian system shown in Fig. 4.7,

$$\dot{q} = \int_{\rho=0}^{R} k\left(\frac{\partial T}{\partial z}\right)_{z=0} 2\pi\rho \, d\rho \tag{4.169}$$

Keeping in mind that the spherical and Cartesian coordinates are connected by

$$\rho = r \sin \theta \tag{4.170}$$

$$z = r \cos \theta \tag{4.171}$$

eqs. (4.168)–(4.171) can be combined to yield the final expression for the heat flux through the base of the droplet. This last algebraic step is left as an exercise for the reader.

## 4.4 THE PRINCIPLE OF SUPERPOSITION IN HEAT CONDUCTION

In the preceding sections of this chapter, two-dimensional steady conduction was discussed in the three coordinate systems commonly used in thermal engineering applications: Cartesian, cylindrical, and spherical. More specifically, a host of conduction examples was presented and solved with the method of separation of variables. An important point made was that care should be exercised before the method is applied. For example, in two-dimensional conduction in Cartesian coordinates three of the four possible boundary conditions as well as the governing equation needed to be homogeneous. In several cases in the previous sections three boundary conditions were converted to homogeneous with the help of a simple variable transformation.

Another useful tool that can be used to make the application of the method of separation of variables in heat conduction possible is the *principle of superposition* [2, 7–10], which constitutes the topic of the present section. For the principle of superposition to apply, the governing conduction equation as well as the boundary conditions need to be linear in the unknown function (temperature). Such is the case for the majority of the problems that fall within the scope of the present book.

The main concept of the principle of superposition is the splitting of a complex problem not solvable by direct application of the method of separation of variables into a number of simpler problems solvable by the method of separation of variables. The solution to the initial complex problems is, then, the sum of the solution of the simpler problems.

The application of the superposition principle in heat conduction problems will be better understood with the help of two illustrative examples. However, before these examples are presented, a sequence of recommended steps to be followed when the principle of superposition is applied is outlined.

■ ***Step 1***

Make sure that the heat conduction equation as well as the accompanying boundary conditions are indeed linear.

■ ***Step 2***

Attempt to reduce the complexity of the problem by introducing a variable transformation that may convert some of the nonhomogeneous boundary conditions to homogeneous. For example, if the boundary condition at a given surface is $T = T_0 =$ const, introducing $\Theta = T - T_0$ transforms this boundary condition to $\Theta = 0$, which is homogeneous.

■ ***Step 3***

Decompose the resulting problem into as small a number as possible of simpler problems solvable with the method of separation of variables. An important remark here is that an attempt should be made to include as many nonhomogeneities as possible with the easiest problem resulting from the decomposition. For example, if the decomposition yields a two-dimensional problem and a one-dimensional problem, it is recommended to include as many nonhomogeneities as possible in the equation and the boundary conditions of the one-dimensional problem whose solution is much easier than the solution of the two-dimensional problem. In addition, in decomposing the original problem one should make sure that *when the respective governing equations and boundary conditions of the resulting "component" problems are added up, the governing equation and boundary conditions of the original problem are recovered.*

■ ***Step 4***

Solve the problems resulting from the decomposition.

■ ***Step 5***

Add up the solutions obtained in step 4 to construct the solution to the original problem.

The sequence of steps presented above is applied next to the solution of two illustrative examples.

### Example 4.5

Consider the fin discussed in Example 4.1, case (c). Assume that the right side of the fin is exposed to a thermal radiation flux $\dot{q}''$, resulting in the configuration shown in Fig. 4.8a. Obtain the temperature distribution in the fin.

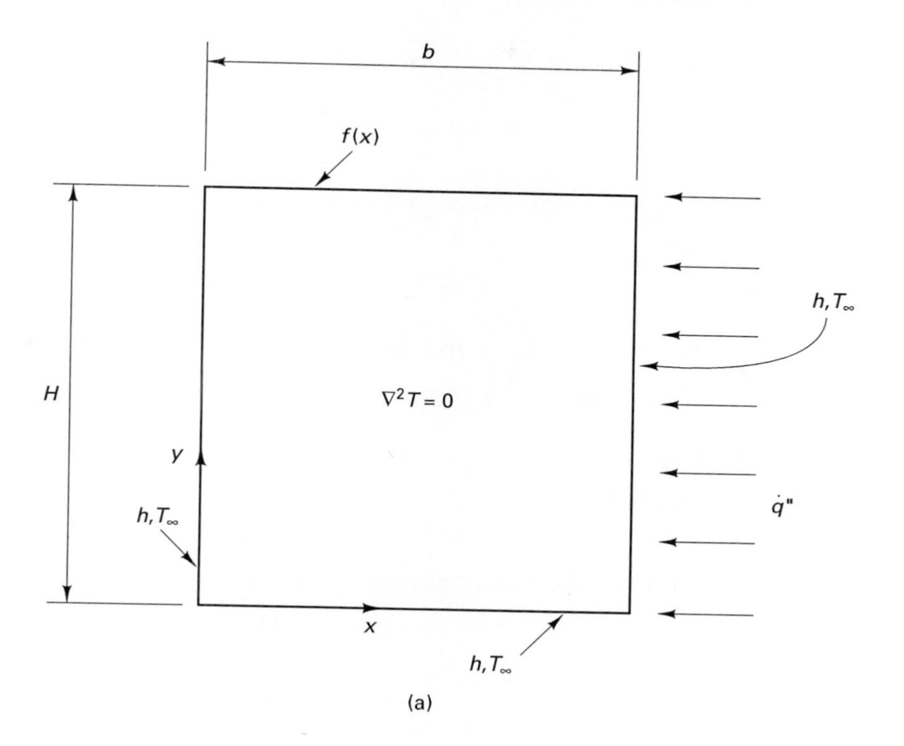


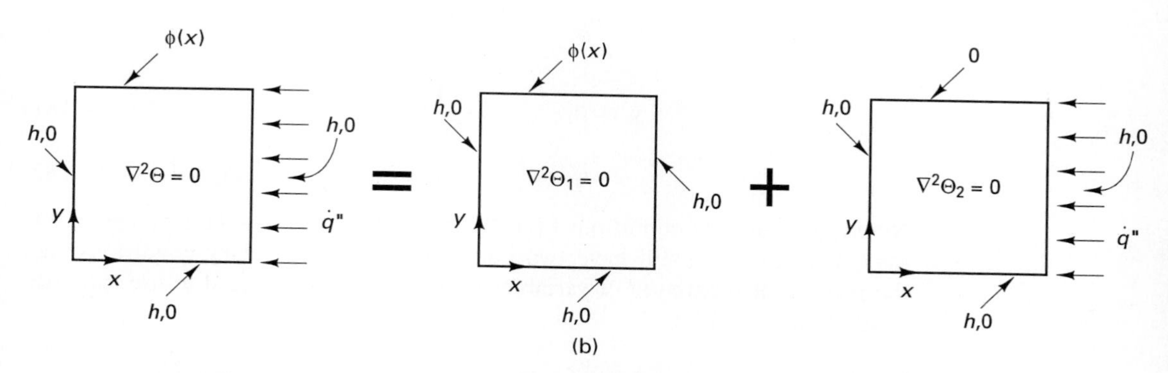


**Figure 4.8** (a) Schematic of a problem to be solved by superposition. (b) Decomposition process.

**Solution**

■ ***Step 1***

The mathematical model (conduction equation and boundary condition) of this problem is

$$\frac{\partial^2 T}{\partial x^2} + \frac{\partial^2 T}{\partial y^2} = 0 \tag{4.172}$$

$$x = 0: \qquad k\frac{\partial T}{\partial x} = h(T - T_\infty) \tag{4.173}$$

$$x = b: \qquad -k\frac{\partial T}{\partial x} = h(T - T_\infty) - \dot{q}'' \tag{4.174}$$

$$y = 0: \qquad k\frac{\partial T}{\partial y} = h(T - T_\infty) \tag{4.175}$$

$$y = H: \qquad T = f(x) \tag{4.176}$$

The set of equations (4.172)–(4.176) is linear in $T$ and its derivatives.

■ ***Step 2***

Introducing

$$\Theta = T - T_\infty \tag{4.177}$$

and rewriting the heat conduction model in terms of $\Theta$ yields

$$\frac{\partial^2 \Theta}{\partial x^2} + \frac{\partial^2 \Theta}{\partial y^2} = 0 \tag{4.178}$$

$$x = 0: \qquad k\frac{\partial \Theta}{\partial x} = h\Theta \tag{4.179}$$

$$x = b: \qquad -k\frac{\partial \Theta}{\partial x} = h\Theta - \dot{q}'' \tag{4.180}$$

$$y = 0: \qquad k\frac{\partial \Theta}{\partial y} = h\Theta \tag{4.181}$$

$$y = H: \qquad \Theta = f(x) - T_\infty = \phi(x) \tag{4.182}$$

Note that boundary conditions (4.179) and (4.181) are now homogeneous. On the other hand, we still have two nonhomogeneous boundary conditions. For the method of separation of variables to be applicable we need to use the principle of superposition.

■ ***Step 3***

We attempt to decompose the problem into two "component" problems as shown in Fig. 4.8b, so that

$$\Theta = \Theta_1 + \Theta_2 \tag{4.183}$$

where $\Theta_1$ and $\Theta_2$ are the temperatures of the first component problem and the second component problem in Fig. 4.8b, respectively. The mathematical model for the first component problem is

$$\frac{\partial^2 \Theta_1}{\partial x^2} + \frac{\partial^2 \Theta_1}{\partial y^2} = 0 \tag{4.184}$$

$$x = 0: \qquad k\frac{\partial \Theta_1}{\partial x} = h\Theta_1 \tag{4.185}$$

$$x = b: \qquad -k\frac{\partial \Theta_1}{\partial x} = h\Theta_1 \tag{4.186}$$

$$y = 0: \qquad k\frac{\partial \Theta_1}{\partial y} = h\Theta_1 \tag{4.187}$$

$$y = H: \qquad \Theta_1 = \phi(x) \tag{4.188}$$

Similarly, the mathematical model for the second component problem reads

$$\frac{\partial^2 \Theta_2}{\partial x^2} + \frac{\partial^2 \Theta_2}{\partial y^2} = 0 \tag{4.189}$$

$$x = 0: \qquad k\frac{\partial \Theta_2}{\partial x} = h\Theta_2 \tag{4.190}$$

$$x = b: \qquad -k\frac{\partial \Theta_2}{\partial x} = h\Theta_2 - \dot{q}'' \tag{4.191}$$

$$y = 0: \qquad k\frac{\partial \Theta_2}{\partial y} = h\Theta_2 \tag{4.192}$$

$$y = H: \qquad \Theta_2 = 0 \tag{4.193}$$

Note that each of the problems above possesses three homogeneous boundary conditions, (4.185)–(4.187), for the first problem and (4.190), (4.192), and (4.193) for the second problem. Therefore, they are both solvable with the method of separation of variables. In addition, the reader can easily show that the sum of the governing equations and the corresponding boundary conditions of the two problems above yields the governing equation and boundary conditions of the original problem. For example, adding eqs. (4.186) and (4.191) side by side yields

$$x = b: \qquad -k\frac{\partial \Theta_1}{\partial x} - k\frac{\partial \Theta_2}{\partial x} = h\Theta_1 + h\Theta_2 - \dot{q}'' \tag{4.194a}$$

or

$$x = b: \qquad -k\frac{\partial}{\partial x}(\Theta_1 + \Theta_2) = h(\Theta_1 + \Theta_2) - \dot{q}'' \tag{4.194b}$$

Invoking eq. (4.183), we realize that eq. (4.194b) is identical to eq. (4.180), which is the boundary condition of the original problem at $x = b$.

### ■ *Step 4*

The solution of the problems for $\Theta_1$ and $\Theta_2$ falls along the lines of the discussion in earlier parts of this chapter. More specifically, the result for $\Theta_1$ can be obtained directly from expression (4.68) since the problem for $\Theta_1$ is identical to the problem solved in Example 4.1, case (c). The problem for $\Theta_2$ is also a standard application of the method of separation of variables. Omitting the details for brevity,

$$\Theta_2 = -\frac{\dot{q}''}{h}\sum_{n=1}^{\infty}\frac{1}{\lambda_n}\left[\frac{1-\cos(\lambda_n H)}{\cos(\lambda_n H)}\right]\left[\frac{k\lambda_n}{h}\cosh(\lambda_n x)+\sinh(\lambda_n x)\right][\sin(\lambda_n y)$$
$$-\tan(\lambda_n H)\cos(\lambda_n y)]\Big/\left(\left\{\left[\left(\frac{k\lambda_n}{h}\right)^2+1\right]\sinh(\lambda_n b)+\frac{2k\lambda_n}{h}\cosh(\lambda_n b)\right\}\right.$$
$$\left.\cdot\left\{\frac{H}{2\cos^2(\lambda_n H)}-\frac{\sin(2\lambda_n H)}{4\lambda_n}[1-\tan^2(\lambda_n H)]-\frac{\tan(\lambda_n H)}{\lambda_n}\sin^2(\lambda_n H)\right\}\right) \tag{4.195}$$

where the values of $\lambda_n$, $n = 1, 2, 3, \ldots$, are obtained from

$$\frac{\lambda_n k}{h} = -\tan(\lambda_n H) \tag{4.196}$$

### ■ *Step 5*

Using eqs. (4.68) and (4.195), it is trivial to carry out this step to obtain

$$\Theta = T - T_\infty$$
$$= \sum_{n=1}^{\infty}\left\{\left[\int_0^b \phi(x)\left[\frac{h}{\alpha_n k}\sin(\alpha_n x)+\cos(\alpha_n x)\right]dx\right\}\left[\frac{h}{\alpha_n k}\sin(\alpha_n x)+\cos(\alpha_n x)\right]\right.$$
$$\cdot\left[\frac{h}{\alpha_n k}\sinh(\alpha_n y)+\cosh(\alpha_n y)\right]\Big/\left(\left[\frac{h}{\alpha_n k}\sinh(\alpha_n H)+\cosh(\alpha_n H)\right]\right.$$
$$\left.\cdot\left\{\int_0^b\left[\frac{h}{\alpha_n k}\sin(\alpha_n x)+\cos(\alpha_n x)\right]^2dx\right\}\right)$$
$$-\frac{q''}{h}\sum_{n=1}^{\infty}\frac{1}{\lambda_n}\left[\frac{1-\cos(\lambda_n H)}{\cos(\lambda_n H)}\right]\left[\frac{k\lambda_n}{h}\cosh(\lambda_n x)+\sinh(\lambda_n x)\right][\sin(\lambda_n y)$$
$$-\tan(\lambda_n H)\cos(\lambda_n y)]\Big/\left(\left\{\left[\left(\frac{k\lambda_n}{h}\right)^2+1\right]\sinh(\lambda_n b)+\frac{2k\lambda_n}{h}\cosh(\lambda_n b)\right\}\right.$$
$$\left.\cdot\left\{\frac{H}{2\cos^2(\lambda_n H)}-\frac{\sin(2\lambda_n H)}{4\lambda_n}[1-\tan^2(\lambda_n H)]-\frac{\tan(\lambda_n H)}{\lambda_n}\sin^2(\lambda_n H)\right\}\right) \tag{4.197}$$

where $\alpha_n$ and $\lambda_n$ ($n = 1, 2, 3, \ldots$) are obtained from eqs. (4.64) and (4.196), respectively, after the numerical values of the various parameters of the problem are prescribed.

Finally, a remark regarding the placement of the coordinate systems in the component problems is of merit. It is not necessary to place the coordinate systems of the two (or more) component problems in the same location (the lower left-hand corner of the rectangles in Fig. 4.8, for example. As discussed earlier in this chapter, the solution for $\Theta_2$ would be easier (algebraically) if the upper left-hand corner had been chosen for placement of the coordinate system for the second component problem of Fig. 4.8. The student is encouraged to re-solve this component problem utilizing the above-mentioned coordinate system. Caution needs to be exercised, however, in step 5 when the solutions are added up: All the solutions of the component problems should be expressed with respect to a unique coordinate system.

**Example 4.6**

The cylindrical capsule of solid radioactive waste shown in Fig. 4.9a generates heat at constant rate $\dot{q}'''(W/m^3)$. The capsule rests on an adiabatic floor and it is being cooled with the help of a forced liquid coolant stream. The heat transfer coefficient between the capsule and the liquid coolant is assumed to be very large. Therefore, the surface temperature of the capsule in contact with the liquid can be set equal to the temperature of the liquid coolant far away from the capsule, $T_\infty$. Obtain the temperature distribution in the capsule.

**Solution**

■ ***Step 1***

The heat conduction equation and boundary conditions for the problem outlined above are

$$\frac{1}{r}\frac{\partial}{\partial r}\left(r\frac{\partial T}{\partial r}\right) + \frac{\partial^2 T}{\partial z^2} + \frac{\dot{q}'''}{k} = 0 \qquad (4.198)$$

$$r = 0: \quad \frac{\partial T}{\partial r} = 0 \qquad (4.199)$$

$$r = R: \quad T = T_\infty \qquad (4.200)$$

$$z = 0: \quad \frac{\partial T}{\partial z} = 0 \qquad (4.201)$$

$$z = L: \quad T = T_\infty \qquad (4.202)$$

The set of equations and boundary conditions above is linear.

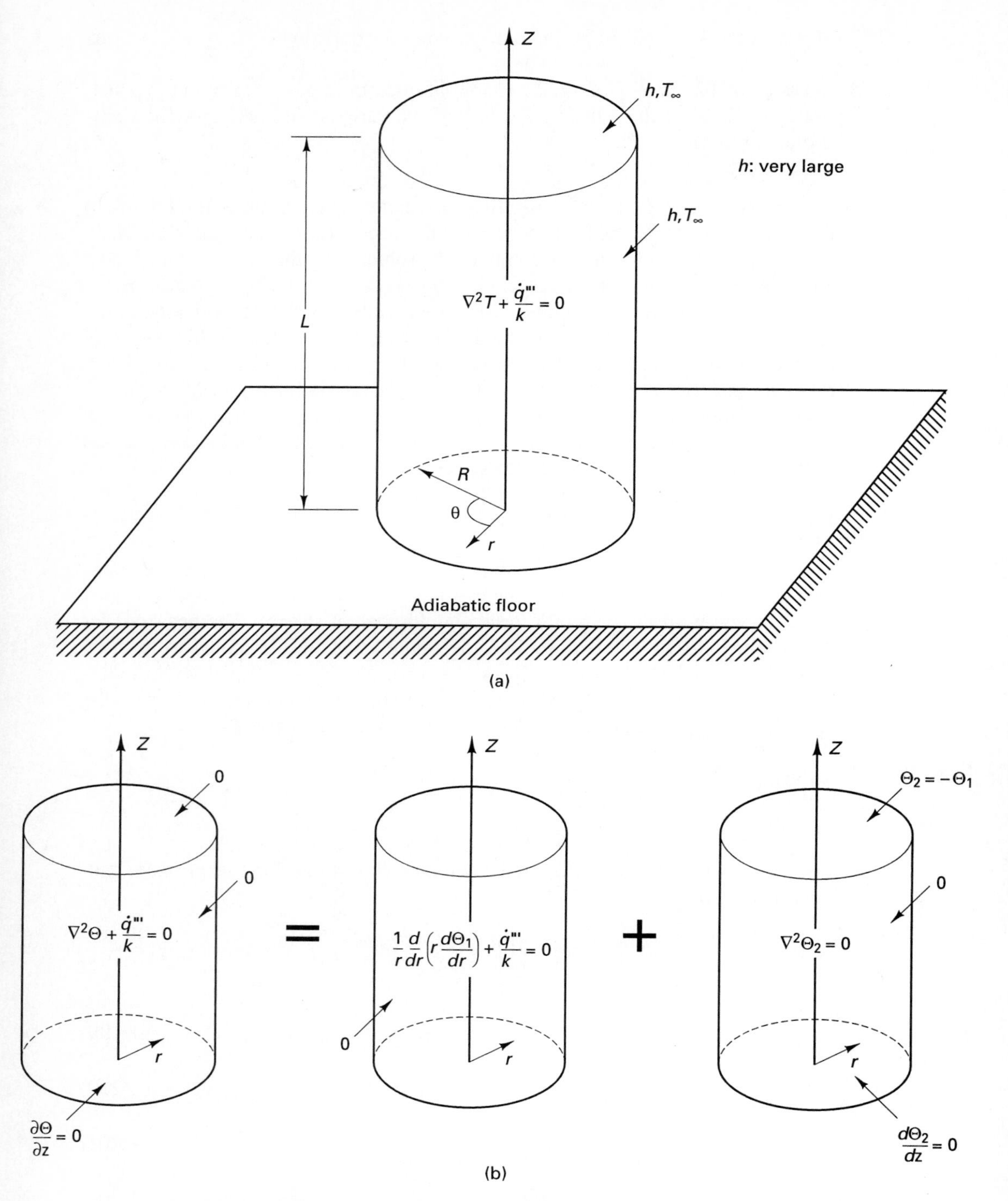


**Figure 4.9** (a) Schematic of heat-generating cylindrical capsule resting on an adiabatic floor. (b) Decomposition process.

■ ***Step 2***

Introducing the variable transformation $\Theta = T - T_\infty$ we cast eqs. (4.198)–(4.202) in the following form:

$$\frac{1}{r}\frac{\partial}{\partial r}\left(r\frac{\partial \Theta}{\partial r}\right) + \frac{\partial^2 \Theta}{\partial z^2} + \frac{\dot{q}'''}{k} = 0 \tag{4.203}$$

$$r = 0: \quad \frac{\partial \Theta}{\partial r} = 0 \tag{4.204}$$

$$r = R: \quad \Theta = 0 \tag{4.205}$$

$$z = 0: \quad \frac{\partial \Theta}{\partial z} = 0 \tag{4.206}$$

$$z = L: \quad \Theta = 0 \tag{4.207}$$

All the boundary conditions are homogeneous, but the governing equation is not. To be able to use the method of separation of variables we invoke the superposition principle.

■ ***Step 3***

The original problem is decomposed into one one-dimensional problem ($\Theta_1$) and one two-dimensional problem ($\Theta_2$) as shown in Fig. 4.9b. The nonhomogeneous (heat generation) part of the original governing equation is included in the governing equation of the one-dimensional problem, which can easily be solved. As before,

$$\Theta = \Theta_1 + \Theta_2 \tag{4.208}$$

The first component problem is described by

$$\frac{1}{r}\frac{d}{dr}\left(r\frac{d\Theta_1}{dr}\right) + \frac{\dot{q}'''}{k} = 0 \tag{4.209}$$

$$r = 0: \quad \frac{d\Theta_1}{dr} = 0 \tag{4.210}$$

$$r = R: \quad \Theta_1 = 0 \tag{4.211}$$

Clearly, no boundary conditions in the $z$-direction are necessary because $\Theta_1$ does not depend on $z$.

Similarly the model of the second component problem reads

$$\frac{1}{r}\frac{\partial}{\partial r}\left(r\frac{\partial \Theta_2}{\partial r}\right) + \frac{\partial^2 \Theta_2}{\partial z^2} = 0 \tag{4.212}$$

$$r = 0: \quad \frac{\partial \Theta_2}{\partial r} = 0 \tag{4.213}$$

$$r = R: \qquad \Theta_2 = 0 \tag{4.214}$$

$$z = 0: \qquad \frac{\partial \Theta_2}{\partial z} = 0 \tag{4.215}$$

$$z = L: \qquad \Theta_2 = -\Theta_1 \tag{4.216}$$

It can be easily shown that if the corresponding boundary conditions and governing equations for the problems for $\Theta_1$ and $\Theta_2$, respectively, are added up side by side, the model for the original problem [eqs. (4.203)–(4.207)] will be recovered. To exemplify this fact, we show how boundary condition (4.206) is recovered. Since $\Theta_1$ is not a function of $z$, $d\Theta_1/dz = 0$ everywhere. Hence

$$z = 0: \qquad \frac{d\Theta_1}{dz} = 0 \tag{4.217}$$

Adding up eqs. (4.215) and (4.217) side by side yields

$$z = 0: \qquad \frac{d}{dz}(\Theta_1 + \Theta_2) = \frac{d\Theta}{dz} = 0 \tag{4.218}$$

which is identical to eq. (4.206).

■ ***Step 4***

It is of interest to note that for the second problem to be solved, knowledge of $\Theta_1$ is necessary since it appears in boundary condition (4.217). The solution for the first problem is obtained by first integrating eq. (4.209) twice and, next, applying the two boundary conditions (4.210) and (4.211). The final result reads

$$\Theta_1 = \frac{\dot{q}'''R^2}{4k}\left(1 - \frac{r^2}{R^2}\right) \tag{4.219}$$

The problem for $\Theta_2$ falls directly along the lines of what was discussed in Examples 4.1 and 4.2 with reference to conduction in cylindrical coordinates. Since the governing equation (4.212) as well as three boundary conditions are homogeneous [eqs. (4.213)–(4.215)] separation of variables can be applied directly. Note that the homogeneous direction is the radial direction and orthogonal functions (Bessel functions) should be sought in that direction. Leaving the solution as an exercise to the reader, we provide here only the final result:

$$\Theta_2 = -\frac{\dot{q}'''}{k} \sum_{n=1}^{\infty} \frac{J_2(\lambda_n R)/J_1(\lambda_n R)}{\lambda_n^2 \cosh(\lambda_n L)} J_0(\lambda_n r) \cosh(\lambda_n z) \tag{4.220}$$

where the values for $\lambda_n$, $n = 1, 2, 3, \ldots$ are the solution of the following algebraic equation:

$$J_0(\lambda_n R) = 0 \tag{4.221}$$

■ ***Step 5***

The temperature field of the original problem is then [eqs. (4.208), (4.219), and (4.220)]

$$\Theta = T - 2T_\infty = \frac{\dot{q}'''R^2}{4k}\left(1 - \frac{r^2}{R^2}\right) - \frac{\dot{q}'''}{k}\sum_{n=1}^{\infty}\frac{J_2(\lambda_n R)/J_1(\lambda_n R)}{\lambda_n^2 \cosh(\lambda_n L)} J_0(\lambda_n r)\cosh(\lambda_n z) \tag{4.222}$$

where $\lambda_n$ is obtained from eq. (4.221) after a numerical value for $R$ is prescribed.

Before closing our discussion on the use of the principle of superposition in heat conduction, we find it important to point out that applying step 3 in the procedure outlined earlier is not always easy since several possibilities exist on how a problem can be decomposed and the most efficient way may be initially elusive. It is the experience gathered by applying the principle of superposition to a host of problems that enables problem solvers to "pick their way" in an optimal manner when decomposing the initial problem.

## 4.5 THREE-DIMENSIONAL STEADY-STATE CONDUCTION

The basic principles of solution of two-dimensional, steady heat conduction problems with the method of separation of variables apply here as well. The principle of superposition can be involved, if needed, following guidelines analogous to those discussed in two-dimensional steady conduction. To illustrate the solution procedure we present the following example.

**Example 4.7**

A cubical fin with sides of length $L$ is mounted on a hot surface as shown in Fig. 4.10. The base temperature distribution of the fin is known (measured) and it is denoted by $f(x, y)$. A fluid coolant flows past the fin. The coolant temperature is denoted by $T_\infty$. For simplicity, it is assumed that the heat transfer coefficient between the fin surface and the coolant is very large ($h \to \infty$). Obtain the temperature distribution in the fin and the overall heat transfer through the fin base.

**Solution**

The assumption of a very large heat transfer coefficient allows for the approximation that the surface temperature of the fin in contact with the fluid equals the fluid temperature. Defining $\Theta = T - T_\infty$ the mathematical model for the conduction process under consideration becomes

$$\frac{\partial^2 \Theta}{\partial x^2} + \frac{\partial^2 \Theta}{\partial y^2} + \frac{\partial^2 \Theta}{\partial z^2} = 0 \tag{4.223}$$

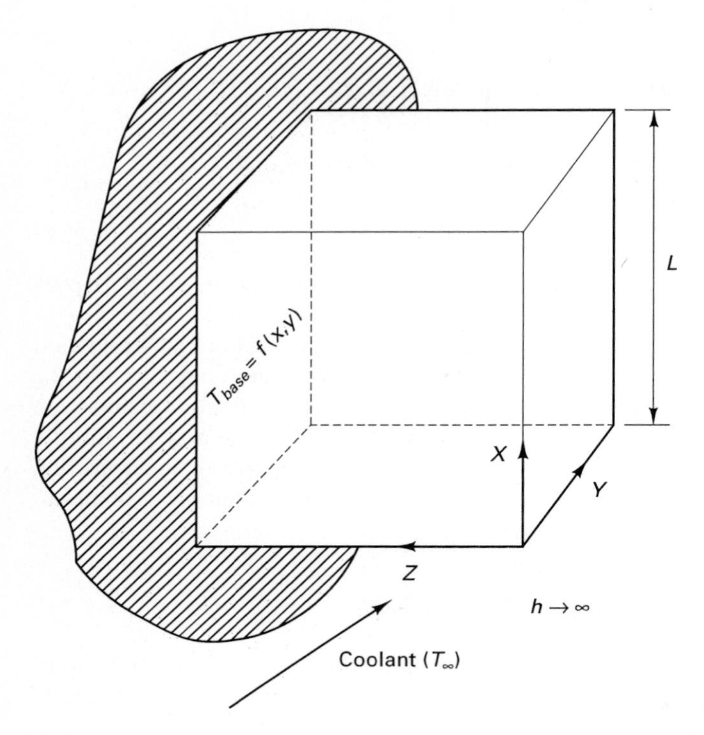


**Figure 4.10** Cubical fin.

$$z = 0: \quad \Theta = 0 \tag{4.224}$$

$$z = L: \quad \Theta = f(x, y) - T_\infty = \phi(x, y) \tag{4.225}$$

$$y = 0: \quad \Theta = 0 \tag{4.226}$$

$$y = L: \quad \Theta = 0 \tag{4.227}$$

$$x = 0: \quad \Theta = 0 \tag{4.228}$$

$$x = L: \quad \Theta = 0 \tag{4.229}$$

It is worth mentioning that for the method of separation of variables to apply directly in three-dimensional heat conduction, *five* out of the six boundary conditions need to be homogeneous. This is clearly the case in the present problem. Assuming that

$$\Theta = X(x)Y(y)Z(z) \tag{4.230}$$

substituting expression (4.230) into the governing equation (4.223), and dividing both sides of the resulting equation by the product $XYZ$ yields

$$\underbrace{\frac{X''}{X}}_{-\lambda^2} + \underbrace{\frac{Y''}{Y} + \frac{Z''}{Z}}_{-\beta^2} = 0 \tag{4.231}$$

Seeking orthogonality in the two homogeneous directions we set the $x$-dependent term of eq. (4.231) equal to $-\lambda^2$ and the $y$-dependent term to $-\beta^2$, to obtain

$$X'' + \lambda^2 X = 0 \tag{4.232}$$

$$Y'' + \beta^2 Y = 0 \tag{4.233}$$

$$Z'' - (\lambda^2 + \beta^2)Z = 0 \tag{4.234}$$

Solving the foregoing ordinary differential equations (Appendix B) and substituting the solutions into expression (4.230) yields

$$\Theta = [A_1 \sin(\lambda x) + A_2 \cos(\lambda x)][A_3 \sin(\beta y) + A_4 \cos(\beta y)] \cdot [A_5 \sinh(\sqrt{\lambda^2 + \beta^2}\, z) + A_6 \cosh(\sqrt{\lambda^2 + \beta^2}\, z)] \tag{4.235}$$

To evaluate the constants $A_1$ to $A_6$, $\lambda$ and $\beta$ we start applying the boundary conditions (4.224)–(4.229). First, boundary conditions (4.224), (4.226), and (4.228) yield

$$A_2 = A_4 = A_6 = 0 \tag{4.236}$$

Second, boundary conditions, (4.227) and (4.229), respectively, yield

$$\lambda_n = \frac{n\pi}{L} \qquad n = 1, 2, 3, \ldots \tag{4.237}$$

$$\beta_m = \frac{m\pi}{L} \qquad m = 1, 2, 3, \ldots \tag{4.238}$$

Combining eqs. (4.235)–(4.238) yields

$$\Theta_{mn} = E_{mn} \sin\left(\frac{n\pi}{L}x\right)\sin\left(\frac{m\pi}{L}y\right) \sinh\left(\pi\sqrt{m^2 + n^2}\,\frac{z}{L}\right) \tag{4.239}$$

The constant $E_{mn}$ was obtained by combining the constants $A_1$, $A_3$, and $A_5$. The subscript $mn$ accounts for the fact that all values of $m = 1, 2, 3, \ldots$ and all values of $n = 1, 2, 3$ yield solutions for the temperature distribution. The general solution is the linear combination of all these solutions. Based on the above, the expression for the general solution for the temperature distribution reads

$$\Theta = \sum_{m=1}^{\infty}\sum_{n=1}^{\infty} E_{mn} \sin\left(\frac{n\pi}{L}x\right)\sin\left(\frac{m\pi}{L}y\right) \sinh\left(\pi\sqrt{m^2 + n^2}\,\frac{z}{L}\right) \tag{4.240}$$

Next, we apply boundary condition (4.225):

$$\phi(x, y) = \sum_{m=1}^{\infty}\sum_{n=1}^{\infty} E_{mn} \sin\left(\frac{n\pi}{L}x\right) \sinh(\pi\sqrt{m^2 + n^2}) \sin\left(\frac{m\pi}{L}y\right) \tag{4.241}$$

To proceed, we notice that the quantity

$$\sum_{n=1}^{\infty} E_{mn} \sin\left(\frac{n\pi}{L}x\right) \sinh(\pi\sqrt{m^2 + n^2})$$

depends only on index $m$ (after, imagine, the summation is carried out). Defining

$$G_m(x) = \sum_{n=1}^{\infty} E_{mn} \sin\left(\frac{n\pi}{L}x\right) \sinh\left(\pi\sqrt{m^2 + n^2}\right) \tag{4.242}$$

eq. (4.241) becomes

$$\phi(x, y) = \sum_{m=1}^{\infty} G_m(x) \sin\left(\frac{m\pi}{L}y\right) \tag{4.243}$$

Utilizing the orthogonality property of the $\sin[(m\pi/L)y]$ function in eq. (4.243) in the usual manner, we obtain

$$G_m(x) = \frac{2}{L}\int_0^L \phi(x, y) \sin\left(\frac{m\pi}{L}y\right) dy \tag{4.244}$$

Recalling the definition of $G_m(x)$ [eq. (4.242)], substituting it in eq. (4.244), and applying the orthogonality property of the functions $\sin[(m\pi/L)x]$ yields

$$E_{mn} = \frac{4}{L^2}\frac{\displaystyle\int_0^L\left[\int_0^L (\phi(x, y) \sin\left(\frac{m\pi}{L}y\right) dy\right] \sin\left(\frac{n\pi}{L}x\right) dx}{\sinh(\pi\sqrt{m^2 + n^2})} \tag{4.245}$$

At this point the solution for the temperature distribution is completed. Combining eqs. (4.240) and (4.245) results in the following final expression for the temperature distribution:

$$T = T_\infty + \frac{4}{L^2}\sum_{m=1}^{\infty}\sum_{n=1}^{\infty}\frac{\displaystyle\int_0^L\left[\int_0^L (f(x, y) - T_\infty) \sin\left(\frac{m\pi}{L}y\right) dy\right] \sin\left(\frac{n\pi}{L}x\right) dx}{\sinh(\pi\sqrt{m^2 + n^2})}$$

$$\cdot\ sin\left(\frac{n\pi}{L}x\right) sin\left(\frac{m\pi}{L}y\right) \sinh\left(\pi\sqrt{m^2 + n^2}\,\frac{z}{L}\right) \tag{4.246}$$

To obtain the total heat transfer rate through the fin base, we apply Fourier's law. The total heat transfer rate through a differential element of area $dx\ dy$ is

$$d\dot{q} = -k\left(\frac{\partial T}{\partial z}\right)_{z=L} dx\ dy \tag{4.247}$$

Therefore, the total heat transfer rate through the fin base is

$$\dot{q} = -k \int_0^L \int_0^L \left(\frac{\partial T}{\partial z}\right)_{z=L} dx\, dy \tag{4.248}$$

Combining eqs. (4.246) and (4.248) and performing the algebra gives

$$\dot{q} = -4\frac{k}{\pi L} \sum_{m=1}^{\infty} \sum_{n=1}^{\infty} \frac{\int_0^L \left[ \int_0^L (f(x, y) - T_\infty) \sin\left(\frac{m\pi}{L} y\right) dy \right] \sin\left(\frac{n\pi}{L} x\right) dx}{\sinh(\pi \sqrt{m^2 + n^2})}$$

$$\cdot \frac{[(-1)^n - 1][(-1)^m - 1]}{mn} \sqrt{m^2 + n^2} \cosh(\pi \sqrt{m^2 + n^2}) \tag{4.249}$$

By inspection, we realize that if either $m$ or $n$ are even, the result of the double series expression (4.249) is zero. Therefore, defining $m = 2k + 1$ and $n = 2j + 1$, we rewrite expression (4.249) as follows:

$$\dot{q} = \frac{16k}{\pi L} \sum_{k=0}^{\infty} \sum_{j=0}^{\infty} \frac{\int_0^L \left[ \int_0^L (f(x, y) - T_\infty) \sin\left[\frac{(2k + 1)\pi}{L} y\right] dy \right] \sin\left[\frac{(2j + 1)\pi}{L} x\right] dx}{(2k + 1)(2j + 1)\sqrt{(2k + 1)^2 + (2j + 1)^2}}$$

$$\cdot \coth[\pi \sqrt{(2k + 1)^2 + (2j + 1)^2}] \tag{4.250}$$

In the special case where the base temperature is constant $[f(x, y) = T_0]$, eq. (4.250) simplifies to

$$\dot{q} = \frac{64kL}{\pi^3}(T_0 - T_\infty) \sum_{k=0}^{\infty} \sum_{j=0}^{\infty} \frac{\coth[\pi \sqrt{(2k + 1)^2 + (2j + 1)^2}]}{(2k + 1)^2(2j + 1)^2\sqrt{(2k + 1)^2 + (2j + 1)^2}} \tag{4.251}$$

The result (4.251) for the present idealized problem does not account for the high local heat fluxes around the perimeter of the fin base that one would anticipate to encounter in a real problem with similar boundary conditions, as discussed earlier after eq. (4.38).

## PROBLEMS

**4.1.** Utilize the results for the temperature field, $T(x, y)$ in eq. (4.70), and graph the dimensionless temperature distribution $[(T - T_0)/(T_0 - T_\infty)]$ at the midheight of the fin cross section ($y = H/2$) for the following values of the Biot number: Bi = 0.1, 1, 10, for a "square" fin of aspect ratio $H/b = 1$. Repeat the foregoing procedure for $H/b = 10$, 50. What do you conclude? [*Hint:* First, the nonlinear algebraic equation (4.64) has to be solved numerically for the eigenvalues $(\alpha_n b)$. Next, nondimensionalize expressions (4.70), utilizing $b$ as the length scale for both the $x$ and $y$ directions such that the only parameters appearing in these expressions are $(a_n b)$, Bi, $\hat{y} = y/b$, $H/b$.]

**4.2.** Utilize the following definitions to nondimensionalize the variables in eqs. (4.97) and (4.98): $\hat{T} = (T - T_\infty)/(T_0 - T_\infty)$, $\hat{r} = r/R$, $\hat{z} = z/R$, $\hat{\dot{q}} = \dot{q}/kL(T_0 - T_\infty)$. Then graph the dimensionless temperature distribution at $z = L/2$ and the heat transfer rate at the base of the cylindrical fin, for $L/R = 0.1, 1, 10, 50$. What are your conclusions?

**4.3.** Re-solve the case (c) of Example 4.1 with the following modification: The heat transfer coefficient at the "tip" of the bar is different from that at the sides of the bar ($h_1 = h_2 = h$, $h_3 \neq h$). Do your results reduce to those of Example 4.1, case (c), in the limit $h_3 \to h$?

**4.4.** Densely populated electronic components mounted on substrate surfaces may result in undesirably high temperatures affecting the operation and reliability of electronic equipment. To assure the reliable operation of this equipment, efficient removal of heat

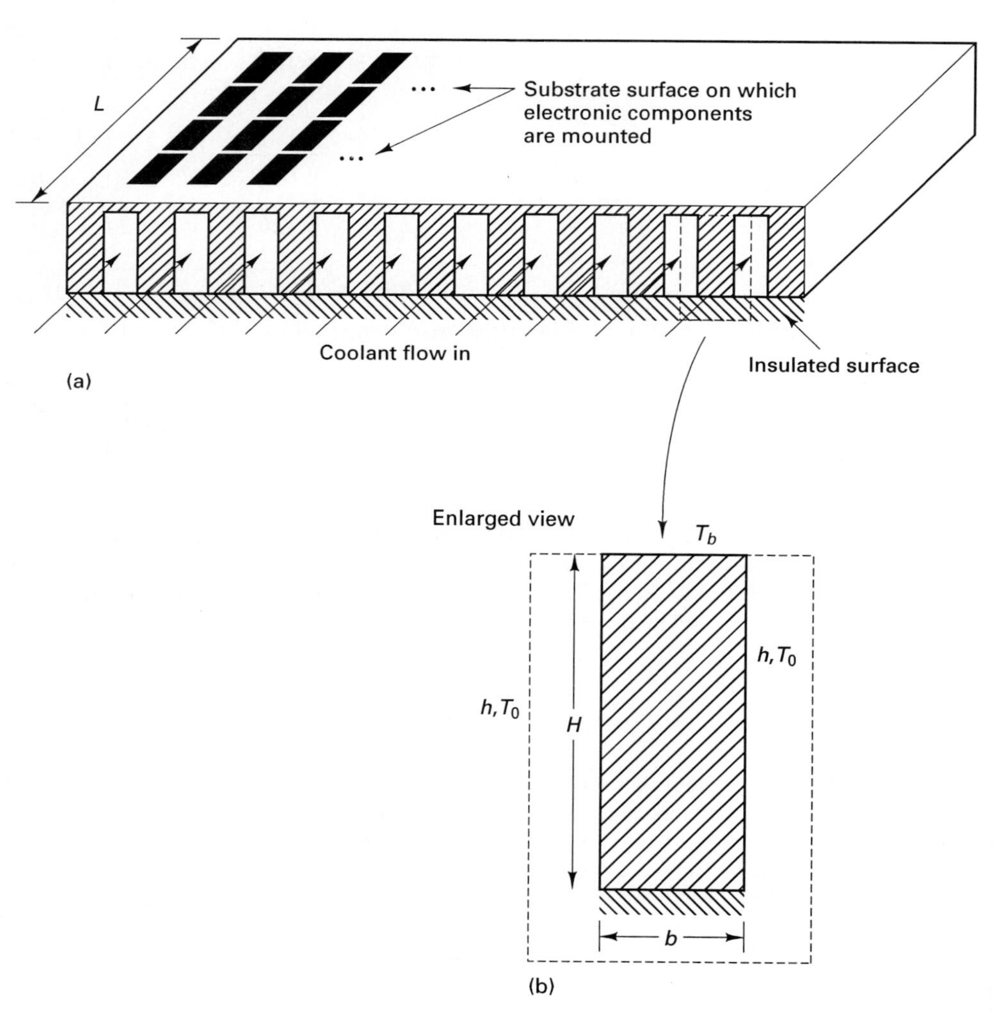


**Figure P4.4** (a) Schematic of a heat exchanger needed for cooling electronic components. (b) Schematic of the region to be modeled.

from the electronic components is desired. The heat exchanger design shown in Fig. P4.4 aims at the cooling of a substrate surface on top of which electronic components are mounted. The lid at the bottom of the devise is assumed to be adiabatic. The heat transfer through a typical fin (rib) between two adjacent coolant channels can be obtained by studying the two-dimensional model shown in the detail of Fig. P4.4. It is assumed that all the fins and coolant channels are identical, that the heat transfer coefficient ($h$) and the bulk fluid temperature ($T_0$) are known, and that the base temperature of the fin ($T_b$) is measured with thermocouples and it is also known and constant. End effects are neglected.

**(a)** Calculate the temperature distribution in the typical fin shown in the detail of Fig. P4.4.

**(b)** Calculate the local and overall heat transfer through the base of the fin.

**(c)** How does the overall heat transfer through the base of the fin vary with the ratio $H/b$? Based on your findings, what are your recommendations to the fin manufacturer for the design of the fin?

**4.5.** The cylindrical bars shown in Fig. P4.5 serve both as structural support for the two side surfaces of a heat exchanger and as a means of enhancing the heat transfer between

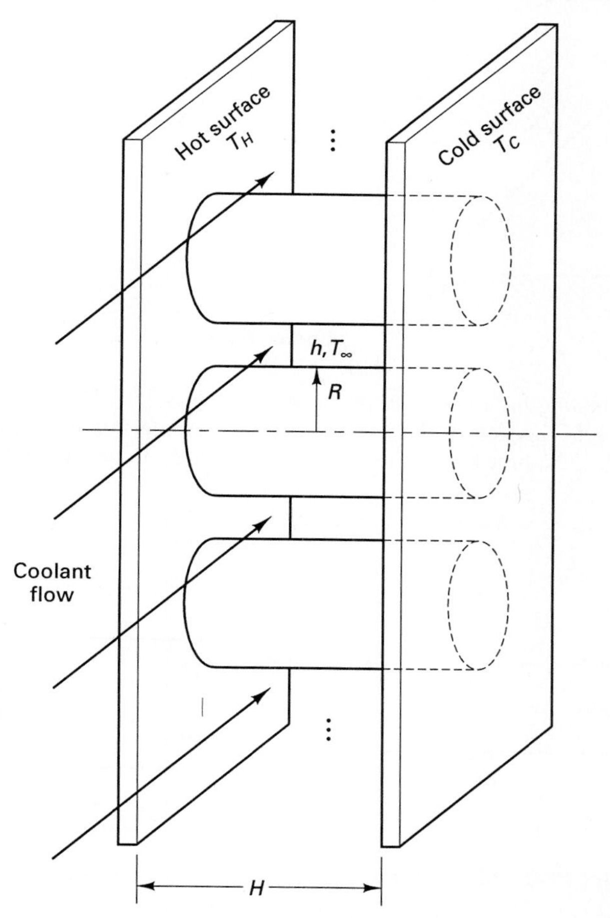


**Figure P4.5**

these surfaces. Assume for simplicity that the temperature of the cold surface is identical to the bulk temperature of the coolant ($T_c = T_\infty$). The temperature of the hot surface is denoted by $T_H$, the height of the bars by $H$, the radius of the bars by $R$, and the heat transfer coefficient by $h$. Assuming that the heat transfer behavior of all bars is identical:

**(a)** Obtain the temperature distribution in one bar.

**(b)** Calculate the overall heat transfer removed through one bar from the hot surface.

**(c)** How does varying the ratio $R/H$ affect the overall heat transfer? Based on your analysis, if the volume of each bar is fixed, what values of $R/H$ (large or small) would you recommend to the manufacturer of the heat exchanger to enhance the overall heat transfer?

**4.6.** Figure P4.6 shows the cross section of a long triangular bar. The hypotenuse of the triangular cross section is perfectly insulated. Find the temperature distribution in the bar if **(a)** the heat transfer coefficient ($h$) is very large and **(b)** if the heat transfer coefficient is finite.

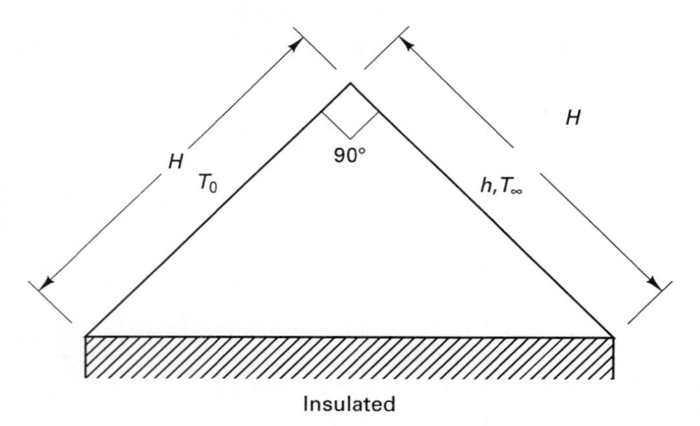


**Figure P4.6**

**4.7.** A cylindrical shaft is rotating inside two sleeves as shown in Fig. P4.7. Because of a failure in the lubrication of the sleeves, heat is generated at the interface between the sleeves and the cylinder. The heat flux at this interface is assumed to be constant, $\dot{q}''$.

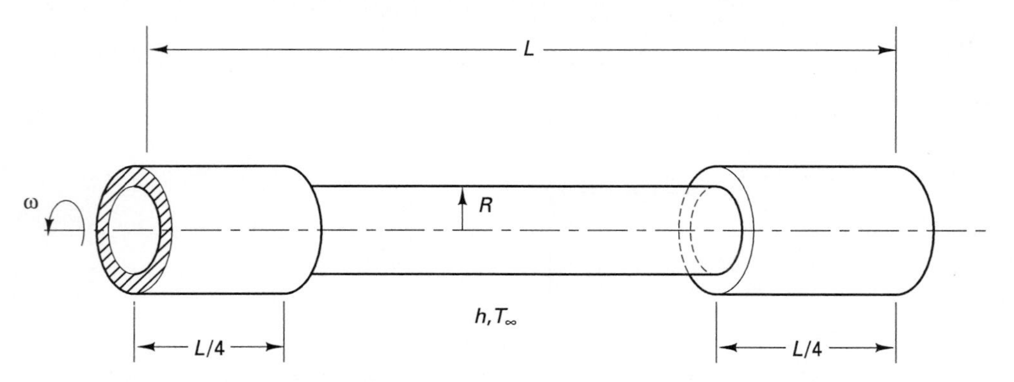


**Figure P4.7** Shaft rotating inside sleeves.

The remainder of the lateral surface of the shaft is cooled convectively ($h$, $T_\infty$). The two ends of the shaft can be approximated as adiabatic. Obtain the steady-state temperature distribution in the shaft.

**4.8.** Reconsider Problem 4.4. The two-dimensional heat transfer model of the rib fin can be improved if radiative cooling is taken into account as well. Such an improvement may be necessary if the coolant is a gas (air, for example). A schematic of an improved model is shown in Fig. P4.8. The radiative cooling is approximated with the help of a constant heat flux $\dot{q}''$ applied at both sides of the fin.

**(a)** Use the principle of superposition and decompose the problem described above into as small a number of problems as possible solvable by direct application of the method of separation of variables.

**(b)** What is the temperature distribution of the fin?

**(c)** What is the local heat flux and overall heat transfer rate at the base of the fin?

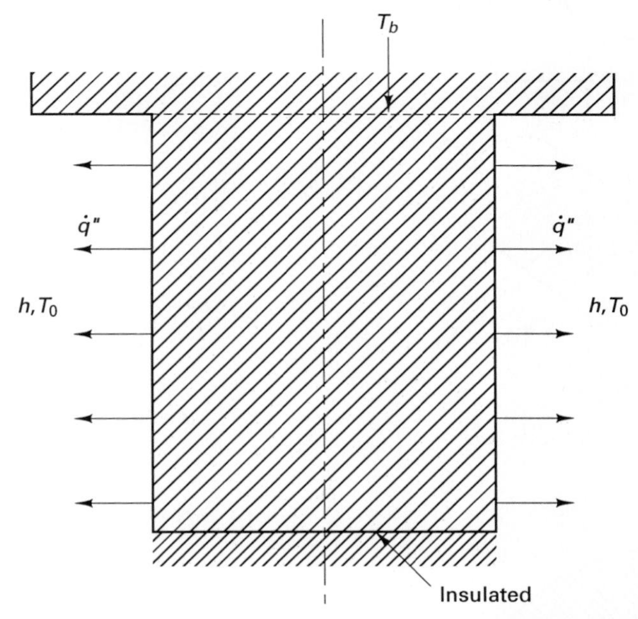


**Figure P4.8**

**4.9.** Reconsider Problem 4.5 but by allowing for $T_c$ to be different than $T_\infty$. Obtain the temperature distribution in the bar as well as the overall heat transfer removed from the hot surface because of the presence of the bar. Do your results reduce to those of Problem 4.5 in the limit $T_c \rightarrow T_\infty$?

**4.10.** A barrel of nuclear waste is represented schematically by the cylinder of Fig. P4.10. The nuclear waste generates heat at a constant rate $\dot{u}'''$ $W/m^3$. It is being cooled convectively ($h$, $T_\infty$) as well as radiatively ($\dot{q}''$) through its lateral and top surfaces. The bottom surface (base of the barrel) is at a constant temperature $T_0$.

**(a)** Using the principle of superposition decompose the problem into as small a number of problems as possible that are solvable by direct application of the method of separation of variables.

**(b)** Obtain the temperature distribution in the barrel of nuclear waste. What are the location and magnitude of the maximum temperature?

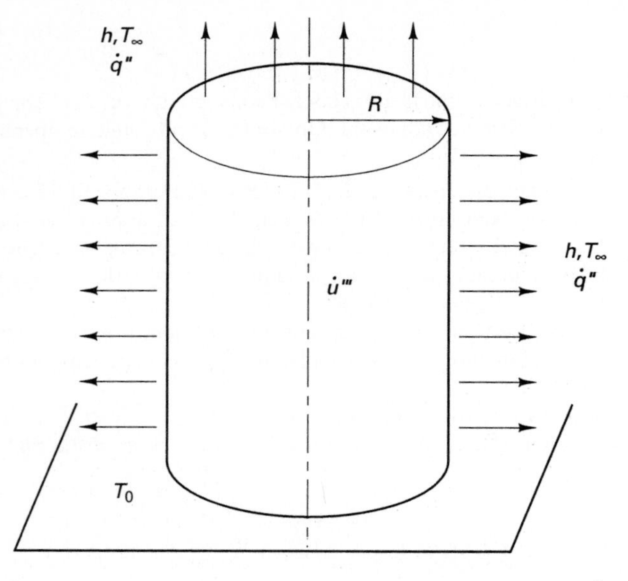


**Figure P4.10** Schematic of a nuclear waste barrel.

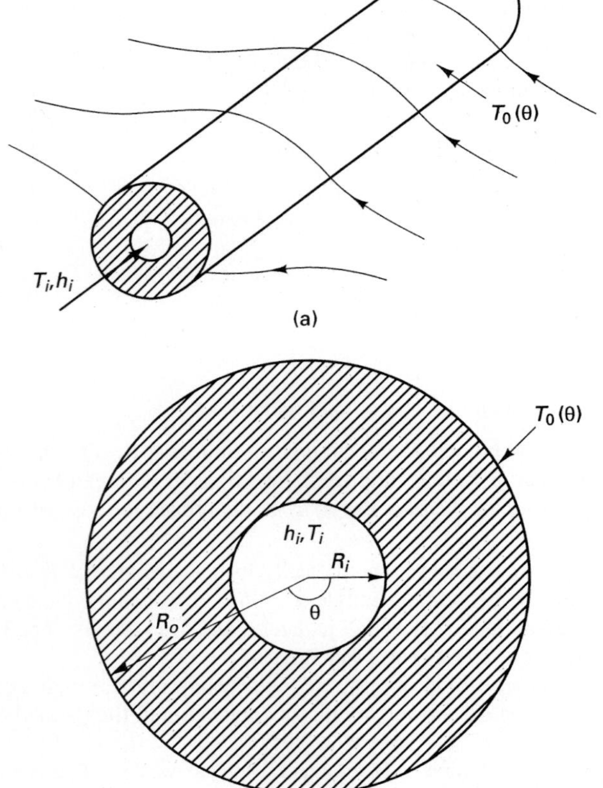


**Figure P4.11** (a) Pipe in cross-flow. (b) Cross section of the pipe.

(c) What is the total heat transfer rate from the barrel of nuclear waste to the environment?

**4.11.** A warm fluid flows outside a long pipe in cross-flow as shown in Fig. P4.11a. Inside the pipe a cold fluid ($T_i$) is circulated. As a result, heat conduction occurs through the pipe wall. The heat transfer coefficient of the inner fluid ($h_i$) is independent of the angular position $\theta$. The temperature of the outer surface of the pipe is measured to be $T_0(\theta)$, a known function of the angular position. Obtain the steady-state temperature distribution in the pipe wall and the heat transfer rate through the wall. A cross section of the pipe is shown in Fig. P4.11b.

**4.12.** Re-solve Problem 4.11 this time by assuming that $h_i \to \infty$ and therefore that the temperature of the inner pipe surface equals $T_i$.

**4.13.** A long and thick electrical cable of radius $R$ is placed in a uniform stream of air for cooling purposes. Heat is generated volumetrically inside the cable ($\dot{q}'''$ in W/m$^3$). The heat transfer rate per unit length between the cable surface and the air is a known function of the angular position $\dot{q}'(\theta)$. Obtain the steady-state temperature distribution inside the cable.

**4.14.** Heat conduction in a tissue is governed by the bioheat equation (Chapter 2). In the case of the thin tissue of Fig. P4.14, an experiment is performed whereby two sides of the tissue are kept at one constant temperature $T_1$ and the other two sides at another constant temperature $T_2$. Neglecting metabolic heat generation the bioheat equation for the steady-state temperature distribution is

$$k_t\left(\frac{\partial^2 T}{\partial x^2} + \frac{\partial^2 T}{\partial y^2}\right) - \dot{m}'''_b c_{pb}(T - T_A) = 0$$

What is the temperature distribution in the tissue?

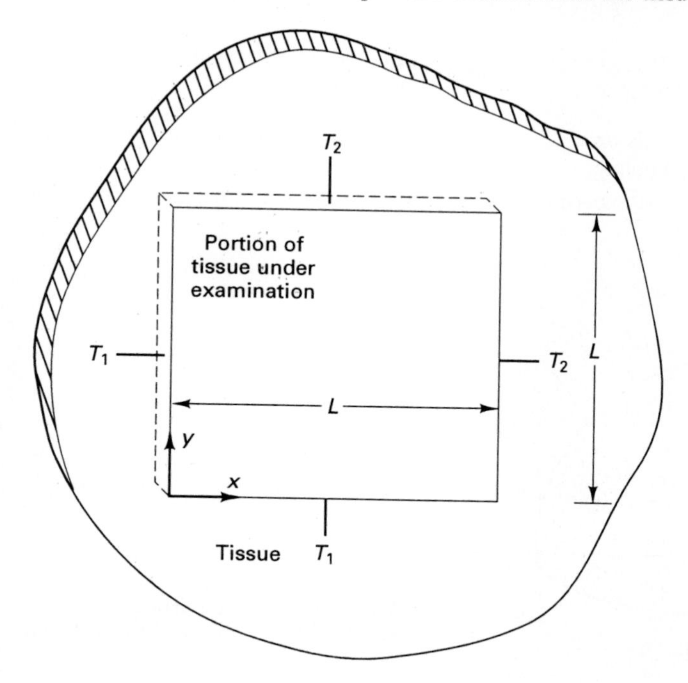


**Figure P4.14**

**4.15.** Re-solve Problem 4.14 by including the metabolic heat generation $\dot{q}''' = \text{const}$ in the model for the heat conduction process in the tissue.

**4.16.** Heat conduction in a standing human leg immersed in flowing warm water ($h$, $T_\infty$) is approximately modeled as heat conduction in the cylinder shown in Fig. P4.16. Using the bioheat equation accounting for blood perfusion and neglecting the metabolic heat generation, obtain the steady-state temperature distribution in the leg. What is the heating rate through the lateral surface of the leg?

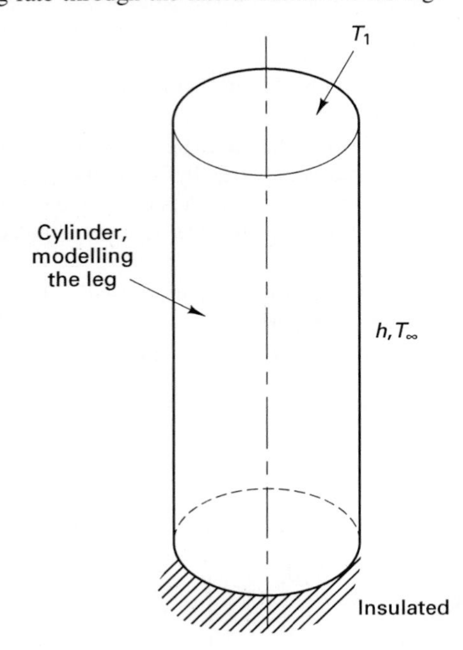


**Figure P4.16** Cylinder modeling a human leg.

**4.17.** Re-solve Problem 4.16 this time accounting for metabolic heat generation $\dot{q}'''$ in addition to blood perfusion cooling.

**4.18.** In a very slow evaporation experiment, a hemispherical droplet of radius $R$ that is situ-

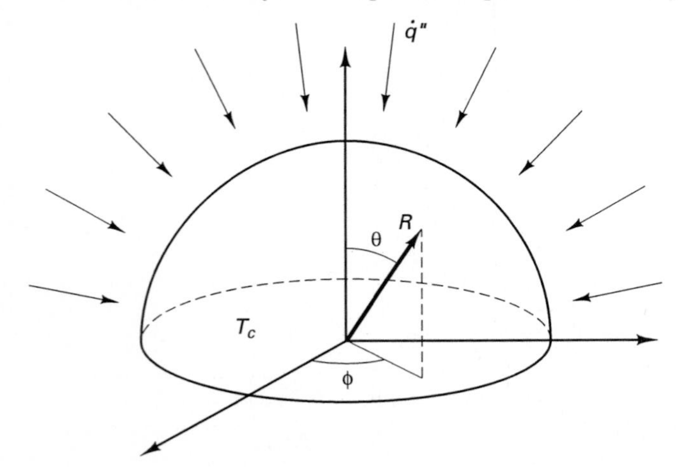


**Figure P4.18** Hemispherical droplet evaporating on an isothermal surface.

ated on an isothermal surface ($T_c$) is subjected to a uniform heat flux at its bottom surface (Fig. P4.18). Since the evaporation process is very slow it can be assumed that for a large time period the droplet radius remains constant and that a steady-state temperature field is established within the droplet. Obtain this temperature field.

**4.19.** A hemispherical droplet of radius $R$ situated on an isothermal hot plate ($T_H$) is placed in an airstream ($h$, $T_\infty$). The evaporation effects at the droplet surface as well as the convection effects within the droplet are negligible. Obtain the steady-state temperature distribution in the droplet.

**4.20.** A cylindrical disk of a composite material is subjected to radiation heat treatment by exposing its top surface to laser radiation as shown in Fig. P4.20. The radiation absorption by the material of the disk decreases exponentially with the distance from its top surface. In a heat conduction model for the heat transfer in the disk the above-mentioned absorbed radiation energy can be modeled as a volumetric heat source $\dot{Q}''' = Ae^{-Bz}$, where $A$ and $B$ are positive constants. The temperature of all the surfaces of the disk is assumed to be constant ($T_0$) for simplicity. Obtain the temperature distribution in the disk. What is the location and the magnitude of the maximum temperature?

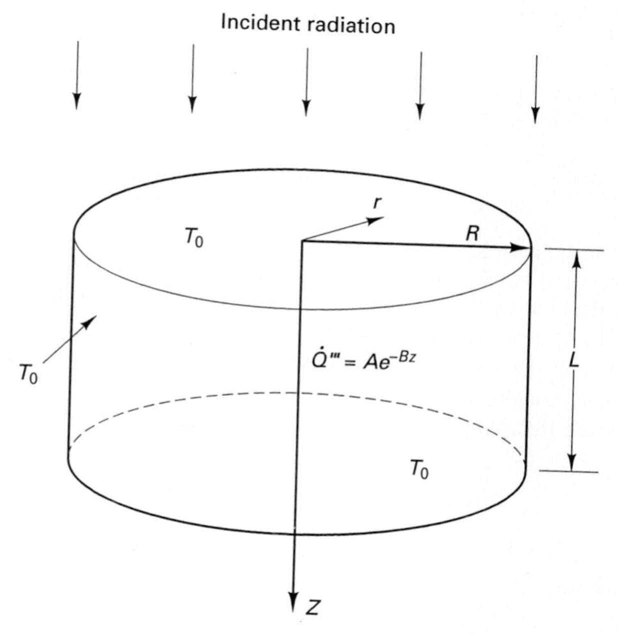


**Figure P4.20**

**4.21.** Reconsider Problem 4.20 with the following modification. Instead of keeping the surface temperature of the disk at a constant value, $T_0$, the disk is placed in a gas stream ($h$, $T_\infty$). As a result, the top, bottom, and lateral surfaces of the disk are cooled by convection. The heat transfer coefficient ($h$) is assumed to be the same for all these surfaces. Obtain the temperature distribution in the disk as well the location and the value of its maximum temperature.

**4.22.** Figure P4.22 shows a cylindrical silo within which corn is stored. The corn can be modeled as a composite medium whose thermophysical properties are assumed known. The floor of the silo is adiabatic. It is a well-known fact that through metabolic reac-

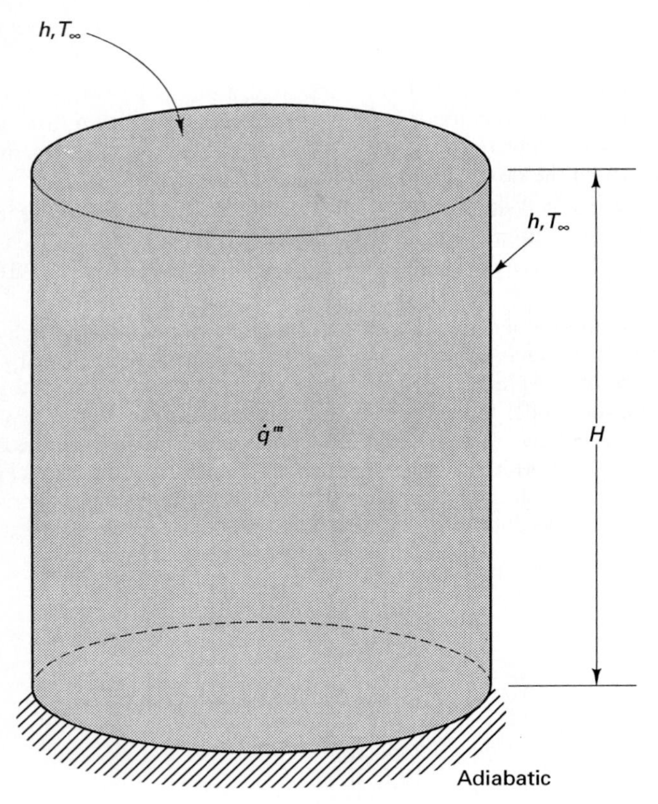

**Figure P4.22** Schematic of a cylindrical corn silo.

tions heat is generated in the corn at a rate $\dot{q}'''$ (W/m$^3$). Cooling of the corn occurs by forced convection of the wind flowing passed the lateral surface and the top of the silo. The air temperature is denoted by $T_\infty$ and the heat transfer coefficient (assumed to be the same for the top and lateral surfaces of the silo) by $h$. Find the temperature distribution in the corn as well as the location and the magnitude of its maximum temperature.

**4.23.** The cross section of an electrically heated heat exchanger wall is shown in Fig. P4.23. Three sides of this wall are thermally insulated and the fourth side is cooled by a fluid

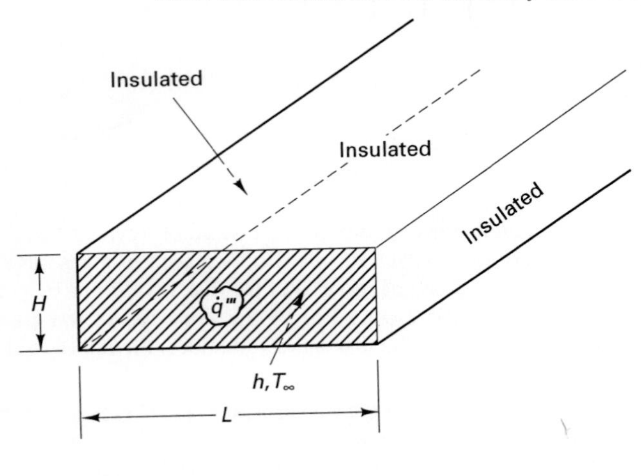

**Figure P4.23**

$(h, T_\infty)$. The electrical heating is modeled as internal heat generation $\dot{q}'''$. Obtain the wall temperature distribution and the location and magnitude of the maximum temperature.

**4.24.** A brick is being baked in an oven as shown in Fig. P4.24. The brick surface in contact with the floor is modeled as insulated. The remaining five surfaces of the brick are being heated radiatively ($\dot{q}''$) and at the same time, cooled somewhat by the action of natural convection $(h, T_\infty)$. Find the steady-state temperature distribution in the brick as well as the net heat input through its lateral surface.

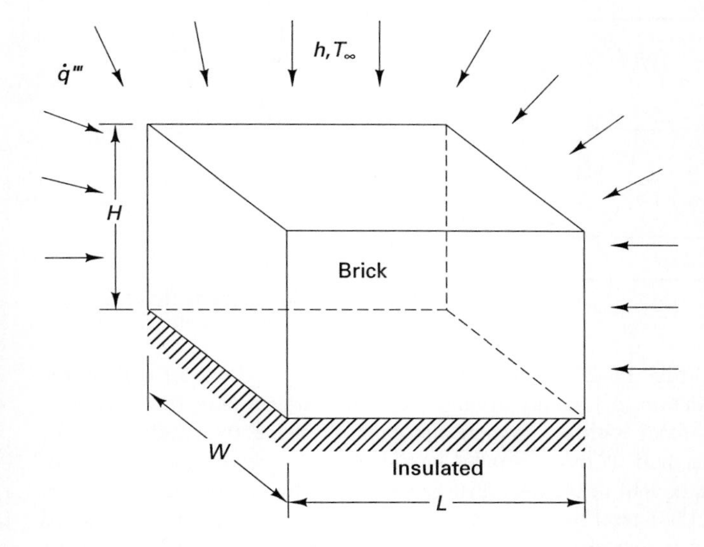


**Figure P4.24** Schematic of brick baked in an oven.

**4.25.** Figure P4.25 shows the cross section of a long semicircular rod. The bottom of the rod is perfectly insulated. The same is true for half of the lateral surface of the rod. The remainder of the lateral surface is kept at a constant temperature, $T_0$. Obtain the temperature distribution in the rod.

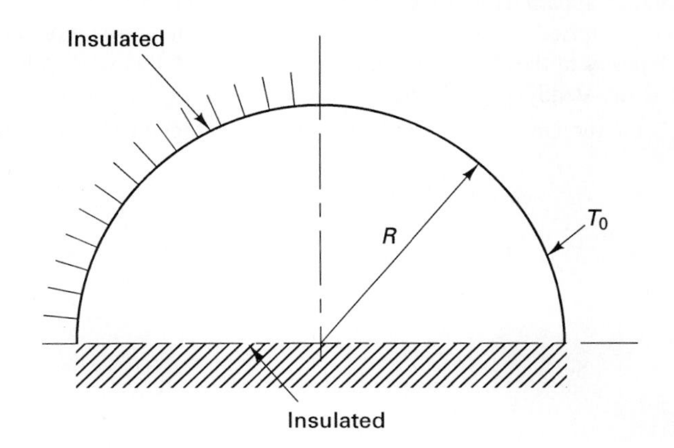


**Figure P4.25**

**4.26.** A short cylindrical fin is shown in Fig. P4.26. Its base temperature is measured carefully and is expressed by the function $T_b = f(r, \theta)$. The remaining surfaces of the fin are in contact with a convecting fluid $(h, T_\infty)$.

**(a)** Obtain the temperature distribution in the fin and the heat transfer rate through its base if $h \to \infty$.

**(b)** Obtain the same results for a finite heat transfer coefficient.

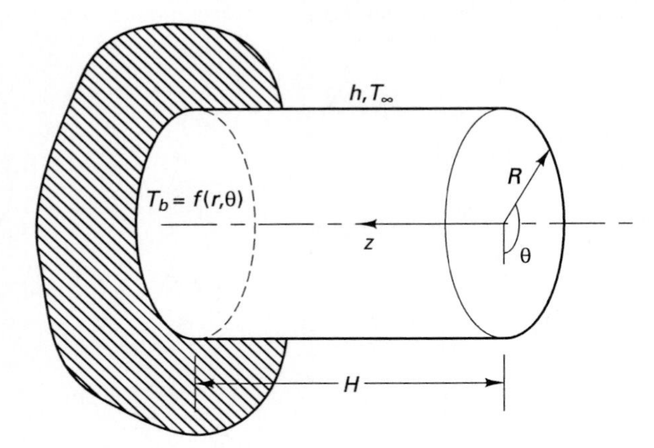


**Figure P4.26** Schematic of a cylindrical fin.

**4.27.** A growing cucumber (Fig. P4.27a) modeled as a cylinder (Fig. P4.27b) still on the vine receives solar radiation ($\dot{q}''$) mainly through its lateral surface. Note that part of its lateral surface is in contact with the ground and cannot receive solar radiation. For simplicity, assume that half of the cucumber lateral surface receives solar radiation. Simultaneously, the same half of the lateral surface is cooled by air convection $(h, T_\infty)$. The remaining half of the lateral surface (that is in contact with the ground) is assumed to be adiabatic. Obtain the steady-state temperature distribution in the cucumber if its end surfaces are adiabatic.

**4.28.** Re-solve Problem 4.27 assuming this time that the solar radiation flux is a known function of the angular position $\dot{q}''(\theta)$.

**4.29.** A ceramic column of rectangular cross section is partially buried in the floor of a test section of an experimental apparatus as shown in Fig. P4.29. The part of the column surface (all sides) that is buried is assumed to be adiabatic. The remainder of the column (all sides) is exposed to the flow of a hot reactive gas $(h, T_g)$ as well as to a radiative flux $\dot{q}''$. What is the steady-state temperature distribution in the column?

**4.30.** Re-solve Example 4.7 for the case where the heat transfer coefficient $(h)$ has a finite value.

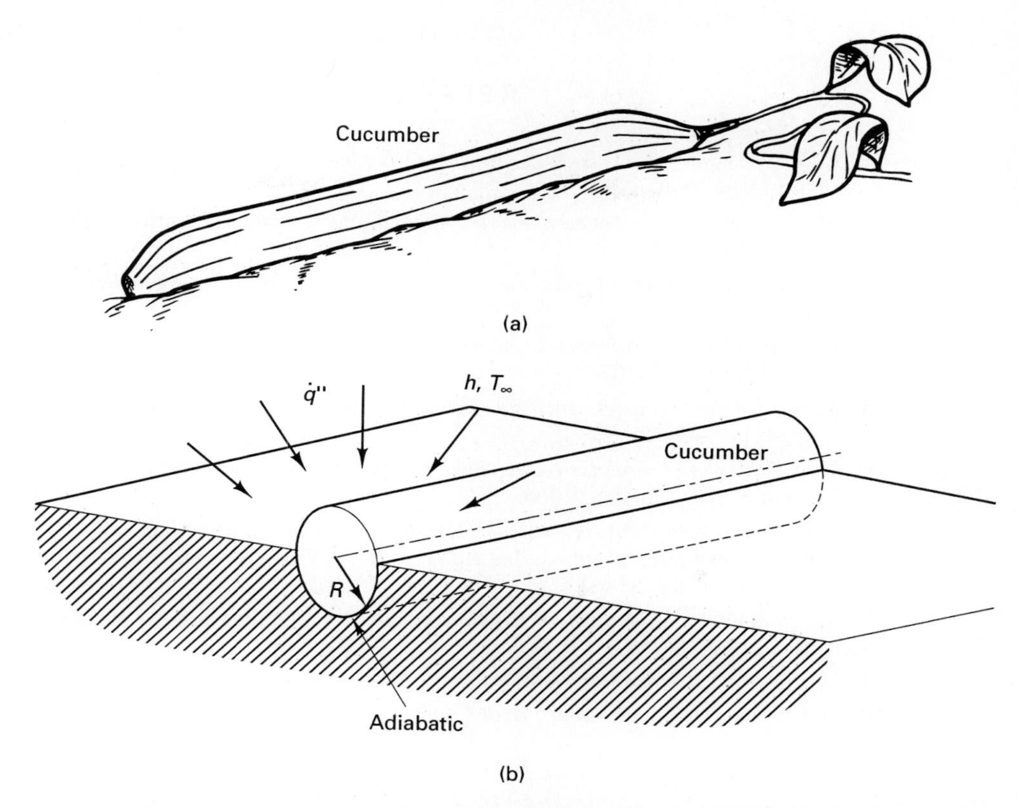


**Figure P4.27** (a) Schematic of a cucumber on the vine, semiburied in the ground. (b) Cylinder modeling the cucumber.

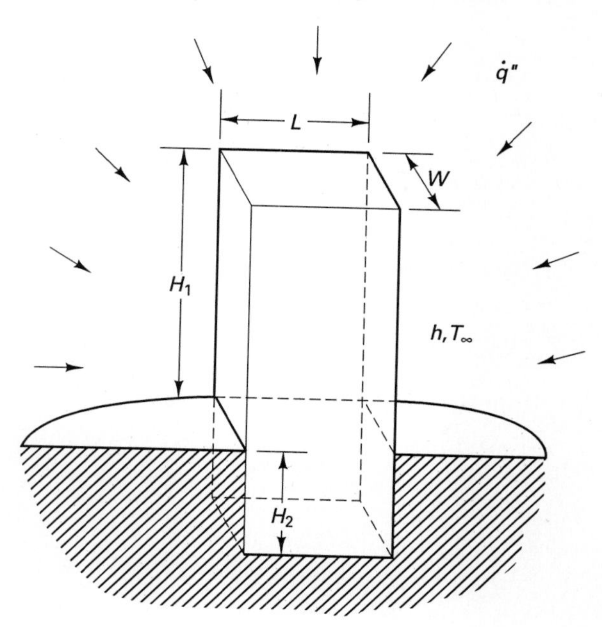


**Figure P4.29** Partially buried rectangular ceramic column.

## REFERENCES

1. C. R. WILIE AND L. C. BARRET, *Advanced Engineering Mathematics,* 5th ed., McGraw-Hill, New York, 1982.
2. M. D. GREENBERG, *Foundations of Applied Mathematics,* Prentice Hall, Englewood Cliffs, NJ, 1978.
3. M. BRAUN, *Differential Equations and Their Applications,* Springer-Verlag, New York, 1975.
4. F. B. HILDEBRANT, *Advanced Calculus for Applications,* 2nd ed., Prentice Hall, Englewood Cliffs, NJ, 1976.
5. M. R. SPIEGEL, *Fourier Analysis,* Schaum's Outline Series, McGraw-Hill, New York, 1974.
6. G. N. WATSON, *A Treatise on the Theory of Bessel Functions,* 2nd ed., Cambridge University Press, London, 1966.
7. M. ABRAMOWITZ AND I. A. STEGUN, *Handbook of Mathematical Functions,* National Bureau of Standards, Applied Mathematics, Series 55, U.S. Government Printing Office, Washington, DC, 1964.
8. M. R. SPIEGEL, *Mathematical Handbook,* Schaum's Outline Series, McGraw-Hill, New York, 1968.
9. V. S. ARPACI, *Conduction Heat Transfer,* Addison-Wesley, Reading, MA, 1966.
10. U. GRIGULL AND H. SANDNER, *Heat Conduction,* Hemisphere, New York, 1984.

CHAPTER **5**

# TRANSIENT CONDUCTION WITH TIME-INDEPENDENT BOUNDARY CONDITIONS

In this chapter the class of heat conduction problems in which the temperature field is a function of time is examined. For the most part, one-dimensional problems in space are considered. Following the style and sequence adopted in Chapter 4, first illustrative examples of one-dimensional transient conduction are presented in the Cartesian, cylindrical, and spherical coordinate systems that are solvable by a rather straightforward application of the method of separation of variables. Next, more complex heat conduction models are analyzed, in which the principle of superposition will aid the solution. Following that, a new powerful tool is introduced and added to the arsenal of solution methods of transient conduction problems: namely, the method of variation of parameters. In addition, a special class of one-dimensional transient conduction problems are presented which possess the feature that with the aid of an appropriate similarity variable, the governing conduction partial differential equation is reduced to an ordinary differential equation and no partial derivatives appear in the boundary conditions. The latter model is markedly easier to handle mathematically. Next, the approximate integral solution method is presented. This method often combines simplicity with acceptable engineering accuracy. Finally, multidimensional transient conduction is discussed with the help of illustrative examples in which the base of knowledge pertaining to one-dimensional conduction, as presented in this chapter, is utilized and new features related to the dependence of the heat conduction process in more than one space coordinates are presented.

## 5.1 ONE-DIMENSIONAL TRANSIENT CONDUCTION IN CARTESIAN COORDINATES

The salient features of the methodology involved in the solution of this class of problems are highlighted with the help of the following example.

### Example 5.1

A long sheet of a composite material, immediately after it is manufactured, is immersed in a cold liquid bath of temperature $T_\infty$. The purpose of this process is gradually to reduce the temperature of the sheet, which has an initial high value, $T_0$. The length of the plate, $L$, is considerably larger than the thickness of the plate, $a$ (i.e., $a/L \ll 1$). Because of the temperature difference between the plate and the liquid bath, a buoyancy-driven (natural convection) flow is established parallel to the plate. The average heat transfer coefficient describing the natural convection phenomenon is assumed constant during the cooling process and is denoted by $h$. The thermal conductivity of the plate is denoted by $k$. A schematic corresponding to the description above is shown in Fig. 5.1. To perform a thermal stress analysis of the plate during this cooling treatment, knowledge of the temperature history of the sheet is needed. It is therefore required to obtain the temperature distribution of the sheet as well as the overall heat transfer rate from each side of the sheet to the liquid bath. End effects occurring near the bottom and the top edges of the plate are negligible. All temperature changes in the direction perpendicular to the $x$–$y$ plane of Fig. 5.1 are also negligible.

**Solution**

After careful consideration of the problem description we realize the following:

**(a)** Since all changes in the direction perpendicular to the $x$–$y$ plane of Fig. 5.1 are negligible, the transient heat conduction phenomenon is "at the most" two-dimensional, occurring in the $x$–$y$ plane. Furthermore, based on scaling arguments, it can easily be shown that

$$\frac{\partial^2 T}{\partial x^2} \sim O\left(\frac{\Delta T}{a^2}\right) \tag{5.1}$$

$$\frac{\partial^2 T}{\partial y^2} \sim O\left(\frac{\Delta T}{L^2}\right) \tag{5.2}$$

where $\Delta T$ is a characteristic temperature difference representative of the heat conduction phenomenon. For example, at early times, $\Delta T = T_0 - T_\infty$. Since $L \gg a$, eqs. (5.1) and (5.2) imply that

$$\frac{\partial^2 T}{\partial x^2} \gg \frac{\partial^2 T}{\partial y^2} \tag{5.3}$$

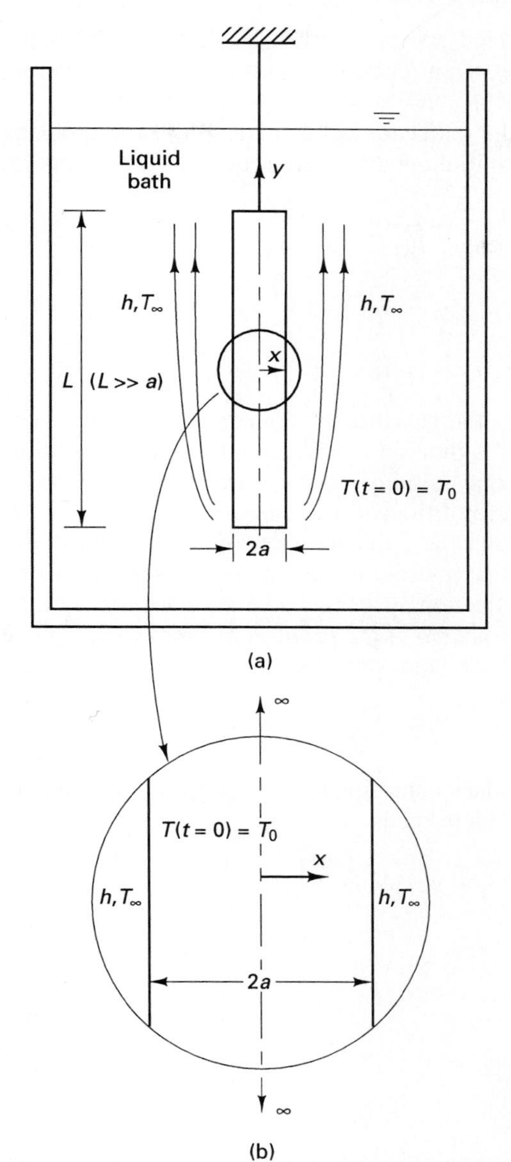


**Figure 5.1** (a) Sheet of composite material immersed in a bath of liquid coolant. (b) Detail of the sheet for the modeling process.

Therefore, the general conduction equation (2.33) reduces to the following one-dimensional transient conduction statement:

$$\frac{1}{\alpha}\frac{\partial T}{\partial t} = \frac{\partial^2 T}{\partial x^2} \tag{5.4}$$

**(b)** Making use of the symmetry of the problem we attached the origin of the $x$- axis at the centerline of the plate. Figure 5.1b enlarges the detail indicated in Fig. 5.1a and represents the schematic of the final model for the present problem (i.e., transient conduction in the $x$-direction in an "infinitely" long plate).

To complete the mathematical formulation of the problem two boundary conditions and one initial condition are required. These conditions are:

$$x = 0: \qquad \frac{\partial T}{\partial x} = 0 \tag{5.5}$$

$$x = a: \qquad -k\frac{\partial T}{\partial x} = h(T - T_\infty) \tag{5.6}$$

$$t = 0: \qquad T = T_0 \tag{5.7}$$

Equation (5.5) states the fact that the temperature field is symmetric about the centerline, eq. (5.6) is the usual convection boundary condition, and eq. (5.7) represents the fact that initially (at the time right before the cooling process starts) the temperature of the sheet is uniform, $T_0$.

Continuing along the guidelines we initiated in the study of steady-state problems, we state that in order for the method of separation of variables to apply, the governing equation as well as *both the boundary conditions in the only space coordinate (direction) of the problem* ($x$) *need to be homogeneous*. While eqs. (5.4) and (5.5) are homogeneous, eq. (5.6) is not. However, much as in Chapter 4, introducing

$$\Theta = T - T_\infty \tag{5.8}$$

transforms this boundary condition to homogeneous. In terms of the variable $\Theta$ the model of the problem reads

$$\frac{1}{\alpha}\frac{\partial \Theta}{\partial t} = \frac{\partial^2 \Theta}{\partial x^2} \tag{5.9}$$

$$x = 0: \qquad \frac{\partial \Theta}{\partial x} = 0 \tag{5.10}$$

$$x = a: \qquad -k\frac{\partial \Theta}{\partial x} = h\Theta \tag{5.11}$$

$$t = 0: \qquad \Theta = T_0 - T_\infty = \Theta_0 \tag{5.12}$$

Clearly, both the governing equation (5.9) and the two boundary conditions (5.10) and (5.11) are homogeneous and the method of separation of variables can be applied directly. The procedure is similar to that used in steady-state problems. We begin by assuming that the temperature can be written as

$$\Theta = X(x)S(t) \tag{5.13}$$

where $X(x)$ is a function of $x$ only and $S(t)$ is a function of $t$ only. Substituting eq. (5.13) into eq. (5.9) and dividing through by the product $XS$ yields

$$\underbrace{\frac{1}{\alpha}\frac{S'}{S}}_{-\beta^2} = \underbrace{\frac{X''}{X}}_{-\beta^2} \tag{5.14}$$

where the primes denote differentiation with respect to $t$ or $x$. Since $S(t)$ depends only on $t$ and $X(x)$ depends only on $x$, the left-side of eq. (5.14) is a function of $t$ only, while the right-hand side is a function of $x$ only. For eq. (5.14) to hold in general, both sides should equal a constant, say, $-\beta^2$. As a result, two ordinary differential equations are recovered from eq. (5.14):

$$X'' + \beta^2 X = 0 \tag{5.15}$$

$$S' + \beta^2 \alpha S = 0 \tag{5.16}$$

The choice of setting each side of eq. (5.14) equal to a *negative* constant $(-\beta^2)$ instead of a positive constant $(\beta^2)$ is guided by the fact that we want the solution of eq. (5.15) to be in terms of *orthogonal* functions. A similar discussion took place in Chapter 4 in connection with steady-state problems. In addition, for the present situation, the physics of the problem requires that after a long time, a steady state will be reached and the temperature of the sheet will equal that of the fluid. The solution of eq. (5.16) should reflect this fact. As will be shown next, it does. Had we chosen to set each side of eq. (5.14) equal to a positive constant, the resulting temperature field would "blow up" at very large times. An inspection of the solution for eq. (5.16) verifies the discussion above. This solution is straightforward to obtain (Appendix B) and reads

$$S = A\exp(-\alpha\beta^2 t) \tag{5.17}$$

On the other hand, the solution of eq. (5.15) is

$$X = B\sin(\beta x) + C\cos(\beta x) \tag{5.18}$$

Combining eqs. (5.13), (5.17), and (5.18) and renaming the constants ($AB = C$, $AC = K$) yields

$$\Theta = [C\sin(\beta x) + K\cos(\beta x)]e^{-\alpha\beta^2 t} \tag{5.19}$$

Next, we apply boundary condition (5.10) and obtain

$$C = 0 \tag{5.20}$$

The value of $\beta$ is determined by substituting eq. (5.19) into eq. (5.11):

$$\text{Bi}\cot(\beta_n a) = \beta_n a \qquad n = 1, 2, 3, \ldots \tag{5.21}$$

where the subscript $n$ denotes the fact that an infinite number of eigenvalues $\beta_n$ satisfy eq. (5.21) and where the Biot number is defined as

$$\text{Bi} = \frac{ah}{k} \tag{5.22}$$

Once numerical values for $a$ and Bi are assigned, eq. (5.21) can be solved to yield the solutions for $\beta_n$, $n = 1, 2, \ldots$. To proceed, we assume that the eigenvalues $\beta_n$ are known. Up to this point, then, our result for the temperature field reads

$$\Theta = \sum_{n=1}^{\infty} K_n \cos(\beta_n x) e^{-\alpha\beta_n^2 t} \tag{5.23}$$

In writing eq. (5.23) we made use of the fact that a set of linearly independent solutions exists, each of which corresponds to a value of $\beta_n$. Therefore, the temperature field, $\Theta$, is the linear superposition of the above-mentioned solutions.

The constant $K_n$ is obtained by applying the initial condition (5.12) together with the orthogonality property of the cosine function. Substituting eq. (5.23) into eq. (5.12) yields

$$\Theta_0 = \sum_{n=1}^{\infty} K_n \cos(\beta_n x) \tag{5.24}$$

To make use of the orthogonality property of the cosine function, we multiply both sides of eq. (5.24) by $\cos(\beta_m x)$ and integrate both sides from $x = 0$ to $x = a$. This procedure is identical to what was discussed in Chapter 4. After the above, we have

$$\Theta_0 \int_0^a \cos(\beta_m x)\, dx = \sum_{n=1}^{\infty} K_n \int_0^a \cos(\beta_m x) \cos(\beta_n x)\, dx \tag{5.25}$$

The integral on the right-hand side of eq. (5.25) is calculated in general to be

$$\int_0^a \cos(\beta_m x) \cos(\beta_n x)\, dx = \begin{cases} \dfrac{\sin[(\beta_m - \beta_n)a]}{2(\beta_m - \beta_n)} + \dfrac{\sin[(\beta_m + \beta_n)a]}{2(\beta_m + \beta_n)} & \text{if } \beta_m \neq \beta_n \\ \left[\dfrac{a\beta_n}{2} + \dfrac{\sin(2\beta_n a)}{4}\right]\dfrac{1}{\beta_n} & \text{if } \beta_m = \beta_n \end{cases} \tag{5.26}$$

Since $\beta_m$ and $\beta_n$ satisfy eq. (5.21), this equation can be used together with the top branch of the right-hand side of eq. (5.26) to prove that the value of the integral on the left-hand side of eq. (5.25) equals zero if $\beta_m \neq \beta_n$. Therefore, all the terms of the series in eq. (5.25) are zero, except for the term for which $n = m$. For this term, the value of the integral in eq. (5.25) is given by the bottom branch of the right-hand side of eq. (5.26). Based on the above,

$$K_n = \frac{\Theta_0 \sin(\beta_n a)}{\dfrac{\beta_n a}{2} + \dfrac{\sin(2\beta_n a)}{4}} \tag{5.27}$$

Substituting eq. (5.27) into eq. (5.23) results in the final expression for the temperature:

$$\Theta = T - T_\infty = (T_0 - T_\infty) \sum_{n=1}^{\infty} \frac{\sin(\beta_n a)}{\dfrac{\beta_n a}{2} + \dfrac{\sin(2\beta_n a)}{4}} \cos(\beta_n x) e^{-\alpha\beta_n^2 t} \quad (5.28)$$

The local heat transfer rate at the right side of the sheet ($x = a$) can be obtained with the help of Fourier's law:

$$\dot{q}'' = -k\left(\frac{\partial T}{\partial x}\right)_{x=a} = k(T_0 - T_\infty) \sum_{n=1}^{\infty} \frac{\beta_n \sin^2(\beta_n a)}{\dfrac{\beta_n a}{2} + \dfrac{\sin(2\beta_n a)}{4}} e^{-\alpha\beta_n^2 t} \quad (5.29)$$

One may also evaluate the total of heat flux transferred from the fluid to the sheet through its right side from $t = 0$ to any specified time simply by integrating eq. (5.29) in time from $t = 0$ to any specified time. Here we perform this integration to calculate the total heat flux from $t = 0$ until steady state is reached ($t \to \infty$). Realizing that

$$\dot{q}'' = \frac{dq''}{dt} \quad (5.30)$$

and utilizing eq. (5.29) yields

$$q''_\infty = k(T_0 - T_\infty) \sum_{n=1}^{\infty} \frac{\beta_n \sin^2(\beta_n a)}{\dfrac{\beta_n a}{2} + \dfrac{\sin(2\beta_n a)}{4}} \int_0^\infty e^{-\alpha\beta_n^2 t}\, dt$$

$$= \frac{k(T_0 - T_\infty)}{\alpha} \sum_{n=1}^{\infty} \frac{\sin^2(\beta_n a)}{\left[\dfrac{\beta_n a}{2} + \dfrac{\sin(2\beta_n a)}{4}\right]\beta_n} \quad (5.31)$$

Clearly, at large times the temperature of the sheet approaches the fluid temperature, $T_\infty$ [eq. (5.28)]. The total amount of heat removed from one side of the sheet during the entire cooling process ($t \to \infty$) is one-half the difference between the energy content of the sheet at $t = 0$ ($2\rho cAaT_0$) and at $t \to \infty$ ($2\rho cAaT_\infty$) [i.e., $q = \rho cAa(T_0 - T_\infty)$, where $A$ is the area of one side of the sheet]. The heat flux through the entire cooling process therefore, is $q''_\infty = q_\infty/A = \rho ca(T_0 - T_\infty)$. It is recommended that the student show that eq. (5.31) reduces to the same result.

## 5.2 ONE-DIMENSIONAL TRANSIENT CONDUCTION IN CYLINDRICAL COORDINATES

Transient heat conduction problems in cylindrical coordinates are exemplified by the cooling of shafts and rods the manufacturing of wires and bars and the operation of pin fins. The solution procedure in cases where the method of separation of variables can be applied almost directly is presented with the help of the following example.

### Example 5.2

Consider the arrangement described in Example 5.1 with the difference that instead of a flat sheet a long cylindrical rod is cooled. The radius of the rod is denoted by $R$. Obtain the temperature field in the rod during the cooling process, the heat flux at any location at the outer surface of the rod, and the total amount of heat transferred from the coolant to the rod during the entire cooling process. A schematic of the cooling process is shown in Fig. 5.2.

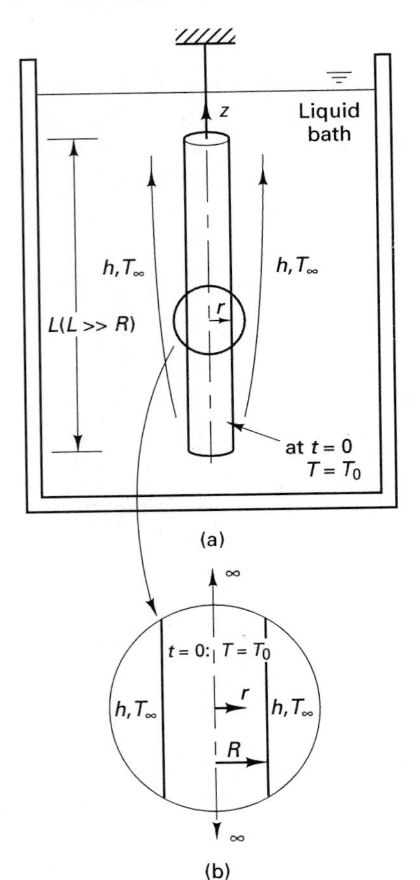


**Figure 5.2** (a) Cylindrical rod immersed in a bath of liquid coolant. (b) Detail of the rod for the modeling process.

### Solution

All the assumptions used in Example 5.1 are adopted here as well but they are not repeated for brevity. The governing equation together with the boundary and initial conditions of the problem are

$$\frac{\partial^2 \Theta}{\partial r^2} + \frac{1}{r}\frac{\partial \Theta}{\partial r} = \frac{1}{\alpha}\frac{\partial \Theta}{\partial t} \tag{5.32}$$

$$r = 0: \qquad \frac{\partial \Theta}{\partial r} = 0 \tag{5.33}$$

$$r = R: \qquad -k\frac{\partial \Theta}{\partial r} = h\Theta \tag{5.34}$$

$$t = 0: \qquad \Theta = T_0 - T_\infty = \Theta_0 \tag{5.35}$$

where, as before,

$$\Theta = T - T_\infty \tag{5.36}$$

Clearly, the governing equation (5.32) as well as the two boundary conditions (5.33) and (5.34) are homogeneous and the method of separation of variables can be applied directly. Before we proceed with the solution, it is worth commenting on the governing equation (5.32). It is a special case of eq. (2.34). It accounts for the fact that there is no angular dependence on the temperature field and that since $R \ll L$,

$$O\left(\frac{\partial^2 T}{\partial z^2}\right) \ll O\left(\frac{\partial^2 T}{\partial r^2}\right) \tag{5.37}$$

Note that

$$\frac{\partial^2 T}{\partial z^2} \sim O\left(\frac{T_0 - T_\infty}{L^2}\right) \tag{5.38}$$

$$\frac{\partial^2 T}{\partial r^2} \sim O\left(\frac{T_0 - T_\infty}{R^2}\right) \tag{5.39}$$

Setting

$$\Theta = G(r)S(t) \tag{5.40}$$

and following the usual steps pertinent to the application of the method of separation of variables yields

$$G'' + \frac{1}{r}G' + \lambda^2 G = 0 \tag{5.41}$$

$$S' + \alpha\lambda^2 S = 0 \tag{5.42}$$

The choice of the sign of the eigenvalue ($\lambda^2$ versus $-\lambda^2$) in obtaining eqs. (5.41) and (5.42) was once again based on the fact that the solution of eq. (5.41) needs to be in terms of orthogonal functions. The solutions to eqs. (5.41) and (5.42) are

$$G = AJ_0(\lambda r) + BY_0(\lambda r) \tag{5.43}$$

$$S = Ce^{-\lambda^2 \alpha t} \tag{5.44}$$

Substituting eqs. (5.43) and (5.44) into eq. (5.40) and combining constants ($K = AC$, $D = BC$) yields

$$\Theta = [KJ_0(\lambda r) + DY_0(\lambda r)]e^{-\lambda^2 \alpha t} \tag{5.45}$$

Proceeding as in the preceding example, we apply the boundary conditions. Equation (5.33) yields

$$D = 0 \tag{5.46}$$

Combining eqs. (5.34) and (5.45) yields the expression for obtaining the eigenvalues $\lambda_n$:

$$(\lambda_n R)J_1(\lambda_n R) = \text{Bi}\, J_0(\lambda_n R) \qquad n = 1, 2, 3, \ldots \tag{5.47}$$

where the Biot number is defined as

$$\text{Bi} = \frac{hR}{k} \tag{5.48}$$

If the values of Bi and $R$ prescribed, eq (5.47) can be solved numerically to yield the eigenvalues $\lambda_n$, $n = 1, 2, 3, \ldots$. We continue by assuming that the values of $\lambda_n$ are known. At this point the solution reads

$$\Theta = \sum_{n=1}^{\infty} K_n J_0(\lambda_n r) e^{-\lambda_n^2 \alpha t} \tag{5.49}$$

The last unknown constant, $K_n$, is determined by applying the initial condition (5.35) together with the orthogonality property of the Bessel function. Substituting eq. (5.49) into eq. (5.35), we obtain

$$\Theta_0 = \sum_{n=1}^{\infty} K_n J_0(\lambda_n r) \tag{5.50}$$

Multiplying both sides of eq. (5.50) by $rJ_0(\lambda_m r)$, integrating both sides from $r = 0$ to $r = R$ and making use of the fact that the family of functions $J_0(\lambda_n r)$ are orthogonal with respect to the weighting factor $r$ (Appendix C) yields

$$K_n = \frac{2\Theta_0}{J_0(\lambda_n R)\left(\text{Bi} + \dfrac{\lambda_n^2 R^2}{\text{Bi}}\right)} \tag{5.51}$$

The final expression for the temperature results simply from combining eqs. (5.49) and (5.51):

$$\Theta = T - T_\infty = 2\text{Bi}\Theta_0 \sum_{n=1}^{\infty} \frac{e^{-\alpha\lambda_n^2 t} J_0(\lambda_n r)}{(\lambda_n^2 R^2 + \text{Bi}^2) J_0(\lambda_n R)} \tag{5.52}$$

The total heat transfer per unit length at any axial location at the surface of the bar over a time period from $t = 0$ and $t = t_f$ is obtained by integrating the instantaneous heat transfer rate over the time period of interest. The instantaneous heat transfer rate per unit length is given by

$$\dot{q}' = -2\pi Rk \left(\frac{\partial T}{\partial r}\right)_{r=R} \tag{5.53}$$

or, by invoking eq. (5.52) to evaluate the temperature gradient,

$$\dot{q}' = 4\pi k \Theta_0 \, \mathrm{Bi} \sum_{n=1}^{\infty} \frac{R\lambda_n e^{-\alpha\lambda_n^2 t} J_1(\lambda_n R)}{\lambda_n^2 R^2 + \mathrm{Bi}^2 J_0(\lambda_n R)} \tag{5.54}$$

Therefore,

$$q'_{t_f} = \int_0^{t_f} \dot{q}' \, dt = 4\pi\rho c R \Theta_0 \, \mathrm{Bi} \sum_{n=1}^{\infty} \frac{1 - e^{-\alpha\lambda_n^2 t_f} J_1(\lambda_n R)}{\lambda_n(\lambda_n^2 R^2 + \mathrm{Bi}^2) J_0(\lambda_n R)} \tag{5.55a}$$

In the limit $t_f \rightarrow \infty$, eq. (5.55) reduces to

$$q'_\infty = 4\pi\rho c \mathrm{R} \Theta_0 \, \mathrm{Bi} \sum_{n=1}^{\infty} \frac{1}{\lambda_n(\lambda_n^2 R^2 + \mathrm{Bi}^2)} \frac{J_1(\lambda_n R)}{J_0(\lambda_n R)} \tag{5.55b}$$

The heat per unit length removed by the coolant through the entire cooling process ($t \rightarrow \infty$) equals the difference in the energy content of the rod per unit length between the initial and final states [i.e., $q'_\infty = q_\infty/L = 2\pi R\rho c(T_0 - T_\infty)$]. The student is encouraged to show that eq. (5.55b) for the total heat removal from the rod per unit length reduces to the above-mentioned result.

## 5.3 ONE-DIMENSIONAL TRANSIENT CONDUCTION IN SPHERICAL COORDINATES

Spherical bodies are commonly encountered in processes involving conduction heat transfer. Such processes are exemplified by heat transfer in packed or fluidized beds consisting of spherical beads, condensation or evaporation of droplets, and the quenching of spheres of metallic or other materials during manufacturing. This last application constitutes the topic of the following example, with the help of which the solution of transient, one-dimensional problems in conduction is illustrated.

### Example 5.3

Consider the arrangement shown in Fig. 5.3. It is identical to the arrangement of Fig. 5.1, with the exception that a hot sphere (instead of a plate) is immersed in a bath of coolant after it is manufactured. The sphere radius is denoted by $R$. The remaining definitions of quantities in Example 5.1 are valid here as well and are not repeated for brevity. It is desired to obtain the temperature field in the sphere as well as the total amount of heat transferred from the sphere to the coolant during the cooling process. Note that the presence of the plume of liquid coolant at the top of the sphere (Fig. 5.3) upsets the use of an average heat transfer coefficient in the modeling of the heat conduction process. However, to facilitate the classroom-level presentation of the solution, the use of an average heat transfer coefficient is recommended.

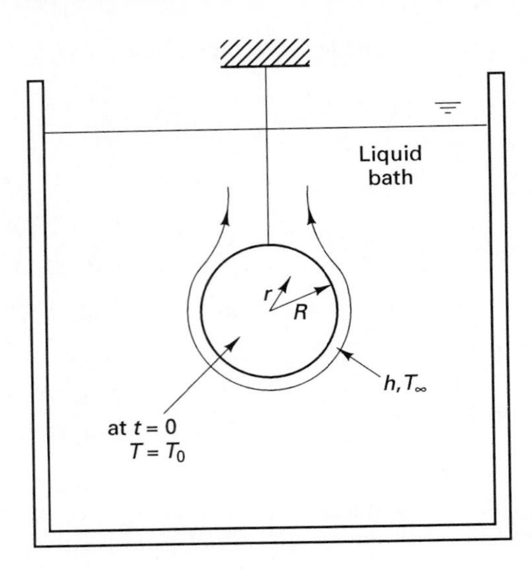


**Figure 5.3** Sphere immersed in a bath of liquid coolant.

**Solution** Since the problem is formulated by using an average heat transfer coefficient, $h$, according to the arrangement of Fig. 5.3, the temperature field is expected to be independent of the coordinates $\phi$ and $\theta$ of a spherical coordinate system. Therefore, the general conduction equation (2.35) reduces to

$$\frac{1}{\alpha}\frac{\partial \Theta}{\partial t} = \frac{1}{r^2}\frac{\partial}{\partial r}\left(r^2 \frac{\partial \Theta}{\partial r}\right) \tag{5.56}$$

The accompanying boundary and initial conditions are

$$r = 0: \qquad \frac{\partial \Theta}{\partial r} = 0 \tag{5.57}$$

$$r = R: \qquad -k\frac{\partial \Theta}{\partial r} = h\Theta \tag{5.58}$$

$$t = 0: \qquad \Theta = T_0 - T_\infty = \Theta_0 \tag{5.59}$$

where again we introduce $\Theta = T - T_\infty$ to render boundary condition (5.58) homogeneous. Boundary condition (5.57) reflects the fact the temperature field is symmetric about the center of the sphere and that the heat flux is continuous everywhere within the sphere. Since the two boundary conditions are homogeneous, we may proceed with the method of separation of variables. However, it is recommended that the following variable transformation be introduced first:

$$u = r\Theta \tag{5.60}$$

In terms of the variable $u$, the model of the problem [eqs. (5.56)–(5.59)] reads

$$\frac{\partial u}{\partial t} = \alpha \frac{\partial^2 u}{\partial r^2} \tag{5.61}$$

$$r = 0: \qquad u = 0 \quad \text{(since } \Theta \text{ is finite)} \tag{5.62}$$

$$r = R: \qquad -k\frac{\partial u}{\partial r} + \left(\frac{k}{r} - h\right)u = 0 \tag{5.63}$$

$$t = 0: \qquad u = r\Theta_0 \tag{5.64}$$

Interestingly, the governing equation in terms of $u$ [eq. (5.61)] is identical to the one-dimensional conduction equation in *Cartesian* coordinates. In addition, boundary conditions (5.62) and (5.63) are homogeneous. Thus, instead of solving the problem for $\Theta$ in spherical coordinates, we proceed with the solution for $u$ in Cartesian coordinates. Once $u$ is unknown, $\Theta$ can be easily recovered through eq. (5.60). Adopting the above-mentioned transformation is recommended when dealing with radial transient conduction in spherical coordinates.

The application of the method of separation of variables is straightforward. Assuming that

$$u = G(r)S(t) \tag{5.65}$$

combining eqs. (5.61) and (5.63), and separating variables in the usual manner yields the following two ordinary differential equations:

$$G'' + \lambda^2 G = 0 \tag{5.66}$$

$$S' + \alpha\lambda^2 S = 0 \tag{5.67}$$

where the primes denote differentiation with respect to $r$ or $t$. After solving eqs. (5.66) and (5.67) the solution for $u$ reads

$$u = [A\sin(\lambda r) + B\cos(\lambda r)]e^{-\alpha\lambda^2 t} \tag{5.68}$$

Applying boundary condition (5.62) yields $B = 0$. The eigenvalues $\lambda_n$, $n = 1, 2, \ldots$ are obtained by substituting eq. (5.68) into boundary condition (5.63):

$$\tan(\lambda_n R) = \frac{\lambda_n R}{1 - \text{Bi}} \qquad n = 1, 2, 3, \ldots \tag{5.69}$$

where $\text{Bi} = hR/k$ is the Biot number. The nonlinear algebraic equation (5.69) can be solved numerically to yield $\lambda_n$ once the values of $R$ and Bi are prescribed. From this point on, the values of $\lambda_n$, $n = 1, 2, \ldots$ are assumed known. Therefore, the expression for $u$ reads

$$u = \sum_{n=1}^{\infty} K_n \sin(\lambda_n r)e^{-\alpha\lambda_n^2 t} \tag{5.70}$$

As in previous examples, the unknown constant $K_n$ is obtained by applying the initial condition and making use of the orthogonality property of $\sin(\lambda_n r)$. Applying the initial condition (5.46) yields

$$r\Theta_0 = \sum_{n=1}^{\infty} K_n \sin(\lambda_n r) \tag{5.71}$$

Next, we multiply both sides of eq. (5.71) by $\sin(\lambda_m r)$ and integrate both sides from $r = 0$ to $r = R$. After noting that the orthogonality of $\sin(\lambda_n r)$ can easily be proven with the help of Appendix C and eq. (5.69) that determines $\lambda_n$, we report the final result for $K_n$:

$$K_n = \frac{\Theta_0 \int_0^R r \sin(\lambda_n r)\, dr}{\int_0^R \sin^2(\lambda_n r)} = 4\Theta_0 \frac{\sin(\lambda_n R) - \lambda_n R \cos(\lambda_n R)}{\lambda_n[2\lambda_n R - \sin(2\lambda_n R)]} \tag{5.72}$$

Substituting eq. (5.72) into eq. (5.70) and recalling the definition of $u$, eq. (5.60), we obtain the final expression for the temperature field in the sphere:

$$\Theta = T - T_\infty = 4\Theta_0 \sum_{n=1}^{\infty} \frac{\sin(\lambda_n R) - \lambda_n R \cos(\lambda_n R)}{2\lambda_n R - \sin(2\lambda_n R)} \frac{\sin(\lambda_n r)}{\lambda_n r} e^{-\alpha\lambda_n^2 t} \tag{5.73}$$

The amount of heat released by the sphere to the coolant from $t = 0$ to any specific final time $t = t_f$ is calculated with the help of Fourier's law:

$$\begin{aligned} \dot{q} = \frac{dq}{dt} &= -k(4\pi R^2)\left(\frac{\partial T}{\partial r}\right)_{r=R} \\ &= 16\pi k \Theta_0 R \sum_{n=1}^{\infty} \frac{[\lambda_n R \cos(\lambda_n R) - \sin(\lambda_n R)]^2}{\lambda_n R[2\lambda_n R - \sin(2\lambda_n R)]} e^{-\alpha\lambda_n^2 t} \end{aligned} \tag{5.74}$$

Integrating over the above-mentioned time span gives

$$\begin{aligned} q_f &= -k(4\pi R^2)\int_0^{t=t_f} \left(\frac{\partial T}{\partial r}\right)_{r=R} dt \\ &= \frac{16\pi k \Theta_0}{\alpha} \sum_{n=1}^{\infty} \frac{[\lambda_n R \cos(\lambda_n R) - \sin(\lambda_n R)]^2}{\lambda_n^3[2\lambda_n R - \sin(2\lambda_n R)]} \left(1 - e^{-\alpha\lambda_n^2 t_f}\right) \end{aligned} \tag{5.75a}$$

In the limit $t \rightarrow \infty$ at which steady state is reached, eq. (5.75a) can be used to provide the result for the total amount of heat released from the sphere during the cooling process until the temperture of the sphere becomes equal to that of the fluid. This result is

$$q_\infty = \frac{16\pi k \Theta_0}{\alpha} \sum_{n=1}^{\infty} \frac{[\lambda_n R \cos(\lambda_n R) - \sin(\lambda_n R)]^2}{\lambda_n^3[2\lambda_n R - \sin(2\lambda_n R)]} \tag{5.75b}$$

The total heat removed from the sphere equals the difference in the energy content of the sphere between the initial ($t = 0$) and the final ($t \rightarrow \infty$) states [i.e., $q_\infty = \frac{4}{3}\pi R^3 \rho c(T_0 - T_\infty)$]. It can be shown that eq. (5.75b) for $q_\infty$ reduces to the above-mentioned result.

## 5.4 THE PRINCIPLE OF SUPERPOSITION

In Sections 5.1 to 5.3, simple one-dimensional transient problems were considered in the three major coordinate systems. These problems were solved with the method of separation of variables. They are termed *simple* because the method of separation of variables was applicable almost directly, without requiring significant preliminary work and transformations or modifications to ensure its applicability. In the present and the following sections more complex problems will be considered in which the method of separation of variables is not directly applicable and the aid of additional tools is necessary. The first of these tools is the *principle of superposition* [1–3]. The general rules that were presented in Chapter 4 to systematize the application of the principle of superposition apply here as well. In addition, it is noted that the principle of superposition can be used for problems in all coordinate systems (Cartesian, cylindrical, or spherical). Two general categories of problems are defined next. The first category contains transient problems in which a steady state exists. In the second category, transient problems in which steady state does not exist are included.

### Problems in Which Steady State Exists

In this category it is recommended that the decomposition of the original problem begins by splitting the problem in two parts: a transient part and a steady-state part. Since the steady-state part is simply one-dimensional steady conduction, it is easily solvable. Therefore, it is advised that as many nonhomogeneities as possible are included with the steady-state part. This simplifies the transient part of the problem. Mathematically, then, it is recommended that the superposition procedure begins as follows:

$$T(\mathbf{r}, t) = T_t(\mathbf{r}, t) + T_s(\mathbf{r}) \tag{5.76}$$

where $\mathbf{r}$ is a position vector. In the case of one-dimensional transient conduction in Cartesian coordinates, if the space coordinate is denoted by $x$, eq. (5.70) becomes

$$T(x, t) = T_t(x, t) + T_s(x) \tag{5.77}$$

In the case of radial conduction in cylindrical coordinates, if the radial coordinate is denoted by $r$, eq. (5.76) becomes

$$T(r, t) = T_t(r, t) + T_s(r) \tag{5.78}$$

Finally, in the case of radial conduction in spherical coordinates, if the radial coordinate is denoted by $r$, eq. (5.76) becomes

$$T(r, t) = T_t(r, t) + T_s(r) \tag{5.79}$$

An example is presented next which illustrates the use of the superposition principle in transient one-dimensional conduction. The example is in cylindrical coordinates. The procedure is identical for problems in Cartesian or spherical coordinates.

**Example 5.4**

Figure 5.4 shows a long electrical wire of radius $a$. When electric current passes through the wire it generates heat (resistance heating). Assume that the volumetric heat generation in the wire is constant, $\dot{q}'''$. The wire is cooled by a cold airstream. The average heat transfer coefficient between the wire and the stream is denoted by $h$. The air temperature far away from the wire is denoted

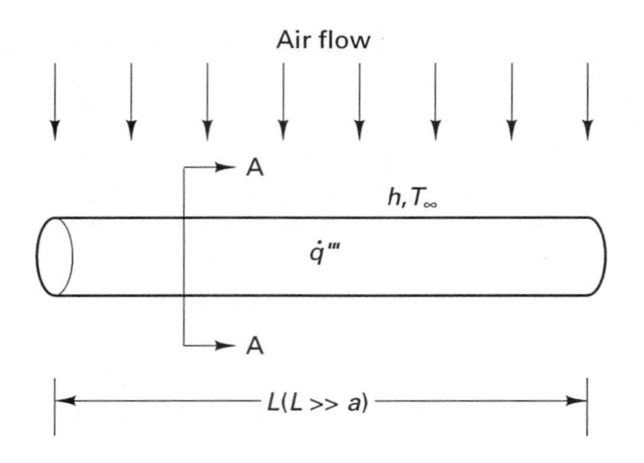


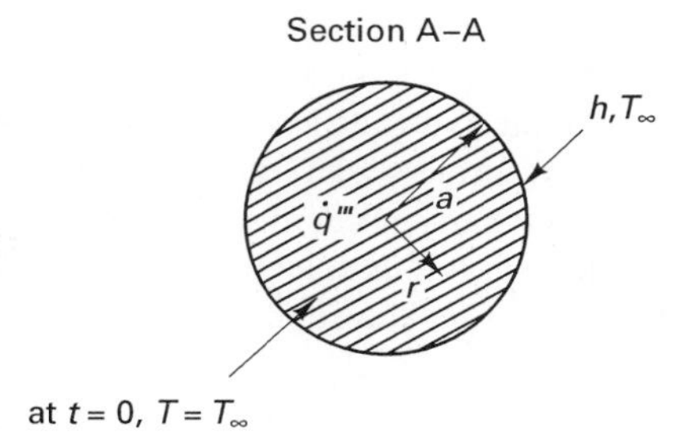


**Figure 5.4** Heat-generating electrical wire in cross-flow.

by $T_\infty$. Initially, no current is passing through the wire and no heat is generated. The initial temperature of the wire equals that of air, $T_\infty$. Suddenly, a constant current, $I$, is passed through the wire, and the resulting constant volumetric heat generation is $\dot{q}''' = I^2R/\pi a^2 L$, where $R$ is the electrical resistance and $L$ the length of the wire. If $\rho$ denotes the resistance of the wire per length, then $\rho = R/L$. We are interested in obtaining the temperature history of the wire as well as the instantaneous and the steady-state heat transfer rates from the wire surface to the fluid. What is the total heat transfer from the wire to the fluid during the transient time period?

## Solution

Before writing down the mathematical model of the problem, several facts need to be discussed. First, a steady-state clearly exists since the heating of the wire caused by the internal heat generation is balanced by the cooling provided by the airstream. Second, because of the symmetry of the problem there is no angular dependence of the temperature field. Third, from scaling arguments it can be easily shown that

$$O\left(\frac{\partial^2 T}{\partial r^2}\right) = \frac{\Delta T}{a^2} \qquad O\left(\frac{1}{r}\frac{\partial T}{\partial r}\right) = \frac{\Delta T}{a^2} \qquad O\left(\frac{\partial^2 T}{\partial z^2}\right) = \frac{\Delta T}{L^2} \tag{5.80}$$

where $\Delta T$ is a characteristic temperature difference. Since $L^2 \gg a^2$, equations (5.80) imply that

$$\frac{\partial^2 T}{\partial z^2} \ll \frac{\partial^2 T}{\partial r^2} \quad \text{and} \quad \frac{\partial^2 T}{\partial z^2} \ll \frac{1}{r}\frac{\partial T}{\partial r} \tag{5.81}$$

Based on the above, the general conduction equation (2.24) when applied to the present problem reduces to

$$\frac{1}{\alpha}\frac{\partial \Theta}{\partial t} = \frac{\partial^2 \Theta}{\partial r^2} + \frac{1}{r}\frac{\partial \Theta}{\partial r} + \frac{\dot{q}'''}{k} \tag{5.82}$$

The corresponding boundary and initial conditions are

$$r = 0: \qquad \frac{\partial \Theta}{\partial r} = 0 \tag{5.83}$$

$$r = a: \qquad -k\frac{\partial \Theta}{\partial r} = h\Theta \tag{5.84}$$

$$t = 0: \qquad \Theta = 0 \tag{5.85}$$

where introducing $\Theta = T - T_\infty$ yielded two homogeneous boundary conditions (5.83) and (5.84). Noting that the energy equation (5.82) is not homogeneous, we use the principle of superposition and decompose the problem above into a transient and a steady-state part as discussed earlier:

$$\Theta(r, t) = \Theta_t(r, t) + \Theta_s(r) \tag{5.86}$$

In creating the boundary conditions for the two "component" problems we attempt to include as many nonhomogeneities as possible with the easy steady-state problem. The various steps in using the superposition principle were outlined in Chapter 4 and are followed (implicitly) here.

The model for the transient problem is

$$\frac{1}{\alpha}\frac{\partial \Theta_t}{\partial t} = \frac{\partial^2 \Theta_t}{\partial r^2} + \frac{1}{r}\frac{\partial \Theta_t}{\partial r} \tag{5.87}$$

$$r = 0: \qquad \frac{\partial \Theta_t}{\partial r} = 0 \tag{5.88}$$

$$r = a: \qquad -k\frac{\partial \Theta_t}{\partial r} = h\Theta_t \tag{5.89}$$

$$t = 0: \qquad \Theta_t = -\Theta_s \tag{5.90}$$

The model for the steady-state problem is

$$\frac{d^2 \Theta_s}{dr^2} + \frac{1}{r}\frac{d\Theta_s}{dr} + \frac{\dot{q}'''}{k} = 0 \tag{5.91}$$

$$r = 0: \qquad \frac{d\Theta_s}{dr} = 0 \tag{5.92}$$

$$r = a: \qquad -k\frac{d\Theta_s}{dr} = h\Theta_s \tag{5.93}$$

Clearly, if we add up the corresponding governing equations and boundary conditions for the transient and steady-state problems side by side and take eq. (5.86) into account, we recover the original governing equation (5.82) and boundary conditions (5.83) and (5.84). In addition, eq. (5.90) is identical to the initial condition (5.85).

Since the solution for the steady-state problem is needed for the solution of the transient problem, we attempt to solve the steady-state problem first. Equation (5.91) can be written as

$$\frac{1}{r}\frac{d}{dr}\left(r\frac{d\Theta_s}{dr}\right) = -\frac{\dot{q}'''}{k} \tag{5.94}$$

Integrating eq. (5.94) directly twice and applying the two boundary conditions (5.92) and (5.93) yields

$$\Theta_s = \frac{\dot{q}'''}{4k}(a^2 - r^2) + \frac{\dot{q}'''a}{2h} \tag{5.95}$$

With the steady-state temperature distribution known, we proceed with the solution of the transient problem.

Assuming that

$$\Theta_t(r, t) = G(r)S(t) \tag{5.96}$$

substituting eq. (5.96) into eq. (5.87), and going through the usual procedure, we derive the following two ordinary differential equations:

$$G'' + \frac{1}{r}G' + \lambda^2 G = 0 \tag{5.97}$$

$$S' + \lambda^2 \alpha S = 0 \tag{5.98}$$

Again, the choice of the sign $\lambda^2$ (versus $-\lambda^2$) in obtaining the equations above was made by taking into account that the solution of eq. (5.97) should be in terms of orthogonal functions. Solving eqs. (5.97) and (5.98) yields (Appendixes A and B)

$$G = AJ_0(\lambda r) + BY_0(\lambda r) \tag{5.99}$$

$$S = Ce^{-\lambda^2 \alpha t} \tag{5.100}$$

Combing eqs. (5.96), (5.99), and (5.100) results in the following expression for the temperature:

$$\Theta_t(r, t) = [KJ_0(\lambda r) + DY_0(\lambda r)]e^{-\lambda^2 \alpha t} \tag{5.101}$$

To proceed, we need to apply boundary conditions (5.92) and (5.93). We note, however that those boundary conditions are identical to the boundary conditions (5.33) and (5.34) used in Example 5.2. In addition, eq. (5.101) is identical to eq. (5.45). Therefore, the steps involving the application of the boundary conditions in the present problem are identical to the analogous steps in Example 5.2 and for brevity, will not be repeated. The expression for the temperature after applying the boundary conditions is

$$\Theta_t = \sum_{n=1}^{\infty} K_n J_0(\lambda_n r) e^{-\lambda_n^2 \alpha t} \tag{5.102}$$

where the eigenvalues $\lambda_n$, $n = 1, 2, 3, \ldots$ are the roots of

$$(\lambda_n a) J_1(\lambda_n a) = \text{Bi}\, J_0(\lambda_n a) \tag{5.103}$$

The Biot number is defined as

$$\text{Bi} = \frac{ha}{k} \tag{5.104}$$

To determine the unknown constant $K_n$ we need to apply eq. (5.90) and use the orthogonality of $J_0(\lambda_n r)$. Applying the initial condition, eq. (5.90), yields

$$-\frac{\dot{q}'''}{4k}(a^2 - r^2) - \frac{\dot{q}'''a}{2h} = \sum_{n=1}^{\infty} K_n J_0(\lambda_n r) \tag{5.105}$$

Multiplying both sides of eq. (5.105) by $rJ_0(\lambda_m r)$, integrating both sides from $r = 0$ to $r = a$, and making use of the fact that $J_0(\lambda_n r)$, $n = 1, 2, \ldots$, are orthogonal with respect to the weighting factor $r$ (Appendix A) yields

$$K_n = -\frac{2}{a^2} \frac{\displaystyle\int_0^a \left[\frac{\dot{q}'''}{4k}(a^2 - r^2) - \frac{\dot{q}'''a}{2h}\right] rJ_0(\lambda_n r)\, dr}{J_0^2(\lambda_n a) + J_1^2(\lambda_n a)} \tag{5.106}$$

To obtain the final expression for $\Theta_t$, we simply substitute eq. (5.106) into eq. (5.102):

$$\Theta_t = \frac{2}{a^2} \sum_{n=1}^{\infty} \frac{\displaystyle\int_0^a \left[\frac{\dot{q}'''}{4k}(r^2 - a^2) - \frac{\dot{q}'''a}{2h}\right] rJ_0(\lambda_n r)\, dr}{J_0^2(\lambda_n a) + J_1^2(\lambda_n a)} J_0(\lambda_n r) e^{-\lambda_n^2 \alpha t} \tag{5.107}$$

The solution for the temperature field of the original problem is constructed by adding eqs. (5.95) and (5.107) side by side:

$$\begin{aligned}
\Theta = T - T_\infty = \Theta_s + \Theta_t &= \frac{\dot{q}'''}{4k}(a^2 - r^2) + \frac{\dot{q}'''a}{2h} \\
&\quad + \frac{2}{a^2} \sum_{n=1}^{\infty} \frac{\displaystyle\int_0^a \left[\frac{\dot{q}'''}{4k}(r^2 - a^2) - \frac{\dot{q}'''a}{2h}\right] rJ_0(\lambda_n r)\, dr}{J_0^2(\lambda_n a) + J_1^2(\lambda_n a)} J_0(\lambda_n r) e^{-\lambda_n^2 \alpha t} \\
&= \frac{\dot{q}'''a^2}{k} \left\{ \frac{1}{4}\left(1 - \frac{r^2}{a^2}\right) + \frac{1}{2\text{Bi}} - 2\text{Bi} \sum_{n=1}^{\infty} \frac{J_0(\lambda_n r) e^{-\lambda_n^2 \alpha t}}{(\lambda_n a)^4 J_0(\lambda_n a)\left[1 + \dfrac{\text{Bi}^2}{(\lambda_n a)^2}\right]} \right\}
\end{aligned} \tag{5.108}$$

The heat transfer rate per unit length of the wire at any axial location is

$$\dot{q}' = -2\pi ak\left(\frac{\partial T}{\partial r}\right)_{r=a} \tag{5.109}$$

Invoking eq. (5.108) to evaluate the temperature gradient in eq. (5.109) yields

$$\dot{q}' = \left\{-2\pi a^3\dot{q}''' \quad -\frac{1}{2a} + 2\frac{\mathrm{Bi}^2}{a}\sum_{n=1}^{\infty}\frac{e^{-\lambda_n^2\alpha t}}{(\lambda_n a)^4\left[1+\dfrac{\mathrm{Bi}^2}{(\lambda_n a)^2}\right]}\right\} \tag{5.110}$$

Recalling that $\dot{q}' = dq'/dt$ in eq. (5.110), the total amount of heat per unit length over a period of time from $t = 0$ to $t = t_f$ can be calculated:

$$q'_{t_f} = \int_0^{t_f}\dot{q}'\,dt = 2\pi a^3\dot{q}'''\left\{\frac{t_f}{2a} + 2\frac{\mathrm{Bi}^2 a}{\alpha}\sum_{n=1}^{\infty}\frac{e^{-\lambda_n^2\alpha t_f}-1}{(\lambda_n a)^6\left[1+\dfrac{\mathrm{Bi}^2}{(\lambda_n a)^2}\right]}\right\} \tag{5.111}$$

### Problems in Which Steady State Does Not Exist

If a steady state does not exist, the use of the superposition principle is less straightforward and some additional creativity is required in decomposing the problem of interest. Even though no general recommendations are made, the following decomposition will be applied in Example 5.5 later in this section:

$$T(\mathbf{r}, t) = T_{et}(\mathbf{r}, t) + T_{lt1}(\mathbf{r}) + T_{lt2}(t) \tag{5.112}$$

In eq. (5.112), $T_{et}(\mathbf{r}, t)$ represents the behavior of the transient conduction temperature field at early times and the *sum* $T_{lt1}(\mathbf{r}) + T_{lt2}(t)$ represents the transient conduction temperature field at late times (since steady state is never reached). In one-dimensional Cartesian coordinates, if the space coordinate is denoted by $x$, eq. (5.112) reads

$$T(x, t) = T_{et}(x, t) + T_{lt1}(x) + T_{lt2}(t) \tag{5.113}$$

In one-dimensional transient conduction in cylindrical or spherical coordinates, if the radial coordinate is denoted by $r$, eq. (5.112) becomes

$$T(r, t) = T_{et}(r, t) + T_{lt1}(r) + T_{lt2}(t) \tag{5.114}$$

In decomposing a problem, an effort should be made to include as many complexities as possible with the easier problems describing the late-time behavior of the temperature field, $T_{lt1}$, $T_{lt2}$. The superposition procedure outlined above is illustrated with the help of the following example.

**Example 5.5**

---

A long cylindrical rod of radius $R$ and length $L(L \gg R)$ is placed inside a furnace to be melted (Fig. 5.5). The initial temperature of the rod is $T_0$. The con-

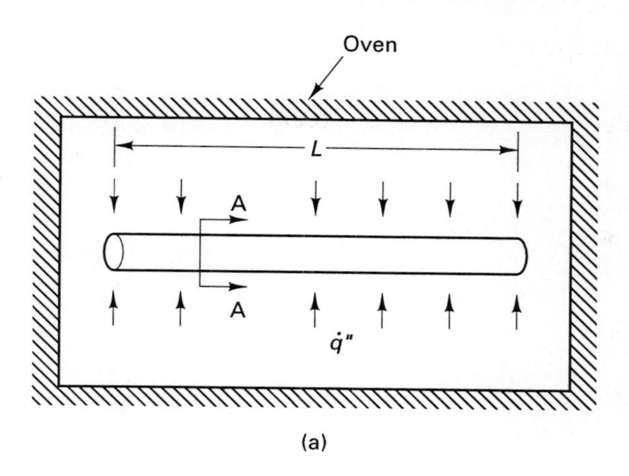


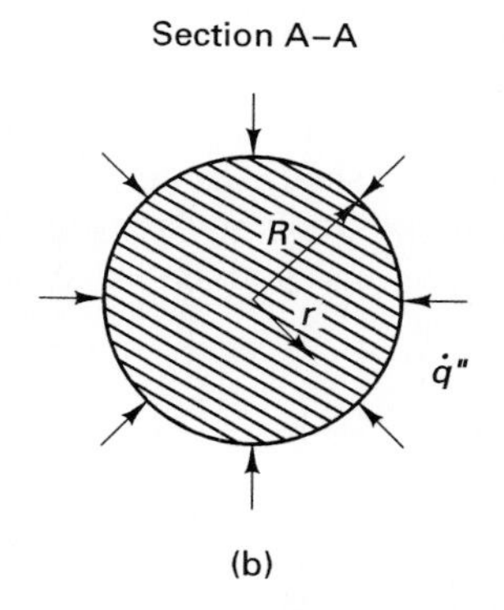


**Figure 5.5** Cylindrical rod being heated in an oven.

vective effects of the furnace are negligible. The heating of the rod occurs by means of a uniform heat flux, $\dot{q}''$, at its surface. It is of interest to obtain the temperature field in the cylinder and to estimate the time required for the melting process to begin if the melting temperature of the cylinder material is denoted by $T_m$.

**Solution** Based on the problem description, the following are clear: First, the temperature field does not depend on the angular coordinate. Second, one may repeat here the scaling arguments of Example 5.4 to show that since $L \gg R$,

$$\frac{\partial^2 T}{\partial z^2} \ll \frac{\partial^2 T}{\partial r^2} \quad \frac{\partial^2 T}{\partial z^2} \ll \frac{1}{r}\frac{\partial T}{\partial r} \tag{5.115}$$

Therefore, the dependence of the temperature field on the axial coordinate, $z$, is negligible. Third, no steady state exists since the cylinder is heated continuously by the oven and no cooling mechanism is provided (i.e., no "escape" for the thermal energy entering the control volume defined by the surface of the rod exists). We are dealing, then, with a transient radial conduction problem with no steady state (Fig. 5.5b).

The mathematical model for the problem of interest after the clarifications is

$$\frac{1}{\alpha}\frac{\partial \Theta}{\partial t} = \frac{1}{r}\frac{\partial}{\partial r}\left(r\frac{\partial \Theta}{\partial r}\right) \tag{5.116}$$

$$r = 0: \qquad \frac{\partial \Theta}{\partial r} = 0 \tag{5.117}$$

$$r = R: \qquad k\frac{\partial \Theta}{\partial r} = \dot{q}'' \tag{5.118}$$

$$t = 0: \qquad \Theta = 0 \tag{5.119}$$

where $\Theta = T - T_0$. As discussed earlier, we attempt to apply the superposition principle as follows:

$$\Theta(r, t) = \Theta_{et}(r, t) + \Theta_{lt1}(r) + \Theta_{lt2}(t) \tag{5.120}$$

In defining the mathematical models for three component problems, one should account for the fact that $\Theta_{et}(r, t)$ should diminish as $t \to \infty$. In addition, the nonhomogeneity associated with the boundary condition at $r = R$[eq. (5.118)] should be included in the easier problem $\Theta_{lt1}(r)$. The mathematical model for $\Theta_{et}(r, t)$ is

$$\frac{1}{\alpha}\frac{\partial \Theta_{et}}{\partial t} = \frac{1}{r}\frac{\partial}{\partial r}\left(r\frac{\partial \Theta_{et}}{\partial r}\right) \tag{5.121}$$

$$r = 0: \qquad \frac{\partial \Theta_{et}}{\partial r} = 0 \tag{5.122}$$

$$r = R: \qquad \frac{\partial \Theta_{et}}{\partial r} = 0 \tag{5.123}$$

$$t = 0: \qquad \Theta_{et} = -\Theta_{lt1} - \Theta_{lt2} \tag{5.124}$$

Note that eq. (5.124) assures already that the original initial condition, eq. (5.119), is satisfied.

The transient radial conduction phenomenon at late times is described by the sum $\Theta_{lt1}(r) + \Theta_{lt2}(t)$. To obtain the mathematical models for $\Theta_{lt1}(r)$ and $\Theta_{lt2}(t)$ we must realize that since at large times $\Theta_{et}(r, t) \to 0$, the energy equation (5.116) *at late times* becomes

$$\frac{1}{\alpha}\frac{\partial}{\partial t}[\Theta_{lt1}(r) + \Theta_{lt2}(t)] = \frac{1}{r}\frac{\partial}{\partial r}\left\{r\frac{\partial}{\partial r}[\Theta_{lt1}(r) + \Theta_{lt2}(t)]\right\} \tag{5.125}$$

or

$$\frac{1}{\alpha}\frac{\partial}{\partial t}[\Theta_{lt2}(t)] = \frac{1}{r}\frac{\partial}{\partial r}\left\{r\frac{\partial}{\partial r}[\Theta_{lt1}(r)]\right\} \tag{5.126}$$

The left-hand side of eq. (5.126) is a function of time only, and the right-hand side is a function of the radial coordinate only. Therefore, each side equals a constant, say $K$. Based on the above, the mathematical model for $\Theta_{lt1}(r)$ is

$$\frac{1}{r}\frac{d}{dr}\left\{r\frac{d}{dr}[\Theta_{lt1}(r)]\right\} = K \tag{5.127}$$

$$r = 0: \qquad \frac{d\Theta_{lt1}}{dr} = 0 \tag{5.128}$$

$$r = R: \qquad k\frac{d\Theta_{lt1}}{dr} = \dot{q}'' \tag{5.129}$$

The governing equation for $\Theta_{lt2}(t)$ is

$$\frac{1}{\alpha}\frac{d}{dt}[\Theta_{lt2}(t)] = K \tag{5.130}$$

Since $\Theta_{lt2}(t)$ represents the late-time dependence of the temperature field on time, we cannot impose an initial condition on eq. (5.130): This equation is not valid at $t = 0$. It will be shown later how we account for the constant of integration of eq. (5.130) despite the lack of an initial condition.

The solutions for $\Theta_{lt1}(r)$ and $\Theta_{lt2}(t)$ are needed for solution of the $\Theta_{et}(r, t)$ model. Integrating eq. (5.127) subject to conditions (5.128) and (5.129) yields

$$\Theta_{lt1} = \frac{\dot{q}''}{2kR}r^2 + C_1 \tag{5.131}$$

Similarly, integrating eq. (5.130), we obtain

$$\Theta_{lt2} = 2\frac{\dot{q}''}{Rk}\alpha t + C_2 \tag{5.132}$$

where $C_1$ and $C_2$ are constants of integration. No additional conditions exist to evaluate $C_1$ and $C_2$. However, since the solution of $\Theta_{et}$ depends on $\Theta_{lt1}$ and $\Theta_{lt2}$ via eq. (5.124), we can arbitrarily set $C_1$ and $C_2$ equal to known values. Clearly, the solution for $\Theta_{et}$ will correct for these arbitrary values and the final expression for $\Theta$ will satisfy all eqs. (5.116)–(5.119). We decide to set

$$C_1 = C_2 = 0 \tag{5.133}$$

Based on eqs. (5.131)–(5.133) the initial condition (5.124) becomes

$$t = 0: \qquad \Theta_{et} = -\frac{\dot{q}''}{2kR}r^2 \tag{5.134}$$

At this point we are ready to solve eqs. (5.121)–(5.123) and (5.134) by direct application of the method of separation of variables to obtain the early-times temperature distribution $\Theta_{et}(r, t)$. Assuming that

$$\Theta_{et} = G(r)S(t) \tag{5.135}$$

and going through the usual separation of variables procedure, we obtain

$$G'' + \frac{1}{r}G' + \lambda^2 G = 0 \tag{5.136}$$

$$S' + \alpha\lambda^2 S = 0 \tag{5.137}$$

The solutions for $G(r)$ and $S(t)$ are

$$G = AJ_0(\lambda r) + BY_0(\lambda r) \tag{5.138}$$

$$S = Ce^{-\lambda^2 \alpha t} \tag{5.139}$$

Combining eqs. (5.135), (5.138) and (5.139) (see also Example 5.4) gives

$$\Theta_{et} = [KJ_0(\lambda r) + DY_0(\lambda r)]e^{-\lambda^2 \alpha t} \tag{5.140}$$

Boundary condition (5.122) yields

$$D = 0 \tag{5.141}$$

The eigenvalues $\lambda_n$ are obtained with the help of eq. (5.123), which becomes

$$J_1(\lambda_n R) = 0 \qquad n = 1, 2, \ldots \tag{5.142}$$

Assuming that $\lambda_n$, $n = 1, 2, \ldots$ are known,

$$\Theta_{et}(r, t) = \sum_{n=1}^{\infty} K_n J_0(\lambda_n r)e^{-\lambda_n^2 \alpha t} \tag{5.143}$$

Next, the initial condition (5.134) when combined with expression (5.143) becomes

$$-\frac{\dot{q}''}{2kR}r^2 = \sum_{n=1}^{\infty} K_n J_0(\lambda_n r) \tag{5.144}$$

Proceeding as in Example 5.4 by utilizing the orthogonality property of $J_0(\lambda_n r)$ in eq. (5.144) results in the expression of the final unknown constant $K_n$:

$$K_n = \frac{-\dfrac{\dot{q}''}{2kR}\displaystyle\int_0^R r^3 J_0(\lambda_n r)\, dr}{\displaystyle\int_0^R rJ_0^2(\lambda_n r)\, dr} \tag{5.145}$$

In performing the integration in eq. (5.145) we need to account for the fact that $\lambda_1 = 0$ is the first root of eq. (5.142). Hence

$$K_1 = -\frac{1}{4}\frac{\dot{q}''R}{k} \tag{5.146}$$

$$K_n = -\frac{2\dot{q}''}{Rk}\frac{1}{\lambda_n^2 J_0(\lambda_n R)} \qquad n = 2, 3, \ldots \tag{5.147}$$

The early-time temperature expression therefore reads

$$\Theta_{et} = -\frac{1}{4}\frac{\dot{q}''R}{k} - \frac{2\dot{q}''}{Rk}\sum_{n=2}^{\infty}\frac{1}{\lambda_n^2}\frac{J_0(\lambda_n r)}{J_0(\lambda_n R)} \tag{5.148}$$

Combining eqs. (5.120), (5.131), (5.132), and (5.148) yields the temperature distribution in the rod:

$$\Theta = T - T_0 = \frac{\dot{q}''}{2Rk}r^2 + \frac{2\dot{q}''}{Rk}\alpha t - \frac{1}{4}\frac{\dot{q}''R}{k} - \frac{2\dot{q}''}{Rk}\sum_{n=2}^{\infty}\frac{1}{\lambda_n^2}\frac{J_0(\lambda_n r)}{J_0(\lambda_n R)}e^{-\alpha\lambda_n^2 t} \tag{5.149}$$

The maximum temperature in the bar occurs at the bar surface. The time at which melting occurs is then obtained by evaluating $T$ in eq. (5.149) at $r = R$, setting it equal to the melting temperature, $T_m$, and solving the resulting expression for the melting time, $t_m$. The expression that needs to be solved is

$$T_m - T_0 = \frac{\dot{q}''R}{4k} + \frac{2\dot{q}''}{Rk}\alpha t_m - \frac{2\dot{q}''}{Rk}\sum_{n=2}^{\infty}\frac{1}{\lambda_n^2}e^{-\alpha\lambda_n^2 t_m} \tag{5.150}$$

Clearly, $t_m$ needs to be evaluated numerically after the values of all parameters are prescribed.

## 5.5 THE METHOD OF VARIATION OF PARAMETERS

As discussed in Section 5.4 the principle of superposition can be used to solve transient nonhomogeneous problems. An alternative is offered via the method of variation of parameters. This is a well-structured mathematical method [1, 4] that can be used for the solution of problems both when steady state exists and when steady state does not exist. It is up to the problem solver to decide whether to adopt the method of variation of parameters or the principle of superposition.

Generally speaking, when a steady state does *not* exist, the application of the principle of superposition contains several subtleties and requires a certain degree of experience or even creativity, as Example 5.5 illustrated. On the other hand, the application of the method of variation of parameters is, almost always, conceptually straightforward.

To aid the reader, we propose that a sequence of steps is followed when the method of variation of parameters is applied:

### ■ *Step 1*

Create the homogeneous problem equivalent to the nonhomogeneous problem in hand simply by setting all the nonhomogeneous terms of the nonhomogeneous problem in its governing equation and/or its boundary conditions equal to zero.

■ ***Step 2***

Determine the eigenfunctions, $\phi_n(\lambda_n x)$, and eigenvalues, $\lambda_n$, of the equivalent homogeneous problem created in step 1. The space variable is denoted by $x$.

■ ***Step 3***

Use the eigenfunctions and eigenvalues from step 2 to construct the solution of the original nonhomogeneous problem by setting

$$T(x, t) = \sum_n A_n(t)\phi_n(\lambda_n x) \tag{5.151}$$

Note that at this point the equivalent homogeneous problem can be forgotten, for it has served its purpose and will not be utilized in the remainder of the solution.

■ ***Step 4***

Evaluate $A_n(t)$ by making use of the orthogonality property of $\phi_n(\lambda_n x)$ (Appendix C). If the space variable varies in the range $0 \leq x \leq a$, then

$$\begin{aligned}\int_0^a T(x, t)w(x)\phi_m(\lambda_m x)\, dx &= \sum_n A_n(t) \int_0^a w(x)\phi_n(\lambda_n x)\phi_m(\lambda_m x)\, dx \\ &= C_m A_m(t)\end{aligned} \tag{5.152}$$

Therefore,

$$A_n(t) = \frac{1}{C_n}\int_0^a T(x, t)w(x)\phi_n(\lambda_n x)\, dx \tag{5.153}$$

where $w(x)$ is the weighting factor and $C_n$ stands for the value of the integral of the middle section of eq. (5.152). Observe that $T(x, t)$ in expression (5.153) for $A_n(t)$ is not known.

■ ***Step 5***

Set up ordinary differential equations and initial conditions for $A_n(t)$ as shown in Example 5.6 (to follow) using eq. (5.153).

■ ***Step 6***

Solve the differential equations of step 5 to obtain $A_n(t)$ and therefore complete the solution process [with both $A_n(t)$ and $\phi_n(\lambda_n x)$ known $T(x, t)$ is known through eq. (5.151)].

The procedure outlined above will be better understood with the help of an illustrative example. To contrast the method of variation of parameters to the principle of superposition, the problem of Example 5.5 is re-solved but this time by using the method of variation of parameters.

**Example 5.6**

Re-solve the problem of Example 5.5 by using the method of variation of parameters.

**Solution**

The procedure outlined above will be followed.

■ ***Step 1***

The mathematical model of the problem of interest consists of eqs. (5.116)–(5.119). To create the equivalent homogeneous problem, we set $\dot{q}'' = 0$ in eq. (5.118). The resulting equivalent homogeneous problem is described by

$$\frac{1}{\alpha}\frac{\partial \Theta_h}{\partial t} = \frac{1}{r}\frac{\partial}{\partial r}\left(r\frac{\partial \Theta_h}{\partial r}\right) \tag{5.154}$$

$$r = 0: \qquad \frac{\partial \Theta_h}{\partial r} = 0 \tag{5.155}$$

$$r = R: \qquad \frac{\partial \Theta_h}{\partial r} = 0 \tag{5.156}$$

$$t = 0: \qquad \Theta_h = 0 \tag{5.157}$$

where subscript $h$ denotes the equivalent homogeneous problem.

■ ***Step 2***

To obtain the eigenfunctions and eigenvalues of the equivalent hmomgeneous problem, we use the method of separation of variables. Assume that

$$\Theta_h(r, t) = G_h(r)S_h(t) \tag{5.158}$$

Then substitute eq. (5.158) into eq. (5.154) and separate variables in the usual manner, to obtain one ordinary differential equation for $G_h(r)$ and another for $S_h(t)$. Consideration should be given to the fact that $r$ is the homogeneous direction and the solution for $G_h(r)$ should be obtained in terms of orthogonal functions. Only the differential equation for $G_h(r)$ is of interest in this step.

$$G_h'' + \frac{1}{r}G_h' + \lambda^2 G_h = 0 \tag{5.159}$$

The solution to eq. (5.159) for $\lambda \neq 0$ is

$$G_h = AJ_0(\lambda r) + BY_0(\lambda r) \tag{5.160}$$

The $\lambda = 0$ case will be examined later. Boundary condition (5.155) yields

$$B = 0 \tag{5.161}$$

Also, the eigenvalues are obtained from boundary condition (5.156), which results in

$$J_1(\lambda_n R) = 0 \qquad n = 2, 3, \ldots \tag{5.162}$$

Based on the discussion above, we then conclude that the eigenfunctions for the equivalent homogeneous problem are $J_0(\lambda_n r)$, $n = 2, 3, \ldots$, where $\lambda_n$ is obtained from eq. (5.162) once a numerical value for the radius of the bar, $R$, is prescribed. Note that the first eigenvalue resulting from eq. (5.162) is $\lambda_1 = 0$ (Appendix A, Fig. A.1) and is excluded from the set of eigenvalues of eq. (5.162) because eq. (5.160) is the solution for $\lambda \neq 0$. It can easily be shown that in the case $\lambda = 0$, eqs. (5.159), (5.155,) and (5.156) yield that $G_h = K$, where $K$ is a constant. At this point, all the tasks needed in step 2 have been performed and we proceed to the following step.

### ■ *Step 3*

Assume that the solution of the *original* problem is

$$\Theta(r, t) = Ka_1(t) + \sum_{n=2}^{\infty} A_n(t)J_0(\lambda_n r) = \sum_{n=1}^{\infty} A_n(t)J_0(\lambda_n r) \tag{5.163}$$

In writing the right-hand side of eq. (5.163), we took into account the fact that $\lambda_1 = 0$ and $J_0(0) = 1$ (Appendix A, Fig. A.1) and we renamed $Ka_1(t)$ to $A_1(t)$ since the product of a constant times on arbitrary function of time is simply another arbitrary function of time.

### ■ *Step 4*

Utilizing the orthogonality property of $J_0(\lambda_n r)$ in eq. (5.163), we obtain

$$\int_0^R \Theta(r, t)rJ_0(\lambda_m r)\, dr = \sum_{n=1}^{\infty} A_n(t) \int_0^R J_0(\lambda_n r)J_0(\lambda_m r)r\, dr \tag{5.164}$$

or

$$A_n(t) = \frac{1}{\int_0^R J_0^2(\lambda_n r)r\, dr} \int_0^R \Theta(r, t)rJ_0(\lambda_n r)\, dr \qquad n = 1, 2, \ldots \tag{5.165}$$

Evaluating the integral in the denominator and accounting for eq. (5.162) yields

$$A_n(t) = \frac{1}{(R^2/2)J_0^2(\lambda_n R)} \int_0^R \Theta(r, t)rJ_0(\lambda_n r)\, dr \qquad n = 1, 2, \ldots \tag{5.166}$$

■ ***Step 5***

To set up ordinary differential equation for $A_n(t)$, we take the derivatives of both sides of eq. (5.166) with respect to time:

$$\frac{dA_n}{dt} = \frac{1}{(R^2/2)J_0^2(\lambda_n R)} \int_0^R \frac{\partial \Theta}{\partial t} r J_0(\lambda_n r)\, dr \tag{5.167}$$

Next, we utilize the energy equation (5.116) to eliminate $\partial\Theta/\partial t$ from the right-hand side of eq. (5.167):

$$\frac{dA_n}{dt} = \frac{\alpha}{(R^2/2)J_0^2(\lambda_n R)} \int_0^R \frac{\partial}{\partial r}\left(r\frac{\partial \Theta}{\partial r}\right) J_0(\lambda_n r)\, dr \tag{5.168}$$

Using integration by parts in eq. (5.168) yields

$$\frac{dA_n}{dt} = \frac{\alpha}{(R^2/2)J_0^2(\lambda_n R)} \left\{ \left[ r\frac{\partial \Theta}{\partial r} J_0(\lambda_n r)\right]_{r=0}^{r=R} - \int_0^R r\frac{\partial \Theta}{\partial r}\frac{d}{dr}[J_0(\lambda_n r)]\, dr \right\} \tag{5.169}$$

Taking account the boundary conditions (5.117) and (5.118) and performing the algebra in the integral of expression (5.169) yields

$$\frac{dA_n}{dt} = \frac{\alpha}{(R^2/2)J_0^2(\lambda_n R)} \left[ R\frac{\dot{q}''}{k} J_0(\lambda_n R) + \int_0^R r\frac{\partial \Theta}{\partial r} \lambda_n J_1(\lambda_n r)\, dr \right] \tag{5.170}$$

To proceed, we integrate by parts in eq. (5.170) again. Omitting the details for brevity, the final result reads

$$\frac{dA_n}{dt} = \frac{\alpha}{(R^2/2)J_0^2(\lambda_n R)} \left[ R\frac{\dot{q}''}{k} J_0(\lambda_n R) - \lambda_n^2 \int_0^R \Theta r J_0(\lambda_n r)\, dr \right] \tag{5.171}$$

To eliminate the integral of the right-hand side of eq. (5.171) in favor of $A_n$, we make use of eq. (5.166). After rearranging, a first-order ordinary differential equation for $A_n$ is obtained:

$$\frac{dA_n}{dt} + \alpha\lambda_n^2 A_n - \frac{2\alpha\dot{q}''}{Rk}\frac{1}{J_0(\lambda_n R)} = 0 \qquad n = 1, 2, 3, \ldots \tag{5.172}$$

The initial condition for $A_n$ accompanying eq. (5.172) results from substituting eq. (5.163) into the initial condition (5.119) and reads

$$t = 0: \qquad A_n = 0 \tag{5.173}$$

### Step 6

The solution of eq. (5.172) subject to condition (5.173) is straightforward. Only the final result is shown:

$$A_n(t) = \begin{cases} \dfrac{2\dot{q}''}{Rk\lambda_n^2} \dfrac{1}{J_0(\lambda_n R)} (1 - e^{-\alpha\lambda_n^2 t}) & \lambda_n \neq 0 \quad (5.174) \\ \dfrac{2\alpha\dot{q}''}{Rk} t & \lambda_n = 0 \quad (5.175) \end{cases}$$

As mentioned earlier, only the smallest eigenvalue of the present problem ($\lambda_1$) resulting from eq. (5.162) is indeed zero. Therefore, $A_1$ is obtained from eq. (5.174), while $A_2$, $A_3$, . . . , (corresponding to $\lambda_2$, $\lambda_3$, . . . ) are obtained from eq. (5.175).

The final expression for the temperature field, therefore, is [eqs. (5.163), (5.174) and (5.175)]

$$\Theta(r, t) = \frac{2\dot{q}''}{Rk} \alpha t + \frac{2\dot{q}''}{Rk} \sum_{n=2}^{\infty} \frac{1}{\lambda_n^2} (1 - e^{-\alpha\lambda_n^2 t}) \frac{J_0(\lambda_n r)}{J_0(\lambda_n R)} \tag{5.176}$$

This expression should be identical to eq. (5.149) obtained with the principle of superposition. At first glance, the aforementioned two equations do not appear to be identical. To investigate this matter further, we consider the large time limit of eq. (5.176). In this limit,

$$\Theta(r, t) = \frac{2\dot{q}''}{Rk} \alpha t + \frac{2\dot{q}''}{Rk} \sum_{n=2}^{\infty} \frac{1}{\lambda_n^2} \frac{J_0(\lambda_n r)}{J_0(\lambda_n R)} \tag{5.177}$$

Denoting the $r$-dependent infinite series of eq. (5.177) by $F(r)$, we obtain

$$\Theta(r, t) = \frac{2\dot{q}''}{Rk} \alpha t + \frac{2\dot{q}''}{Rk} F(r) \tag{5.178}$$

To determine $F(r)$ we substitute eq. (5.178) into the energy equation (5.116). This results in a differential equation for $F(r)$:

$$\frac{1}{r} \frac{d}{dr} \left[ r \frac{dF(r)}{dr} \right] = 1 \tag{5.179}$$

Solving (5.179) subject to boundary condition (5.117) yields

$$F(r) = \frac{r^2}{4} + C \tag{5.180}$$

Note that the constant of integration $C$ cannot be obtained from the boundary condition (5.118). To determine $C$ we expand $F(r)$ in terms of the following infinite series:

$$\frac{r^2}{4} + C = \sum_{n=1}^{\infty} B_n J_0(\lambda_n r) \tag{5.181}$$

Next, we make use of the orthogonality property of $J_0(\lambda_n r)$ in the usual manner. To this end, eq. (5.181) yields

$$\frac{1}{4}\int_0^R r^3 J_0(\lambda_m r)\,dr + C\int_0^R rJ_0(\lambda_m r)\,dr$$
$$= B_1\int_0^R rJ_0(\lambda_1 r)J_0(\lambda_m r)\,dr + \sum_{n=2}^{\infty} B_n \int_0^R J_0(\lambda_n r)J_0(\lambda_m r)\,r\,dr \qquad (5.182)$$

Since the first root of eq. (5.162) is $\lambda_1 = 0$, we investigate the case where $\lambda_m = 0$. Performing the algebra in eq. (5.182) for $\lambda_m = 0$ yields

$$C - B_1 = -\frac{R^2}{8} \qquad (5.183)$$

If $\lambda_m \neq 0$, on the other hand, eq. (5.182) yields

$$B_n = \frac{1}{\lambda_n^2 J_0(\lambda_n R)} \qquad n = 2, 3, 4, \ldots \qquad (5.184)$$

Combining eqs. (5.181), (5.183), and (5.184), we obtain

$$\frac{r^2}{4} - \frac{R^2}{8} = \sum_{n=2}^{\infty} \frac{J_0(\lambda_n r)}{\lambda_n^2 J_0(\lambda_n R)} \qquad (5.185)$$

Invoking eq. (5.185), the expression for the temperature distribution in the rod [eq. (5.176)] becomes

$$\Theta(r, t) = \frac{2\dot{q}''}{Rk}\alpha t + \frac{\dot{q}''}{2Rk}r^2 - \frac{\dot{q}''R}{4k} - \frac{2\dot{q}''}{Rk}\sum_{n=2}^{\infty}\frac{J_0(\lambda_n r)e^{-\alpha\lambda_n^2 t}}{\lambda_n^2 J_0(\lambda_n R)} \qquad (5.186)$$

This alternative expression for $\Theta(r, t)$ is indeed identical to eq. (5.149) obtained with the method of superposition. From a physical standpoint this equation makes sense: It yields a temperature field that at large times increases monotonically with time (no steady state is reached). Regarding the spatial dependence of the temperature field, one clearly distinguishes the series term corresponding to the early time behavior. This term decays exponentially with time. At late times the temperature dependence of the spatial coordinate is proportional to $r^2$.

## 5.6 THE SIMILARITY METHOD IN ONE-DIMENSIONAL TRANSIENT CONDUCTION

It is sometimes possible that the solutions of certain types of heat conduction problems possess the property of self-similarity [5–7]. This property is also a feature of problems in other areas of heat transfer, most notably in certain types of velocity and thermal boundary layers in convection [8–11]. In problems possessing the self-similarity property it is possible to introduce a variable transformation and reduce the governing energy equation from a partial differential equation to an ordinary differential equation. The same transformation needs to be applied to the boundary conditions, if needed, to reduce partial derivatives to ordinary derivatives. The operation prescribed above decreases the difficulty involved in obtaining the problem solution by an *order of magnitude* (mathematical model involving partial differential equations versus mathematical model involving ordinary differential equations). The solution method that takes advantage of the similarity property is termed the *similarity solution method*. In the following example, the similarity method will be applied to obtain the temperature field in a one-dimensional transient conduction problem. Two important requirements should be satisfied for the similarity method to be applicable.

1. *All* partial derivatives in *both* the governing equation *and* the boundary conditions need to be transformed to ordinary derivatives.
2. The space dimension in the problem of interest should be semi-infinite (extend, say, from zero to infinity). If the physical extent of the body is finite, the statement above implies that the time interval for which the temperature field is sought should be such that the portion of the body in which the thermal effect (heating or cooling) is felt is small compared to the physical extent of the body. In this case the body appears to be infinite relative to the size of the thermally affected region.

**Example 5.7**

A thick wall, initially at uniform temperature $T_i$, is suddenly brought into contact with a stream of warm fluid of temperature $T_0$, as shown in Fig. 5.6. Perfect thermal contact is assumed to exist between the wall and the fluid. As

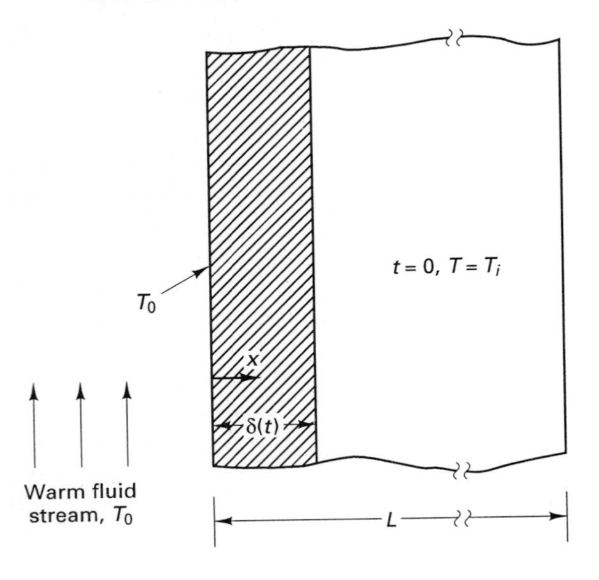


**Figure 5.6**

time progresses, the part of the wall adjacent to the side in contact with the fluid warms up (Fig. 5.6).

**(a)** Obtain the temperature field in the wall for the time interval during which the heating effect of the fluid has not penetrated far inside the wall [i.e., the wall region affected by the heating is very thin compared to the total wall thickness ($\delta(t)/L \ll 1$, Fig. 5.6)].

**(b)** Obtain the variation of the heat flux at the left side of the wall ($x = 0$) with time, during the above-mentioned time interval.

**Solution**

**(a)** The mathematical model for this problem is

$$\frac{1}{\alpha}\frac{\partial T}{\partial t} = \frac{\partial^2 T}{\partial x^2} \tag{5.187}$$

$$T(0, t) = T_0 \tag{5.188}$$

$$T(\infty, t) = T_i \tag{5.189}$$

$$T(x, 0) = T_i \tag{5.190}$$

Note that boundary condition (5.188) accounts for the perfect thermal contact between the fluid and the wall. Note further that boundary condition (5.189) is evaluated at infinity since the region outside $\delta(t)$ is of infinite extent compared to $\delta(t)[L/\delta(t) \to \infty]$.

The first and crucial step in the similarity method is to define the *similarity variable*. It is a combination of the natural independent variables of the problem $(x, t)$, and it is responsible for the transformation of the partial differential equations in the problem model to ordinary differential equations. To define the similarity variable one may resort to purely mathematical methods or utilize the physics of the problem in hand. Here, we opt for the latter: If the scale for the temperature difference driving the heat transfer in the wall is $\Delta T = T_0 - T_i$, and the scale for the thickness of the thermally affected region of the wall at time $t$ is $\Delta x = \delta(t)$, the governing conduction equation yields the following scaling statement:

$$\frac{1}{\alpha}\frac{\Delta T}{t} \sim \frac{\Delta T}{[\delta(t)]^2} \tag{5.191}$$

or, after solving for $\delta(t)$,

$$\delta(t) \sim (\alpha t)^{1/2} \tag{5.192}$$

Noting that the ratio $x/\delta(t) = x/(\alpha t)^{1/2}$ is a dimensionless combination of the two independent variables of the problem $(x, t)$ obtained with the help of the governing equation (5.187), we define the following similarity variable:

$$\eta = A\frac{x}{(\alpha t)^{1/2}} \tag{5.193}$$

where the constant $A$ will be defined later in a manner that will simplify the algebra.

The next step in the solution process is to cast the model of the problem [eqs. (5.187)–(5.190)] in terms of the similarity variable. To carry out this step, we obtain expressions of the partial derivatives involved in terms of $\eta$:

$$\frac{\partial T}{\partial t} = \frac{dT}{d\eta}\frac{\partial \eta}{\partial t} = \frac{dT}{d\eta}\left(-\frac{A}{2}\frac{t^{-3/2}}{\alpha^{1/2}}x\right) \tag{5.194}$$

$$\frac{\partial T}{\partial x} = \frac{dT}{d\eta}\frac{\partial \eta}{\partial x} = \frac{dT}{d\eta}\frac{A}{(\alpha t)^{1/2}} \tag{5.195}$$

$$\frac{\partial^2 T}{\partial x^2} = \frac{d}{d\eta}\left(\frac{\partial T}{\partial x}\right)\frac{\partial \eta}{\partial x} = \frac{d}{d\eta}\left[\frac{dT}{d\eta}\frac{A}{(\alpha t)^{1/2}}\right]\frac{A}{(\alpha t)^{1/2}} = \frac{d^2 T}{d\eta^2}\frac{A^2}{t\alpha} \tag{5.196}$$

Substituting eqs. (5.194) and (5.196) into eq. (5.187) yields

$$-\frac{1}{2A^2}\eta\frac{dT}{d\eta} = \frac{d^2 T}{d\eta^2} \tag{5.197}$$

At this point we assign a value to the constant $A$:

$$A = \tfrac{1}{2} \tag{5.198}$$

This choice for $A$ will simplify somewhat the following algebraic steps. Note, however, that the problem solution can be obtained by assigning any known value to $A$.

Combining eqs. (5.197) and (5.198) yields the similarity form of the energy equation:

$$\frac{d^2 T}{d\eta^2} + 2\eta\frac{dT}{d\eta} = 0 \tag{5.199}$$

Next, we observe that the boundary conditions (5.188) and (5.189) do not contain explicitly the original independent variables. Therefore, these boundary conditions do need to be transformed in terms of the similarity variable $\eta$. If $x$ or $t$ appeared in the boundary conditions (in a derivative, for example) a procedure similar to that applied to the governing equation would be necessary. For the similarity method to work, the original independent variables $(x, t)$ should not appear *anywhere* in the mathematical model after the similarity transformation is applied. Realizing that as $x \to 0$, $\eta \to 0$ and as $x \to \infty$, $\eta \to \infty$ [eq. (5.193)] the boundary conditions (5.188)–(5.189) state that

$$\eta = 0: \qquad T = T_0 \tag{5.200}$$

$$\eta \to \infty: \qquad T \to T_i \tag{5.201}$$

The ordinary differential equation (5.199), subject to the conditions (5.200) and (5.201) can easily be solved as follows: Introducing the variable

$$P = \frac{dT}{d\eta} \tag{5.202}$$

### Example 5.8

In a slab of a heat storage material, a hot fluid is circulated through the channels shown in Fig. 5.7a. The top and the bottom surfaces of the slab are insulated. Neglecting entrance effects, it may be assumed that the heat conduction phenomenon within the wall separating two neighboring flow channels is two-dimensional. If at the starting point of operation of the heat storage process, the temperature distribution of this wall is known, $T(x, y, 0) = f(x, y)$, and the fluid circulation pump is suddenly turned on, obtain the temperature distribution of the wall. You may assume that the heat transfer coefficient between the surfaces of the wall and the fluid is very large ($h \to \infty$). How does the heat transfer rate from each side of the channel wall depend on time?

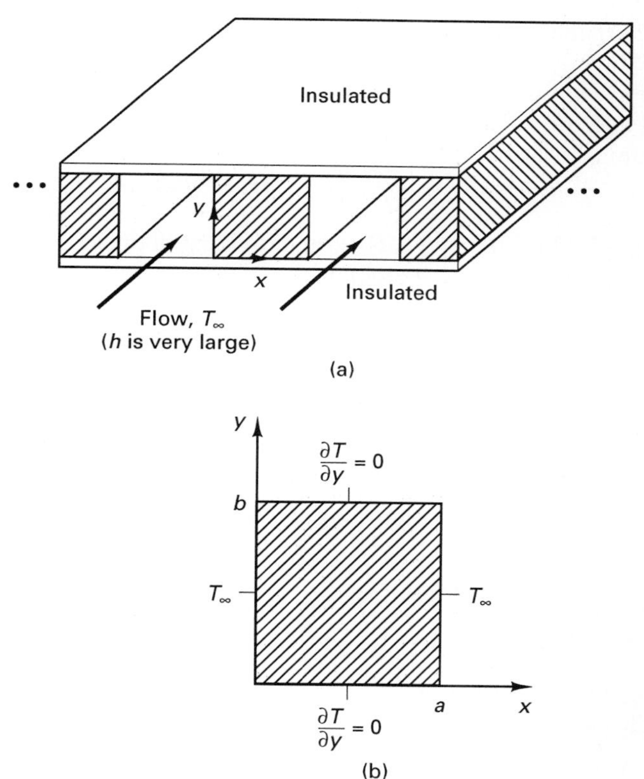


**Figure 5.7** (a) Flow passages in heat storage material. (b) Cross section of the wall separating two flow passages.

### Solution

Taking into account the multidimensionality of the transient problem of interest as well as the fact that the sides of the wall in contact with the fluid are of temperature $T_\infty$ (it is stated in the problem that $h \to \infty$) the mathematical model for the heat conduction process in the wall is (see also Fig. 5.7b)

$$\frac{\partial^2 \Theta}{\partial x^2} + \frac{\partial^2 \Theta}{\partial y^2} = \frac{1}{\alpha}\frac{\partial \Theta}{\partial t} \tag{5.230}$$

$$x = 0: \quad \Theta = 0 \tag{5.231}$$

$$x = a: \quad \Theta = 0 \tag{5.232}$$

$$y = 0: \quad \frac{\partial \Theta}{\partial y} = 0 \tag{5.233}$$

$$y = b: \quad \frac{\partial \Theta}{\partial y} = 0 \tag{5.234}$$

$$t = 0: \quad \Theta_0 = f(x, y) - T_\infty = \phi(x, y) \tag{5.235}$$

In eqs. (5.230)–(5.235) the usual variable transformation $\Theta = T - T_\infty$ was utilized. To proceed directly with the method of separation of variables, the boundary conditions in *both* directions $x$ and $y$ need to be homogeneous. By simple inspection of eqs. (5.214)–(5.218) we realize that this is indeed the case in the present example. Defining

$$\Theta = X(x)Y(y)S(t) \tag{5.236}$$

substituting expression (5.236) into eq. (5.214), and dividing both sides of the resulting equation by the product $XYS$, we obtain

$$\underbrace{\frac{X''}{X}}_{-\lambda^2} + \underbrace{\frac{Y''}{Y}}_{-\beta^2} = \frac{1}{\alpha}\frac{S'}{S} \tag{5.237}$$

Seeking orthogonality in the two homogeneous directions, we set the $x$-dependent term in eq. (5.237) equal to $-\lambda^2$ and the $y$-dependent term equal to $-\beta^2$, to obtain

$$X'' + \lambda^2 X = 0 \tag{5.238}$$

$$Y'' + \beta^2 Y = 0 \tag{5.239}$$

$$S' + \alpha(\lambda^2 + \beta^2)S = 0 \tag{5.240}$$

Solving equations (5.238)–(5.240) (Appendix B) and substituting the solutions into expression (5.236) yields

$$\Theta = [A_1 \sin(\lambda x) + A_2 \cos(\lambda x)][A_3 \sin(\beta y) + A_4 \cos(\beta y)] \exp[-a(\lambda^2 + \beta^2)t] \tag{5.241}$$

To evaluate the constants $A_1$ to $A_4$, $\lambda$, and $\beta$, we will make use of the conditions (5.231)–(5.235) as well as the orthogonality property of the sine and cosine functions. One immediately notices that boundary conditions (5.231) and (5.233) yield

$$A_2 = A_3 = 0 \tag{5.242}$$

Next, applying boundary conditions (5.232) and (5.234) results in

$$\lambda_n = \frac{n\pi}{a} \qquad n = 0, 1, 2, \ldots \tag{5.243}$$

$$\beta_m = \frac{m\pi}{b} \qquad m = 0, 1, 2, \ldots \tag{5.244}$$

Combining eqs. (5.241)–(5.244), we summarize the expression for the temperature distribution as follows:

$$\Theta_{mn} = A_{mn} \sin\left(\frac{n\pi}{a}x\right) \cos\left(\frac{m\pi}{b}y\right) \exp\left[-\alpha\pi^2\left(\frac{n^2}{a^2} + \frac{m^2}{b^2}\right)t\right] \tag{5.245}$$

The constant $A_{mn}$ was obtained by combining the constants $A_1$ and $A_4$. The subscript $mn$ accounts for the fact that all values of $m(m = 0, 1, 2, 3, \ldots)$ and all values of $n(n = 0, 1, 2, 3, \ldots)$ yield solutions for the temperature distribution. Based on the above, the general solution for the temperature distribution in the wall is the linear combination of all the above-mentioned solutions.

$$\Theta = \sum_{m=0}^{\infty} \sum_{n=0}^{\infty} A_{mn} \sin\left(\frac{n\pi}{a}x\right) \exp\left(-\alpha\pi^2\frac{n^2}{a^2}t\right) \cos\left(\frac{m\pi}{b}y\right) \exp\left(-\alpha\pi^2\frac{m^2}{b^2}t\right) \tag{5.246}$$

where the exponential term was conveniently split into two parts. To proceed, we notice that the quantity $\Sigma_{n=0}^{\infty} A_{mn} \sin[(n\pi/a)x] \exp[-\alpha\pi^2(n^2/a^2)t]$ depends on only $m$ (once the summation is carried out the only general index appearing in all terms of the summation is $m$ since specific values have been assigned on the index $n$). With this in mind, we define

$$G_m(x, t) = \sum_{n=0}^{\infty} A_{mn} \sin\left(\frac{n\pi}{a}x\right) \exp\left(-\alpha\pi^2\frac{n^2}{a^2}t\right) \tag{5.247}$$

Combining eqs. (5.246) and (5.247) yields

$$\Theta = \sum_{m=0}^{\infty} G_m(x, t) \cos\left(\frac{m\pi}{b}y\right) \exp\left(-\alpha\pi^2\frac{m^2}{b^2}t\right) \tag{5.248}$$

Next, we apply the initial condition (5.235):

$$\phi(x, y) = \sum_{m=0}^{\infty} G_m(x, 0) \cos\left(\frac{m\pi}{b}y\right) \tag{5.249}$$

Utilizing the orthogonality property of the cosine function in the usual manner in eq. (5.235), we obtain

$$G_m(x, 0) = \frac{2}{b}\int_0^b \phi(x, y) \cos\left(\frac{m\pi}{b}y\right) dy \tag{5.250}$$

Combining eqs. (5.247) and (5.250) and applying the orthogonality property of the sine function yields the final expression for $A_{mn}$:

$$A_{mn} = \frac{4}{ab}\int_0^a \left[\int_0^b \phi(x, y)\cos\left(\frac{m\pi}{b}y\right)dy\right]\sin\left(\frac{n\pi}{a}x\right)dx \tag{5.251}$$

Substituting eq. (5.251) into eq. (5.246) results in the temperature distribution in the wall:

$$T = T_\infty + \frac{4}{ab}\sum_{m=0}^{\infty}\sum_{n=0}^{\infty}\left\{\int_0^a \left[\int_0^b (f(x, y) - T_\infty)\cos\left(\frac{m\pi}{b}y\right)dy\right]\sin\left(\frac{n\pi}{a}x\right)dx\right\}$$
$$\cdot \sin\left(\frac{n\pi}{a}x\right)\cos\left(\frac{m\pi}{b}y\right)\exp\left[-\alpha\pi^2\left(\frac{n^2}{a^2} + \frac{m^2}{b^2}\right)t\right] \tag{5.252}$$

Before proceeding with the calculation of the heat transfer rate between the wall and the fluid, it is constructive to consider the special case where the initial wall temperature is uniform, $f(x) = T_0$. In this case it is expected that the temperature distribution of the wall will be one-dimensional (i.e., it will only depend on the $x$-coordinate and on time). Note that by inspection of Fig. 5.7b, one realizes that if the wall is initially at $T_0$ and the side surfaces of this wall are suddenly, say, cooled down to $T_\infty$, this cooling effect will propagate into the wall uniformly (in the same manner at all points in the $y$-direction for a specific $x$-location since the top and the bottom surfaces are insulated). Indeed, for $f(x, y) = T_0 =$ const, eq. (5.250) yields

$$G_m(x, 0) = \begin{cases} 0 & m \neq 0 \\ 2\Theta_0 & m = 0 \end{cases} \tag{5.253}$$

Next, invoking eq. (5.251), we obtain

$$A_{0n} = -2(T_0 - T_\infty)\frac{a}{n\pi}[(-1)^n - 1] \tag{5.254}$$

$$A_{mn} = 0 \qquad \text{for } m = 1, 2, 3, \ldots \tag{5.255}$$

Since eq. (5.254) states that $A_{0n}$ is nonzero only for the odd values of $n$ and since $A_{0n}$ depends only on one recurrence index, $n$, we rename $A_{0n} = A_n$ and $n = 2k + 1$, $k = 1, 2, 3, \ldots$ to obtain

$$A_{2k+1} = 4(T_0 - T_\infty)\frac{a}{(2k + 1)\pi} \tag{5.256}$$

Based on the above, the expression for the temperature distribution (keeping in mind that $m = 0$ for all nonzero $A_{mn}$) reads

$$T = T_\infty + \frac{4(T_0 - T_\infty)a}{\pi}\sum_{k=0}^{\infty}\frac{\sin\left[\dfrac{(2k + 1)\pi}{a}x\right]}{2k + 1}\exp\left[-\alpha\pi^2\frac{(2k + 1)^2 t}{a^2}\right] \tag{5.257}$$

Going back to the calculation of the heat transfer rate, we apply Fourier's law to obtain the local heat transfer rate through the left vertical surface of the wall and integrate over the length of that surface to obtain the overall heat transfer rate:

$$\dot{q}' = -k \int_0^b \left(\frac{\partial T}{\partial x}\right)_{x=0} dy \tag{5.258}$$

Substituting eq. (5.252) into eq. (5.258) and performing the algebra gives

$$\dot{q}' = -\frac{4}{a^2} \sum_{n=0}^{\infty} n \left\{ \int_0^a \left[ \int_0^b (f(x, y) - T_\infty)\, dy \right] \sin\left(\frac{n\pi}{a} x\right) dx \right\} \exp\left(-\alpha \pi^2 \frac{n^2}{a^2} t\right) \tag{5.259}$$

In the special case of an initially isothermal wall $[f(x, y) = T_0]$, eq. (5.259) reduces to

$$\dot{q}' = -4k(T_0 - T_\infty) \sum_{k=0}^{\infty} \exp\left[-\alpha \pi^2 \frac{(2k + 1)^2}{a^2} t\right] \tag{5.260}$$

Finally, it is worth noting that expressions (5.252) and (5.257) imply that at large times the wall approaches the fluid temperature ($t \to \infty$, $T \to \infty$). Similarly, eqs. (5.259) and (5.260) show that the heat transfer rate diminishes at large times ($t \to \infty$, $\dot{q}' \to 0$).

## PROBLEMS

**5.1.** In a laboratory experiment a plane wall (Fig. P5.1) is heated electrically. To this end, a current $I$ (amperes) is passed through it. As a result, heat is generated within the wall. The volume of the wall is known and is denoted by $V$ ($m^3$) and its electrical resistance by $R$ (ohms). Cooling is provided by a convecting fluid ($h$, $T_\infty$) flowing on both sides of the wall. Only temperature variations in the $x$-direction (along the thickness of the wall) are significant. The experiment proceeds as follows: Initially ($t \leq 0$), no current is passed through the wall and its temperature is identical to that of the coolant ($T_\infty$). Suddenly, the electrical heating begins and stays on thereafter. The temperature history of selected points of the wall is monitored.

**(a)** You are asked to construct and solve a theoretical model that will provide the temperature change of any point in the wall with time so that future comparisons with the experimental measurements can be made.

**(b)** What are the location and magnitude of the maximum temperature in the wall?

**5.2.** A solid slab initially at temperature $T_0$ is placed in an oven and it is heated continuously by the incident constant radiative heat flux $\dot{q}''$ as shown in Fig. P5.2. The thickness of the slab is much smaller than its length ($a \ll L$). Obtain the temperature distribution in the slab. If the melting temperature of the slab material is $T_m$, derive a general expression for the time it takes for the melting process to be initiated.

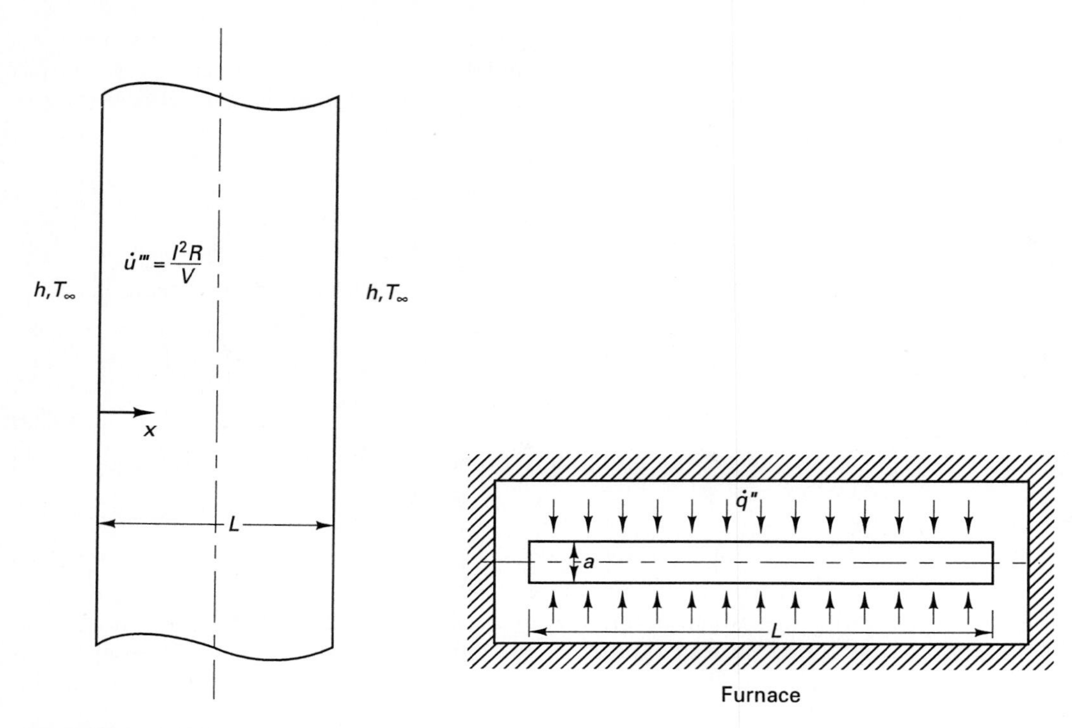


**Figure P5.1** Heat-generating wall cooled convectively.

**Figure P5.2** Slab heated radiatively in an oven.

**5.3.** Re-solve Problem 5.2 by assuming that convection ($h$, $T_\infty$) takes place simultaneously with radiation ($\dot{q}''$) on both sides of the slab.

**5.4.** During the operation of a slider, a constant heat flux ($\dot{q}''$) is generated at the interface between a moving and a stationary plate (Fig. P5.4). The conductivity of the moving plate is much smaller than the conductivity of the stationary plate ($k_2 \ll k_1$). The initial temperature of the stationary plate (before the upper plate in Fig. P5.4 is set into motion) is denoted by $T_0$. Obtain the temperature distribution in this plate if its bottom side is cooled convectively ($h$, $T_\infty$). What is the temperature distribution in the limit $h \rightarrow \infty$?

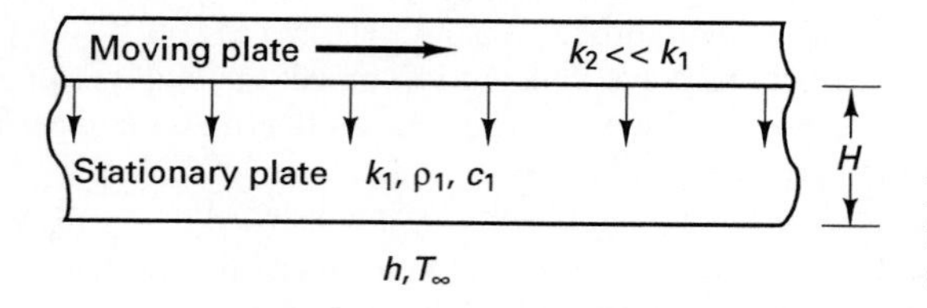


**Figure P5.4** Heat generation at the interface between a moving and a stationary plate.

**5.5.** Re-solve Problem 5.4, but this time by assuming that the lower surface of the stationary plate is insulated.

**5.6.** Figure P5.6 shows the flat wall of a heat exchanger. This wall is initially isothermal at $T = T_0$. Suddenly, the operation of the heat exchanger is initiated. As a result, two dif-

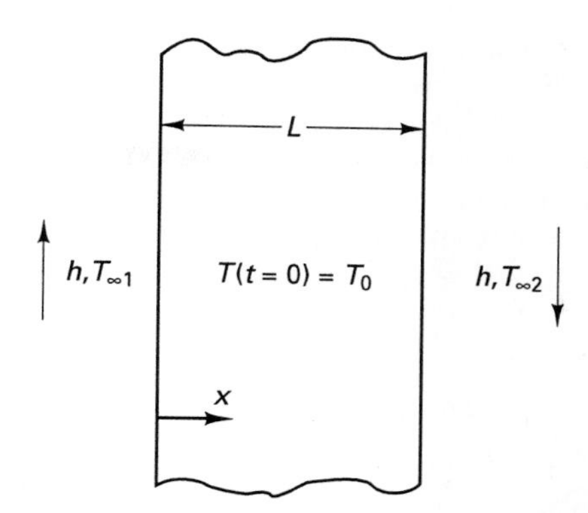


**Figure P5.6** Heat exchanger wall.

ferent fluids ($T_{\infty 1}$, $T_{\infty 2}$) are flowing along the sides of this wall. Obtain the temperature history of the wall as well as the heat flux variation with time at both its sides, if the heat transfer coefficient $h$ is very large ($h \to \infty$).

**5.7.** Re-solve Problem 5.7 assuming this time that the value of the heat transfer coefficient is finite.

**5.8.** Reconsider Problem 5.6 but instead of a flat wall imagine that a circular pipe is part of a heat exchanger (Fig. P5.8). The initial pipe wall temperature is denoted by $T_0$. Obtain the pipe wall temperature history and the heat transfer rate at the inner and the outer pipe wall surfaces.

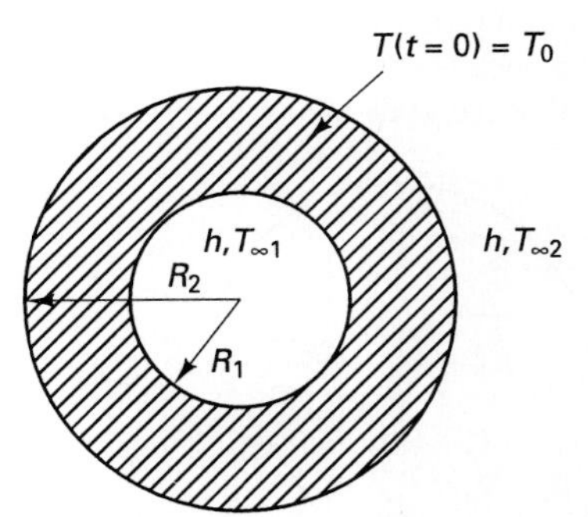


**Figure P5.8** Pipe cross section.

**5.9.** In a heat treatment process a slab of a composite material of thickness $L$ that is initially isothermal at temperature $T_0$ is placed in a furnace. The slab material absorbs the furnace radiation in a manner that decreases exponentially with distance away from the slab surface $\dot{q}'' = Ae^{-Bx}$, where $A$ and $B$ are experimental constants that depend on the slab material. The sides of the slab are cooled by natural convection ($h$, $T_\infty$). Obtain the temperature history in the slab.

**5.10.** Re-solve Problem 5.9 this time for a cylindrical rod that absorbs the furnace radiation in a manner that decreases exponentially with distance away from the cylinder surface $\dot{q}'' = A \exp[-B(R - r)]$, where $R$ is the radius of the cylinder and $r$ is the radial coordinate.

**5.11.** Re-solve Problem 5.10 for the case of a sphere of radius $R$ absorbing the radiation energy in a manner identical to that of the cylinder of Problem 5.10.

**5.12.** A stuffed turkey ready to be baked is placed in an oven, preheated to a temperature $T_\infty$. The initial temperature of the turkey is denoted by $T_i$. The heating is performed by radiation ($\dot{q}''$) and convection ($h$, $T_\infty$) at the surface of the turkey. The bottom of the pan in which the turkey is placed is insulated. Model the turkey as a hemisphere of known constant thermophysical properties, as shown in Fig. P5.12, and obtain its temperature distribution. It is said that the turkey is baked and ready to be consumed when its temperature at $r = R/2$ reaches the value $T = T_B$. Can you derive a criterion for the time it takes for the turkey to be ready to eat?

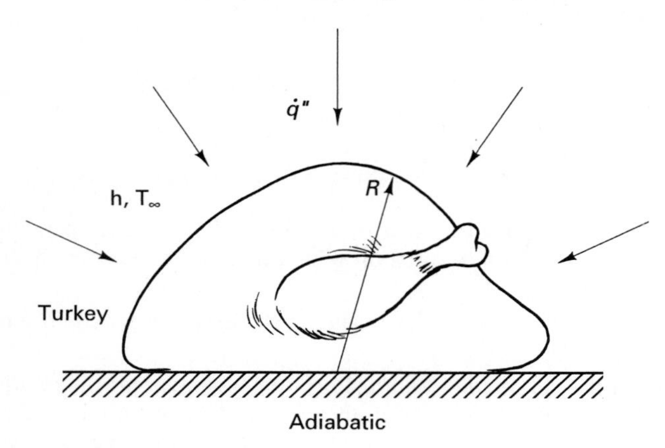


**Figure P5.12** Schematic of a turkey roasting in an oven.

**5.13** Reconsider Problem 5.12. Instead of a turkey, a meat loaf is to be placed in the oven. The meat loaf is shaped as a rather long semicylinder (Fig. P5.13). Obtain the temperature distribution in the meat loaf and derive a criterion for the necessary cooking time. You may assume that the ends of the meat loaf are adiabatic.

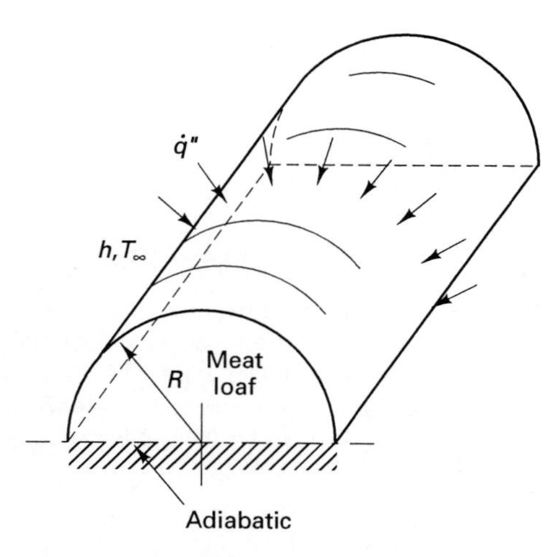


**Figure P5.13** Semicylinder modeling a meat loaf baking in an oven.

**5.14.** A radiative flux ($\dot{q}''$), originating from an expanded laser beam, is received on one side of an isothermal ($T_i$) *thick* metal slab for a rather short time period. As a result, the temperature of the slab is raised. Obtain the temperature history of the slab during the heating process. How long does it take for the slab to start melting if its melting temperature is denoted by $T_m$? [*Hint:* The slab can be modeled as a semi-infinite body for the time period during which it is heated by the laser beam. To make the similarity method applicable, define an artificial temperature $\Theta = \dot{q}'' + k(\partial T/\partial x)$ and obtain a conduction model (governing equation and boundary conditions) for $\Theta$. After solving the model for $\Theta$ use the definition for $\Theta$ given above to obtain $T$.]

**5.15.** A bubble of a chemically reactive gas is trapped inside a large mass of explosive material. The bubble radius is $R_0$. The explosive material is initially at temperature $T_i$. Due to exothermic reactions in the bubble, the temperature of the explosive material is raised to $T_0$ at $r = R_0$ and remains at this value for a relatively long time thereafter. Obtain the temperature distribution in the explosive material. (*Hint:* Model the explosive material as a spherical shell of inner radius $R_0$ and outer radius infinity. After appropriate manipulations, you may use the similarity method to obtain the solution. Note that if before you apply the similarity method you make the coordinate transformation $\tilde{r} = r - R_0$ and work with $\tilde{r}$, the problem simplifies considerably.)

**5.16.** A thick cold slab acts as a heat sink in a cooling device (Fig. P5.16). The heating of the slab occurs by forced convection after one side of the slab comes into contact with a warm fluid of temperature $T_\infty$. The duration of the heating process is such that the other side of the slab is unaffected by it (i.e., the heating effect of the fluid penetrates over a region which is thinner than the thickness of the slab). If $h$ denotes the heat transfer coefficient and $T_i$ the initial temperature of the slab (constant) use the similarity method to obtain the temperature field in the slab. [*Hint:* To make the similarity method applicable, define an artificial temperature $\Theta = (T - T_\infty) - (k/h)(\partial T/\partial x)$ and obtain a conduction model (governing equation and boundary conditions) for $\Theta$. After solving this model for $\Theta$ use the definition of $\Theta$ given above to obtain $T$.]

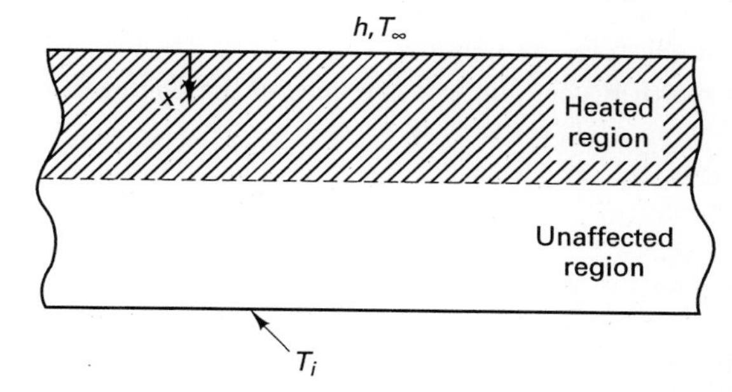


**Figure P5.16**

**5.17.** In a heating process a thin foil heater is installed between two thick slabs of different materials (Fig. P5.17). The slabs are initially isothermal at $T_0$. Suddenly, the electric heater is connected to a power supply and starts producing heat at a rate $\dot{q}''$ (W/m$^2$).

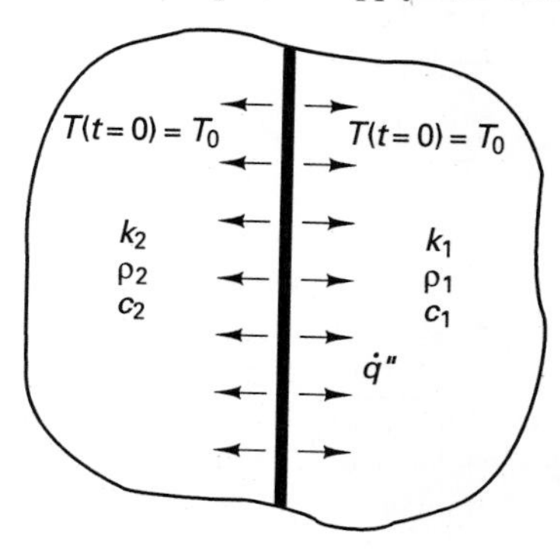


**Figure P5.17** Foil heater placed at the interface between two thick slabs.

Modeling the slabs as semi-infinite bodies with known thermophysical properties ($k_1$, $\rho_1$, $c_1$, $k_2$, $\rho_2$, $c_2$) obtain their temperature history.

**5.18.** During the processing of canned food in a plant, a can of food is to be immersed in a hot water bath of temperature $T_\infty$ for a period of time. The temperature at the center of the can needs to reach the value $T_\infty/2$ in as short a time ($t_a$) as possible. Two can geometries are considered in the canning process. A spherical can of diameter $D$ and a cylindrical can of diameter $D$ and height $D$. Which geometry will yield the shortest heating time $t_a$? You may assume that the heat transfer coefficient at the can surface is very large.

**5.19.** A buried thin electrical wire produces heat per unit length at a rate $\dot{q}'$. If the thermophysical properties of the ground are known and if the initial temperature of the ground (prior to the heat production by the wire) is denoted by $T_0$, find the temperature distribution in the ground.

**5.20.** It is a common (even though not necessarily correct) practice for a person to estimate the temperature of an object by touching the surface of the object with his or her hand. Assuming that both the person's hand and the object are semi-infinite, obtain the temperature distribution in both the object and the hand during the above touching event. For simplicity, assume that the temperature of the object prior to touching is $T_0$ and the temperature of the hand is $T_h$. The thermophysical properties of the hand and the object are known. Obviously, by touching the object the person "feels" the temperature of the contact surface, and not the real temperature of the object ($T_0$) far from that. Obtain an expression for the temperature of the contact surface. How do the thermophysical properties of the object affect this temperature and therefore the process of estimating the temperature of an object by touching it?

**5.21** Re-solve Problem 5.20 approximately, utilizing the integral solution method described in Section 5.7.

**5.22** Utilizing the integral method described in Section 5.7, re-solve Example 5.7 for the following temperature profile:

$$\frac{T - T_i}{T_0 - T_i} = A + B\frac{x}{\delta} + C\left(\frac{x}{\delta}\right)^2 + D\left(\frac{x}{\delta}\right)^3 + E\left(\frac{x}{\delta}\right)^4$$

**5.23.** The short cylindrical fin shown in Fig. P5.23 [$O(R) = O(L)$] is initially isothermal at temperature $T_i$. Suddenly, it is exposed to a hot gas ($h$, $T_\infty$). The fin base temperature is

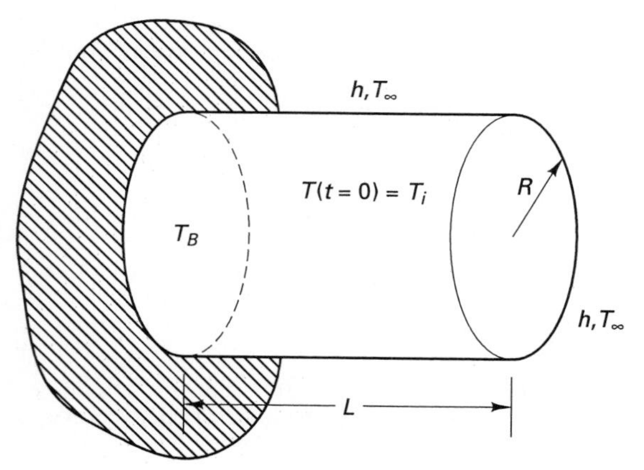


**Figure P5.23** Short cylindrical fin.

known ($T_B$) and constant. Obtain the temperature variation in the fin and the heat transfer rate through its base.

**5.24.** Re-solve Problem 5.23 for a short rectangular fin of cross section $L \times H[O(L) = O(H)]$ as shown in Fig. P5.24.

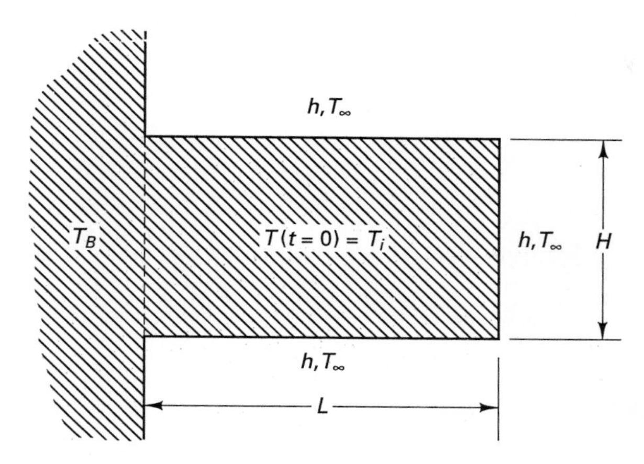


**Figure P5.24** Short rectangular fin.

**5.25.** An isothermal ($T_i$) hemispherical ($R$) solidified metal droplet is resting on a flat heater. Suddenly, the heater is turned on imposing a constant heat flux ($\dot{q}''$) heating condition at the bottom of the droplet (hemisphere). Obtain the temperature history of droplet. If the melting temperature of the metal is denoted by $T_m$, obtain a criterion for estimating the time it takes for the droplet to start melting.

## REFERENCES

1. S. J. FARLOW, *Partial Differential Equations for Scientists and Engineers,* Wiley, New York, 1982.
2. M. D. GREENBERG, *Foundations of Applied Mathematics,* Prentice Hall, Englewood Cliffs, NJ, 1978.
3. V. S. ARPACI, *Conduction Heat Transfer,* Addison Wesley, Reading, MA, 1966.
4. G. E. MYERS, *Analytical Methods in Conduction Heat Transfer,* McGraw-Hill, New York, 1971.
5. H. S. CARSLAW AND J. C. JAEGER, *Conduction of Heat in Solids,* Clarendon Press, London, 1959
6. V. GRIGULL AND H. SANDNER, *Heat Conduction,* Hemisphere, New York, 1984.
7. M. N. ÖZISIK, *Heat Conduction,* Wiley, New York, 1980.
8. H. SCHLICHTING, *Boundary Layer Theory,* 7th ed., McGraw-Hill, New York, 1979.
9. A. BEJAN, *Convection Heat Transfer,* Wiley, New York, 1984.
10. V. S. ARPACI AND P. S. LARSEN, *Convection Heat Transfer,* Prentice Hall, Englewood Cliffs, NJ, 1983.

11. L. C. BURMEISTER, *Convective Heat Transfer,* Wiley, New York, 1983.
12. T. R. GOODMAN, *Trans. Am. Soc. Mech. Eng.,* Vol. 80, pp. 335–342, 1958.
13. T. R. GOODMAN, *J. Heat Transfer,* Vol. 83c, pp. 83–86, 1961.
14. T. R. GOODMAN AND J. SHEA, *J. Appl. Mech.,* Vol. 32, pp. 16–24, 1960.
15. C. R. WYLIE AND L. C. BARRET, *Advanced Engineering Mathematics,* 5th ed., McGraw-Hill, New York, 1982.

CHAPTER 6

# TRANSIENT CONDUCTION WITH TIME-DEPENDENT BOUNDARY CONDITIONS

In this chapter, transient heat conduction problems in which the boundary conditions are time dependent are considered. The presence of the time-dependent boundary conditions calls for solution methodologies that are distinctly different from what was discussed in Chapter 5, which featured problems with time-independent boundary conditions. Two classes of boundary conditions are studied sequentially. First, time-*periodic* boundary conditions proportional to a sine or cosine function are considered. The corresponding solution methodology (the method of complex temperatures) will provide results for the temperature field *after* an initial period during which the system adjusts to the imposed boundary condition. The solution for the temperature field is the steady-periodic or sustained response of the system to the periodic "information" (temperature or heat flux, for example) imposed by the boundary condition, after the initial adjustment period mentioned above. Second, boundary conditions that vary *arbitrarily* with time are studied. Here, the solution methodology is based on Duhamel's theorem, which is presented before the discussion of illustrative heat conduction examples.

## 6.1 TRANSIENT HEAT CONDUCTION WITH TIME-PERIODIC BOUNDARY CONDITIONS PROPORTIONAL TO A SINE OR COSINE FUNCTION

Many heat conduction engineering applications possess mathematical models with time-periodic boundary conditions. For example, the heating of the wall of an engine cylinder is subjected to the periodic behavior of the combustion cycle in the cylinder. The heating of the earth by the sun is also periodic. Here, two cycles can be identified: a daily cycle of a period of 24 hours containing a day and a night, and a yearly cycle of a period spanning over 12 months. The heating or cooling of various structures is also periodic, as dictated by the on–off cycle on which the heating–cooling system operates. The heating and cooling of electronic components is often periodic, depending on the operation of the apparatus in which these components are installed. In the examples above, as well as in other applications, the very early stages of the transient conduction phenomenon, during which the system adjusts to the imposed periodic boundary conditions, are rather short and often are not of importance to the thermal design or operation of the application under consideration. It is the sustained thermal behavior (temperature distribution and heat transfer rates) that needs to be studied carefully. A well-defined method that can be used to study the sustained thermal behavior of various systems when the time-periodic boundary condition is proportional to a sine or a cosine function [$\sin(A\omega t)$ or $\cos(A\omega t)$, for example] is the method of complex temperatures [1–2]. The method is summarized in the following sequence of steps.

### ■ *Step 1*

Construct the mathematical model of the problem of interest (governing equation, and boundary and initial conditions). The unknown temperature is denoted by $T_1$. Before proceeding, make sure that the only nonhomogeneity in the model is the time-periodic boundary condition. Assurance of this fact is necessary for the success of the method. Variable transformations or other manipulations may be necessary to yield a model in which the only nonhomogeneity is the time-periodic boundary condition. In addition, it is worth noting that it is not necessary to include the initial condition in the mathematical model. Since the sustained solution obtained with the method of complex variables does not hold at early times, the initial condition is not satisfied and it is not utilized in the solution process.

### ■ *Step 2*

Construct the mathematical model for a problem that is identical to the one of step 1 except for the fact that its periodic boundary condition is out of phase by $\pi/2$ with the periodic boundary condition of the problem in step 1. For example, if the periodic boundary condition in the problem of step 1 states that the temperature or the heat flux at one boundary is proportional to $\cos(\omega t)$, the corresponding boundary condition in the problem of step 2 should state that the above-mentioned temperature or heat flux is proportional to $\cos(\omega t + \pi/2) = \sin \omega t$. Let the temperature in the problem of step 2 be denoted by $T_2$.

### ■ *Step 3*

Construct the mathematical model for the complex temperature problem $T = T_1 + iT_2$. The governing equation and boundary conditions for this prob-

6.3c. The contribution of each discontinuity $\Delta F_i$ on the temperature distribution in the body is

$$\Delta\Theta_i(x, t) = \Delta F_i S(x, t - \tau_i) \tag{6.32}$$

The contribution of all $N + 1$ discontinuities by application of the principle of superposition is, then,

$$\Delta\Theta = \sum_{i=0}^{N} \Delta F_i S(x, t - \tau_i) \tag{6.33}$$

Combining eqs. (6.31) and (6.33), we obtain the general form of Duhamel's theorem, which yields the temperature distribution in the body in the presence of an arbitrary time-dependent boundary condition of the type shown in Fig. 6.3c, which includes discontinuities:

$$\Theta(x, t) = \int_{\tau=0}^{\tau=t} \frac{dF(\tau)}{d\tau} S(x, t - \tau)\, d\tau + \sum_{i=0}^{N} \Delta F_i S(x, t - \tau_i) \tag{6.34}$$

Note that the jumps $\Delta F_i$ at each discontinuity are obtained by subtracting the final value of $F(\tau)$ at the end of the discontinuity from the initial value of $F(\tau)$, that is,

$$\Delta F_{-i} = F(\tau_i^+) - F(\tau_i^-) \qquad i = 0, 1, \ldots, N \tag{6.35}$$

An alternative form of Duhamel's theorem can be obtained if we first rewrite eq. (6.34) as

$$\begin{aligned} \Theta(x, t) = {} & \int_{\tau=0}^{\tau=\tau_1^-} \frac{dF(\tau)}{d\tau} S(x, t - \tau)\, d\tau + \int_{\tau=\tau_1^+}^{\tau=\tau_2^-} \frac{dF(\tau)}{d\tau} S(x, t - \tau)\, d\tau \\ & + \cdots + \int_{\tau=\tau_{N-1}^+}^{\tau=\tau_N^-} \frac{dF(\tau)}{d\tau} S(x, t - \tau)\, d\tau + \int_{\tau=\tau_N^+}^{\tau=t} \frac{dF(\tau)}{d\tau} S(x, t - \tau)\, d\tau \\ & + \sum_{i=0}^{N} \Delta F_i S(x, t - \tau_i) \end{aligned} \tag{6.36}$$

Integrating by parts the $N + 1$ integrals appearing in eq. (6.36) yields

$$\begin{aligned} \Theta(x, t) = {} & [F(\tau)S(x, t - \tau)]_{\tau=0}^{\tau=\tau_1^-} + [F(\tau)S(x, t - \tau)]_{\tau=\tau_1^+}^{\tau=\tau_2^-} \\ & + \cdots + [F(\tau)S(x, t - \tau)]_{\tau=\tau_{N-1}^+}^{\tau=\tau_N^-} + [F(\tau)S(x, t - \tau)]_{\tau=\tau_N^+}^{\tau=t} - \\ & \left[ \int_{\tau=0}^{\tau=\tau_1^-} \frac{\partial S(x, t - \tau)}{\partial \tau} F(\tau)\, d\tau + \int_{\tau=\tau_1^+}^{\tau=\tau_2^-} \frac{\partial S(x, t - \tau)}{\partial \tau} F(\tau)\, d\tau \right. \\ & \left. + \cdots + \int_{\tau=\tau_{N-1}^+}^{\tau=\tau_N^-} \frac{\partial S(x, t - \tau)}{\partial \tau} F(\tau)\, d\tau + \int_{\tau=\tau_N^+}^{\tau=t} \frac{\partial S(x, t - \tau)}{\partial \tau} F(\tau)\, d\tau \right] \\ & + \sum_{i=0}^{N} \Delta F_i S(x, t - \tau_i) \end{aligned} \tag{6.37}$$

Next, we note that in the integrals in eq. (6.37)

$$\frac{\partial S(x, t-\tau)}{\partial \tau} = \frac{\partial S(x, t-\tau)}{\partial (t-\tau)} \frac{\partial (t-\tau)}{\partial \tau} = -\frac{\partial S(x, t-\tau)}{\partial (t-\tau)} = -\frac{\partial S(x, t-\tau)}{\partial t} \tag{6.38}$$

In addition, the temperature distribution in the body caused by a *unit step* input is continuous, that is,

$$S(x, t - \tau_i^-) = S(x, t - \tau_i^+) \qquad i = 1, 2, \ldots, N \tag{6.39}$$

Combining eqs. (6.37)–(6.39) yields

$$\begin{aligned}\Theta(x, t) = {} & -S(x, t-0)\Delta F_0 - S(x, t-\tau_1)\,\Delta F_1 - S(x, t-\tau_2)\,\Delta F_2 - \cdots \\ & - S(x, t-\tau_N)\,\Delta F_N + F(t)S(x, 0) + \int_{\tau=0}^{\tau=t} \frac{\partial S(x, t-\tau)}{\partial t} F(\tau)\, d\tau \\ & + \sum_{i=0}^{N} \Delta F_i\, S(x, t-\tau_i)\end{aligned} \tag{6.40}$$

where the notation (6.35) was utilized. Observing that $S(x, 0) = 0$ and performing the algebra in eq. (6.40), we obtain the alternative form of Duhamel's theorem:

$$\Theta(x, t) = \int_{\tau=0}^{\tau=t} \frac{\partial S(x, t-\tau)}{\partial t} F(\tau)\, d\tau \tag{6.41}$$

*Since the zero initial condition* [$S(x, 0) = 0$] *was utilized in deriving eq. (6.41), care should be taken in nondimensionalizing the problem of interest such that the resulting dimensionless conduction model has a zero initial condition.*

It is worth stressing that even though it was assumed that $x$ is a Cartesian coordinate for illustrative purposes, eqs. (6.34) and (6.41) hold identically in the cylindrical and spherical coordinate systems. One only needs to replace $x$ in the above-mentioned expressions with the pertinent coordinate notation ($r$, for example, in cylindrical or spherical coordinates when conduction occurs only in the radial direction).

Before proceeding with the presentation of examples in which Duhamel's theorem is utilized to obtain the temperature distribution in a body, we present a sequence of steps that summarize the process of the application of this theorem.

### ■ *Step 1*

Obtain the mathematical model of the conduction problem of interest and cast it in dimensionless form. In nondimensionalizing, care should be taken such that the resulting nondimensional model is homogeneous everywhere except for the time-dependent boundary condition. In addition, the initial dimensionless temperature needs to be zero.

■ ***Step 2***

Create the corresponding unit step problem simply by replacing the time-dependent boundary condition by a unit step boundary condition.

■ ***Step 3***

Solve the corresponding unit step problem and obtain the expression for $S(x, t)$.

■ ***Step 4***

With $S(x, t)$ known from step 3, utilize eq. (6.34) or (6.41) to obtain the temperature distribution for the problem of interest with the time-dependent boundary condition.

The procedure outlined above is applied in the following example.

**Example 6.2**

Resolve Example 6.1 by utilizing Duhamel's theorem. Obtain the solution for the temperature distribution for the entire time domain, including early times during which the system "adjusts" to the time-periodic boundary condition.

**Solution**

■ ***Step 1***

The mathematical model for the problem of interest is

$$\frac{\partial^2 T}{\partial x^2} = \frac{1}{\alpha}\frac{\partial T}{\partial t} \tag{6.42}$$

$$t = 0: \quad T = T_\infty \tag{6.43}$$

$$x = 0: \quad T = T_\infty + T_0 \sin(\omega t) \tag{6.44}$$

$$x \to \infty: \quad T = T_\infty \tag{6.45}$$

To nondimensionalize eqs. (6.42)–(6.45), we define the following variables:

$$\Theta = \frac{T - T_\infty}{T_0} \qquad \hat{t} = \frac{t}{t_p} \qquad \hat{x} = \frac{x}{\sqrt{\alpha t_p}} \tag{6.46}$$

The dimensionless model is obtained by substituting eqs. (6.46) into eqs. (6.42)–(6.45):

$$\frac{\partial^2 \Theta}{\partial \hat{x}^2} = \frac{\partial \Theta}{\partial \hat{t}} \tag{6.47}$$

$$\hat{t} = 0: \quad \Theta = 0 \tag{6.48}$$

$$\hat{x} = 0: \quad \Theta = \sin(2\pi\hat{t}) \tag{6.49}$$

$$\hat{x} \to \infty: \quad \Theta = 0 \tag{6.50}$$

Care was taken in defining the quantities in eq. (6.46) such that the dimensionless temperature is initially zero and that the model is initially homogeneous everywhere except for the time-dependent boundary condition. This boundary condition is shown graphically in Fig. 6.4.

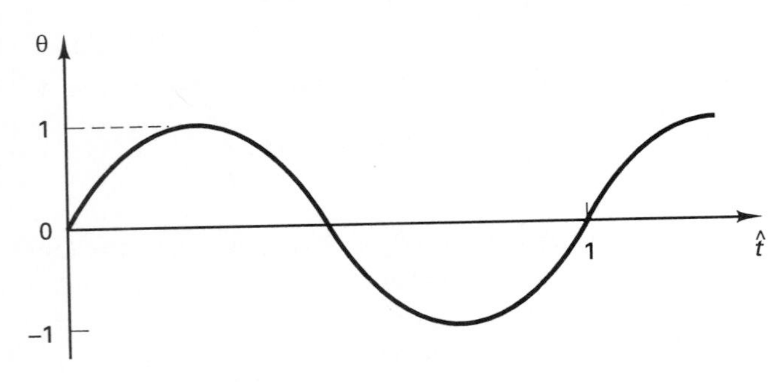


**Figure 6.4** Dimensionless boundary temperature.

### ■ *Step 2*

To create the mathematical model for the corresponding unit step problem, we simply replace the boundary condition (6.49) by a unit step boundary condition. Denoting the temperature for the unit step problem by $S(\hat{x}, \hat{t})$, we obtain

$$\frac{\partial^2 S}{\partial \hat{x}^2} = \frac{\partial S}{\partial \hat{t}} \tag{6.51}$$

$$\hat{t} = 0: \quad S = 0 \tag{6.52}$$

$$\hat{x} = 0: \quad S = 1 \tag{6.53}$$

$$\hat{x} \to \infty: \quad S = 0 \tag{6.54}$$

### ■ *Step 3*

To solve the unit step problem formulated in step 2, we utilize the results of Example 5.7, in particular eq. (5.196). Based on this equation, by inspection and after we replace $T_i$ by $T_\infty$ to accommodate the notation of the present example,

$$S(\hat{x}, \hat{t}) = \frac{T - T_\infty}{T_0 - T_\infty} = 1 - \operatorname{erf}\left(\frac{1}{2}\frac{\hat{x}}{\sqrt{\hat{t}}}\right) \tag{6.55}$$

### ■ *Step 4*

With $S(\hat{x}, \hat{t})$ known via eq. (6.55), we choose to utilize the first version of Duhamel's theorem given in eq. (6.34) to obtain the temperature distribution inside the soil. Note that no discontinuities exist in the transient boundary condition (6.49). Therefore, the second term on the right-hand side of eq. (6.34) is set equal to zero. Also,

$$F(\tau) = \sin(2\pi\tau) \tag{6.56}$$

Combining eqs. (6.34), (6.55), and (6.56), we finally obtain

$$\Theta(\hat{x}, \hat{t}) = \int_0^{\hat{t}} S(\hat{x}, \hat{t} - \tau) \frac{dF}{d\tau} \, d\tau$$

$$= \int_0^{\hat{t}} \left[ 1 - \operatorname{erf}\left( \frac{1}{2} \frac{\hat{x}}{\sqrt{\hat{t} - \tau}} \right) \right] \frac{d[\sin(2\pi\tau)]}{d\tau} \, d\tau \tag{6.57}$$

Integration eq. (6.57) by parts yields

$$\Theta(\hat{x}, \hat{t}) = \left\{ \left[ 1 - \operatorname{erf}\left( \frac{1}{2} \frac{\hat{x}}{\sqrt{\hat{t} - \tau}} \right) \right] \sin(2\pi\tau) \right\}_{\tau=0}^{\tau=\hat{t}}$$

$$+ \frac{\hat{x}}{2\sqrt{\pi}} \int_0^{\hat{t}} (\hat{t} - \tau)^{-3/2} e^{-\hat{x}^2/4(\hat{t}-\tau)} \sin(2\pi\tau) \, d\tau$$

$$= \frac{\hat{x}}{2\sqrt{\pi}} \int_0^{\hat{t}} (\hat{t} - \tau)^{-3/2} e^{-\hat{x}^2/4(\hat{t}-\tau)} \sin(2\pi\tau) \, d\tau \tag{6.57a}$$

The reader may want to show that in the limit of large times the temperature in eq. (6.57a) reduces to the sustained temperature result of eq. (6.26).

**Example 6.3**

The schematic in Fig. 6.5a shows the flat wall of an experimental apparatus. On one side of the wall a heater has been installed such that when it is turned on, the temperature of that side ($x = 0$) increases almost linearly with time. When the heater is turned off, the wall temperature at $x = 0$ attains its original value quickly. Based on the above, and if we denote the initial temperature of the wall by $T_0$, a reasonably accurate representation of the temperature of the wall at $x = 0$ for $t < t_1$ (the heater is turned on at $t = 0$ and off at $t = t_1$) is given by the following equation:

$$T(0, t) = (T_1 - T_0)\frac{t}{t_1} + T_0 \qquad \text{for} \quad t < t_1 \tag{6.58}$$

where $T_1$ is the wall temperature at $x = 0$ at $t = t_1$, right before the heater is turned off. The temperature of the other side of the wall ($x = L$)is kept constant at $T_0$. Equation (6.58) is graphed in Fig. 6.5b. Obtain the temperature distribution inside the wall as well as the heat flux at $x = L$. Utilize both forms of Duhamel's theorem and compare the solutions they yield.

**Solution**

■ ***Step 1***

The mathematical model for the problem of interest is

$$\frac{\partial^2 T}{\partial x^2} = \frac{1}{\alpha} \frac{\partial T}{\partial t} \tag{6.59}$$

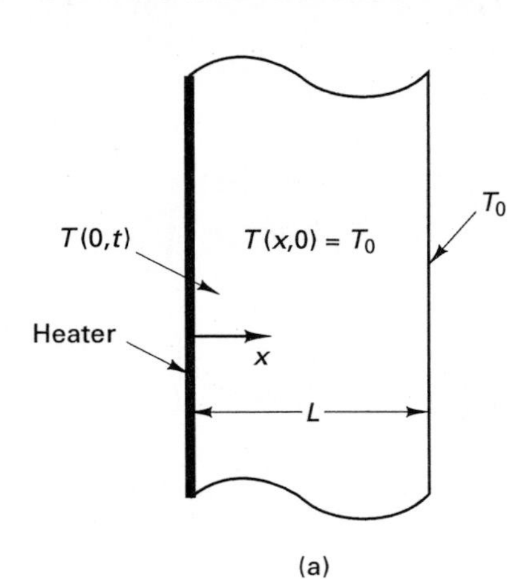


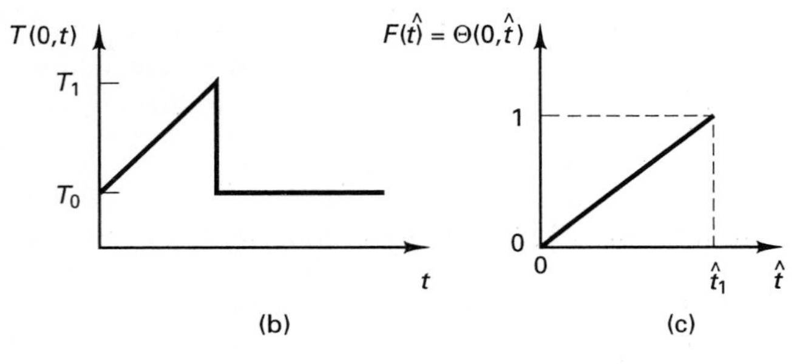


**Figure 6.5** (a) Flat wall with heater attached to its side. (b) Surface temperature. (c) Dimensionless surface temperature.

$$t = 0: \qquad T = T_0 \tag{6.60}$$

$$x = 0: \qquad \begin{cases} T = (T_1 - T_0)\dfrac{t}{t_1} + T_0 & t < t_1 \\ T = T_0 & t > t_1 \end{cases} \tag{6.61}$$

$$x = L: \qquad T = T_0 \tag{6.62}$$

To nondimensionalize the model above, we define the following variables:

$$\Theta = \frac{T - T_0}{T_1 - T_0} \qquad \hat{t} = \frac{t}{L^2/\alpha} \qquad \hat{x} = \frac{x}{L} \tag{6.63}$$

The mathematical model in terms of the dimensionless variables above reads

$$\frac{\partial^2 \Theta}{\partial \hat{x}^2} = \frac{\partial \Theta}{\partial \hat{t}} \tag{6.64}$$

$$\hat{t} = 0: \qquad \Theta = 0 \tag{6.65}$$

$$\hat{x} = 0: \qquad \begin{cases} \Theta = A\hat{t} & \hat{t} < \hat{t}_1 \\ \Theta = 0 & \hat{t} > \hat{t}_1 \end{cases} \tag{6.66}$$

$$\hat{x} = 1: \qquad \Theta = 0 \tag{6.67}$$

where the dimensionless group

$$A = \frac{L^2/\alpha}{t_1} \tag{6.68}$$

is the ratio of the characteristic heat diffusion time of the system to the duration of the temperature pulse. Note that the initial condition for the dimensionless model states that the dimensionless temperature is zero. The model is homogeneous everywhere except for the time-dependent boundary condition (shown graphically in Fig. 6.5c).

### ■ *Step 2*

The mathematical model for the corresponding unit step problem is obtained by replacing boundary condition (6.66) in the model of step 1 by a unit step boundary condition. If, as before, we denote the temperature for the unit step problem by $S(x, t)$, we obtain

$$\frac{\partial S}{\partial \hat{t}} = \frac{\partial^2 S}{\partial \hat{x}^2} \tag{6.69}$$

$$\hat{t} = 0: \quad S = 0 \tag{6.70}$$

$$\hat{x} = 0: \quad S = 1 \tag{6.71}$$

$$\hat{x} = 1: \quad S = 0 \tag{6.72}$$

### ■ *Step 3*

The solution for the unit step problem can be derived based on the methodologies presented in Chapter 5. Both the principle of superposition and the method of variation of parameters are suitable for solving this problem. The details of the solution are left as an exercise for the reader and only the final result is presented:

$$S(\hat{x}, \hat{t}) = 1 - \hat{x} - \frac{2}{\pi} \sum_{n=1}^{\infty} \frac{\sin(n\pi\hat{x})}{n} e^{-n^2\pi^2\hat{t}} \tag{6.73}$$

### ■ *Step 4*

First we solve the problem utilizing the first form of Duhamel's theorem given by eq. (6.34). Since a discontinuity exists in the transient boundary condition at $\hat{t} = \hat{t}_1$, two expressions for the temperature will be obtained: one before the discontinuity and one after it.

**(a)** $\hat{t} < \hat{t}_1$. In this time domain, no discontinuity exists in the system. Therefore, the summation term in eq. (6.34) is set equal to zero. In addition,

$F(\tau) = A\tau$. Based on the above, we have

$$\Theta(\hat{x}, \hat{t}) = \int_0^{\hat{t}} \left[ 1 - \hat{x} - \frac{2}{\pi} \sum_{n=1}^{\infty} \frac{\sin(n\pi\hat{x})}{n} e^{-n^2\pi^2(\hat{t}-\tau)} \right] A\, d\tau$$

$$= A\left[ (1 - \hat{x})\hat{t} - \frac{2}{\pi^3} \sum_{n=1}^{\infty} \frac{\sin(n\pi\hat{x})}{n^3} (1 - e^{-n^2\pi^2\hat{t}}) \right] \tag{6.74}$$

**(b)** $\hat{t} > \hat{t}_1$. Taking into account the discontinuity at $t_1$, eq. (6.34) yields

$$\Theta(\hat{x}, \hat{t}) = \int_0^{\hat{t}_1} S(\hat{x}, \hat{t} - \tau) \frac{dF}{d\tau}\, d\tau + \int_{\hat{t}_1}^{\hat{t}} S(\hat{x}, \hat{t} - \tau) \frac{dF}{d\tau}\, d\tau + S(\hat{x}, \hat{t} - \hat{t}_1)\, \Delta F \tag{6.75}$$

In the first integral on the right-hand side of eq. (6.75), $dF/d\tau = A$. In the second integral, $dF/d\tau = 0$ (Fig. 6.5c). Also in eq. (6.75), $\Delta F = -1$. Taking the above into account and performing the algebra yields

$$\Theta(\hat{x}, \hat{t}) = (A\hat{t}_1 - 1)(1 - \hat{x}) - \frac{2A}{\pi^3} \sum_{n=1}^{\infty} \frac{\sin(n\pi\hat{x})}{n^3} e^{-n^2\pi^2\hat{t}} (e^{n^2\pi^2\hat{t}_1} - 1)$$
$$+ \frac{2}{\pi} \sum_{n=1}^{\infty} \frac{\sin(n\pi\hat{x})}{n} e^{-n^2\pi^2(\hat{t}-\hat{t}_1)} \tag{6.76}$$

To summarize at this point, we have derived expressions for the temperature distribution in the system both before the temperature jump [eq. (6.74)] and after the temperature jump [eq. (6.76)].

Next, we repeat step 4 utilizing the second form of Duhamel's theorem [eq. (6.41)].

**(a)** $t < t_1$. Combining eqs. (6.41), (6.66) and (6.73) and after performing the algebra, we find that

$$\Theta(\hat{x}, \hat{t}) = \frac{2A}{\pi^3} \sum_{n=1}^{\infty} \frac{\sin(n\pi\hat{x})}{n} (n^2\pi^2\hat{t} - 1 + e^{-n^2\pi^2\hat{t}}) \tag{6.77}$$

Taking into account that the sine Fourier series expansion of the function $1 - \hat{x}$ in the region $0 < \hat{x} < 1$ is given by [5]

$$1 - \hat{x} = \frac{2}{\pi} \sum_{n=1}^{\infty} \frac{\sin(n\pi\hat{x})}{n} \tag{6.78}$$

it is obvious that eq. (6.77) is identical to the temperature distribution in eq. (6.74) obtained with the first version of Duhamel's theorem.

**(b)** $t > t_1$. Noting that in eq. (6.41),

$$\int_0^t = \int_0^{t_1} + \int_{t_1}^t \tag{6.79}$$

and that $F(\tau) = 2\tau$ in the domain $0 < \tau < \hat{t}_1$ while $F(\tau) = 0$ in the domain $\tau > \hat{t}_1$, we can easily show that $\Theta(\hat{x}, t)$ is given by an expression identical to eq. (6.76).

To calculate the heat flux at either side of the wall we need to apply Fourier's law. For the side at $\hat{x} = 1$,

$$\dot{q}'' = -k\left(\frac{\partial T}{\partial x}\right)_{x=L} = -\frac{k(T_1 - T_0)}{L}\left(\frac{\partial \Theta}{\partial \hat{x}}\right)_{\hat{x}=1} \tag{6.80}$$

Combining eqs. (6.74) and (6.80) yields the following expression for the heat flux for $\hat{t} < t_1$:

$$\dot{q}'' = \frac{k(T_1 - T_0)A}{L}\left[\hat{t} + \frac{2}{\pi^2}\sum_{n=1}^{\infty}\frac{(-1)^n}{n^2}(1 - e^{-n^2\pi^2\hat{t}})\right] \tag{6.81}$$

Similarly, combining eqs. (6.76) and (6.80) yields the heat flux at $\hat{x} = 1$ for $\hat{t} > \hat{t}_1$:

$$\dot{q}'' = \frac{k(T_1 - T_0)}{L}\left[(A\hat{t}_1 - 1) + \frac{2A}{\pi^2}\sum_{n=1}^{\infty}\frac{(-1)^n}{n^2}e^{-n^2\pi^2\hat{t}}(e^{n^2\pi^2\hat{t}_1} - 1) + 2\sum_{n=1}^{\infty}(-1)^n e^{-n^2\pi^2(\hat{t}-\hat{t}_1)}\right] \tag{6.82}$$

## PROBLEMS

**6.1.** During a therapeutic treatment a specially designed heating–cooling device is attached on the back of a patient such that the heat flux provided to the patient is a periodic function of time:

$$\dot{Q}''_{\text{back}} = \dot{Q}''_0 \cos(\omega t)$$

where $\dot{Q}''_0$ is the amplitude of the heat flux oscillation. Modeling the patient's body as a flat wall of large (infinite) thickness and known properties, find its sustained temperature distribution. The initial temperature of the body is denoted by $T_i$. Comment on the heater design from the perspective of avoiding burning the patient's back where the heater is attached.

**6.2.** The heating device described in Problem 6.1 is now attached around the leg of a patient. Model the leg as a long cylinder of large radius $R \to \infty$, known thermophysical properties, and initial temperature $T_i$ and obtain the sustained temperature distribution in the leg.

**6.3.** Ambient air is used to cool the outer surface of the wall of a furnace that operates continuously. Due to the daily cycle, the temperature of the ambient air varies periodically with time, $T_a = T_\infty - T_0 \cos(2\pi/t_p)t$, where $T_\infty$ is the average ambient air temperature, $T_0$ is the amplitude of the oscillation, and $t_p = 24$ hours is the period of the oscillation. The inner surface of the furnace wall is maintained at a constant temperature $T_i$. The thickness of the wall is known and so are its thermophysical properties and the heat transfer coefficient between the outer surface of the wall and the ambient air. Obtain the sustained temperature distribution in the wall. [*Hint:* Note that the boundary condition at the outer surface of the wall ($x = 0$) is $k(\partial T/\partial x) = h(T - (T_\infty - T_0 \cos \omega t))$.]

**6.4.** Ambient air whose temperature dependence on time was described in Problem 6.3 is used to cool a flat surface of a thick wall in a heat exchanger. Assuming that the wall was initially at a temperature $T_i$ and that it is sufficiently thick such that the cooling effect of the air never penetrates to the opposite side of the wall, obtain its sustained temperature field utilizing the method of complex temperatures. Note that the convection boundary condition applies in this case as well.

**6.5.** Right after it is manufactured, a long, thick cylindrical rod of radius $R$ is sprayed with a coolant of temperature $T_\infty$ (Fig. P6.5). Experiments have shown that the temperature of the outer surface of the rod decreases exponentially with time, $T(R, t) = (T_i - T_\infty)e^{-at} + T_\infty$, where $T_i$ is the initial temperature of the rod before the cooling process is initiated and where $a$ is an experimental constant that depends on the method utilized to spray the coolant upon the rod. Using Duhamel's theorem, obtain the temperature distribution in the rod and the heat flux at the surface of the rod.

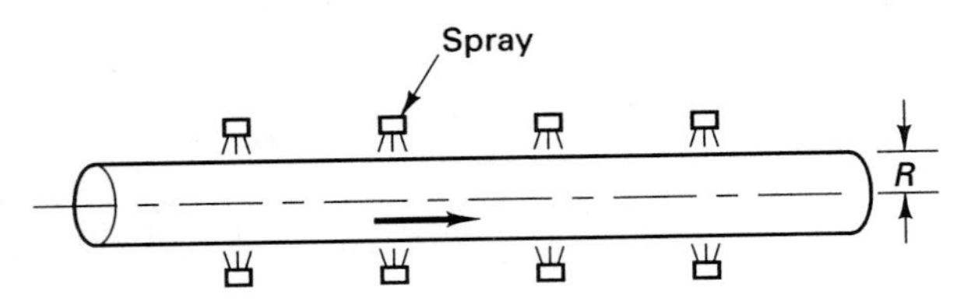


**Figure P6.5** Spray-cooled rod.

**6.6.** Reconsider Problem 6.5. Instead of a rod, a long metal slab is now cooled (Fig. P6.6). The spray cooling process is identical on both sides of the slab and the surface temperatures are given by the expression of Problem 6.5. Obtain the temperature distribution in the slab and the heat flux at the surface of the slab.

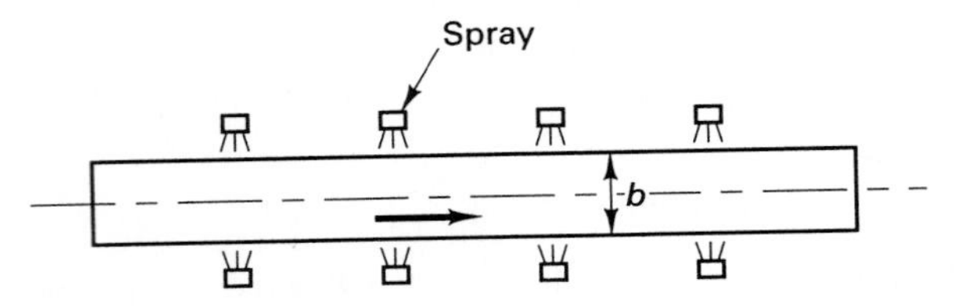


**Figure P6.6** Spray-cooled slab.

**6.7.** Consider the problem of one-dimensional transient heat conduction in a plane wall. Initially, the wall is isothermal at $T_i$ and its right side (at $x = L$) is insulated. Obtain the temperature distribution in the wall as well as the heat flux at $x = 0$ for $0 < t < t_0$ and for $3t_0 < t < 4t_0$ if the left side of the wall (at $x = 0$) undergoes the series of temperature pulses shown in Fig. P6.7.

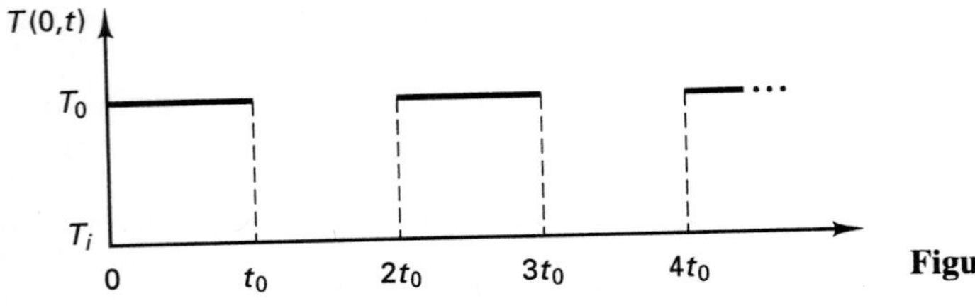


**Figure P6.7**

**6.8.** Re-solve Problem 6.7 for the series of temperature pulses shown in Fig. P6.8.

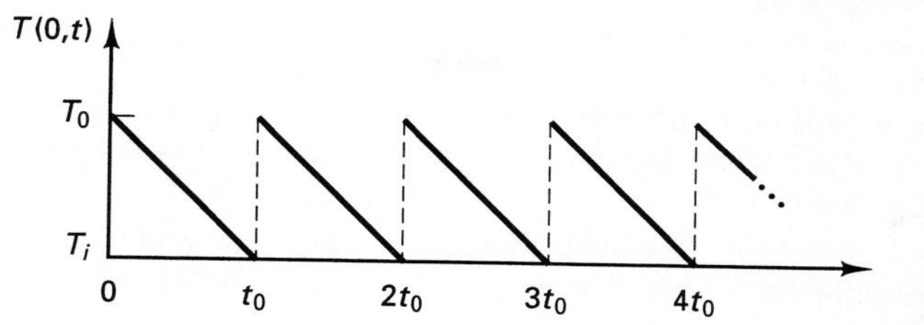


**Figure P6.8**

**6.9.** Consider a very thick (semi-infinite) wall initially isothermal at temperature $T_i$. Suddenly, the temperature of its surface is undergoing the temperature pulses defined earlier in Problem 6.7 (Fig. P6.7). Obtain the transient temperature distribution in the wall and the heat flux at its surface.

**6.10.** Re-solve Problem 6.9 for the temperature pulses defined in Problem 6.8 (Fig. P6.8).

**6.11.** Heat is generated at the interface between two initially isothermal ($T_0$) reciprocating parts of a machine as shown in Fig. P6.11. The reciprocation is such that the parts move with frequency $\omega$, always in opposite directions. The heat generation is time dependent $\dot{q}'' = A\cos(\omega t)$, where $A$ is a constant. Assume that the thermophysical properties of the above-mentioned machine parts are known and that these parts are "thick" enough to be modeled as semi-infinite bodies. Obtain their sustained temperature distribution.

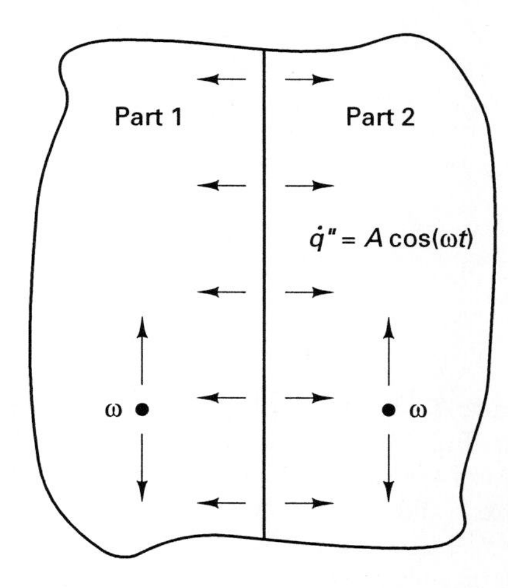


**Figure P6.11** Heat generation at the interface of two reciprocating parts of a machine.

**6.12.** During the operation of a heater (modeled as a slab of thickness $b$ initially at temperature $T_i$) an electric current is passed through the heater generating heat $\dot{q}'''$ for a time period $t_0$. Then the current is turned off for a time period $t_0$. The foregoing scenario is repeated continuously. The sides of the slab are cooled convectively ($h$, $T_i$). Obtain the temperature distribution in the heater and the heat flux through its sides **(a)** for $0 < t < t_0$ and **(b)** for $3t_0 < t < 4t_0$.

**6.13.** Re-solve Problem 6.12 but this time for a long cylindrical (wire) heater of radius $R$.

**6.14.** Figure P6.14 shows the wall of a house. Convection ($h$) takes place on both sides of this wall. The temperature of the air inside the house is kept constant ($T_i$) at all times. The air temperature outside the house varies according to the daily cycle and it is described by $T_a = T_\infty - T_0 \cos(\omega t)$, where $T_\infty$ is the average ambient air temperature and $T_0$ is the amplitude of oscillation of the daily cycle. Obtain the sustained temperature distribution in the wall as well as the heat flux through both its sides. Assume for simplicity that $T_i = T_\infty$ and that $h \to \infty$.

**6.15.** Re-solve Problem 6.14 by utilizing Duhamel's theorem to obtain the wall temperature distribution over the entire time domain. You may assume that the initial wall temperature is $T_\infty$.

**6.16.** Re-solve Problem 6.14 for a finite (realistic) heat transfer coefficient, $h$.

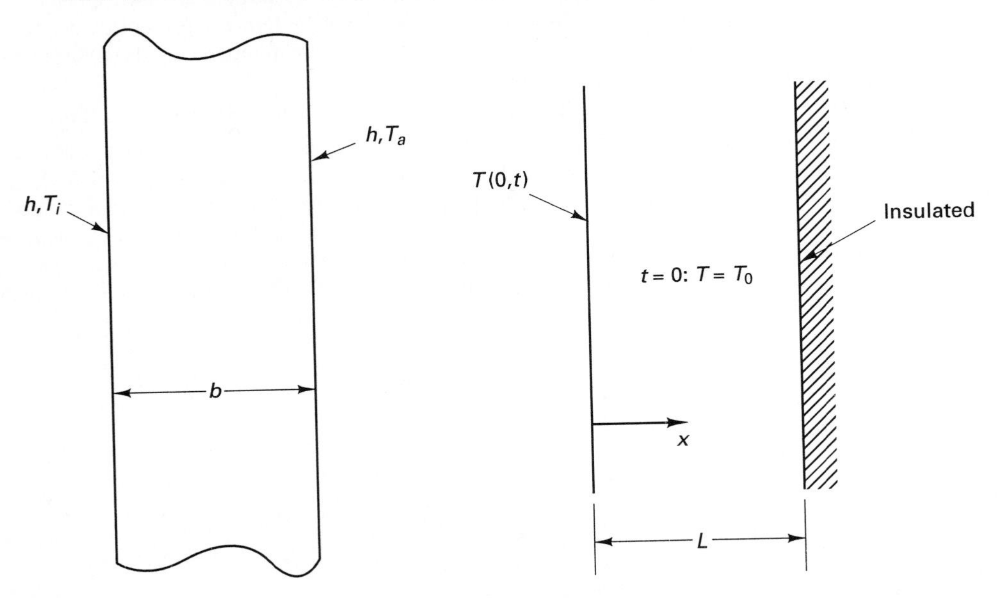


**Figure P6.14**

**Figure P6.17**

**6.17.** A coolant of temperature $T_\infty$ is circulated on one side of the wall of an experimental apparatus shown in Fig. P6.17. The initial temperature of this wall is denoted by $T_0$. The temperature of the side of the wall in contact with the coolant is monitored with thermocouples and it is shown to decrease with time as follows: $T(0, t) = (T_0 - T_\infty)e^{-At} + T_\infty$, where $A$ is a known experimental constant. The opposite side of the wall is insulated. Obtain the transient temperature distribution in the wall.

**6.18.** A coolant of temperature $T_\infty$ is circulated inside a pipe of inner radius $R_i$ and outer radius $R_o$, initially at temperature $T_0$. The outer surface of the pipe is insulated. The temperature of its inner surface is measured to vary with time as follows: $T(R_i, t) = (T_0 - T_\infty)e^{-At} + T_\infty$, where $A$ is a known experimental constant. Obtain the temperature distribution in the pipe wall.

## REFERENCES

1. V. S. ARPACI, *Conduction Heat Transfer,* Addison-Wesley, Reading, MA, 1966.
2. G. E. MYERS, *Analytical Methods in Conduction Heat Transfer,* McGraw-Hill, New York, 1971.
3. S. J. FARLOW, *Partial Differential Equations for Scientists and Engineers,* Wiley, New York, 1982.
4. C. WYLIE, *Differential Equations,* McGraw-Hill, New York, 1979.
5. M. R. SPIEGEL, *Fourier Analysis,* Schaum's Outline Series, McGraw-Hill, New York, 1974.

CHAPTER 7

# TRANSIENT CONDUCTION INVOLVING HEAT SOURCES OR SINKS

In this chapter, a special class of problems is considered with the defining feature that heat is liberated or absorbed instantaneously or continuously in a region of the system of interest that is infinitesimally small compared to the dimensions of the system. Two types of such regions are a point and a straight line. In the former case, we are dealing with a point source (sink) and in the latter case with a line source (sink). In the case of an instantaneous source the time over which the heat is released is minimal compared to the characteristic heat conduction time and the phenomenon of heat release from the source is commonly termed an *explosion*.

Finally, the heat conduction process in several thermal applications (welding, machining, laser processing, tribology [1–3]) can be modeled as one caused by an appropriate moving heat source. Therefore, basic problems involving moving sources in a stationary body are also considered in this chapter.

The discussion is centered primarily around the heat source application, with the understanding that the study of heat sinks involves the identical procedure. Usually, the results for heat sources after simple manipulations can be used directly to study problems involving heat sinks.

## 7.1 HEAT SOURCES PRODUCING HEAT CONTINUOUSLY

Three basic heat sources of this type are considered: the point source, the line source, and the plane source (Fig. 7.1).

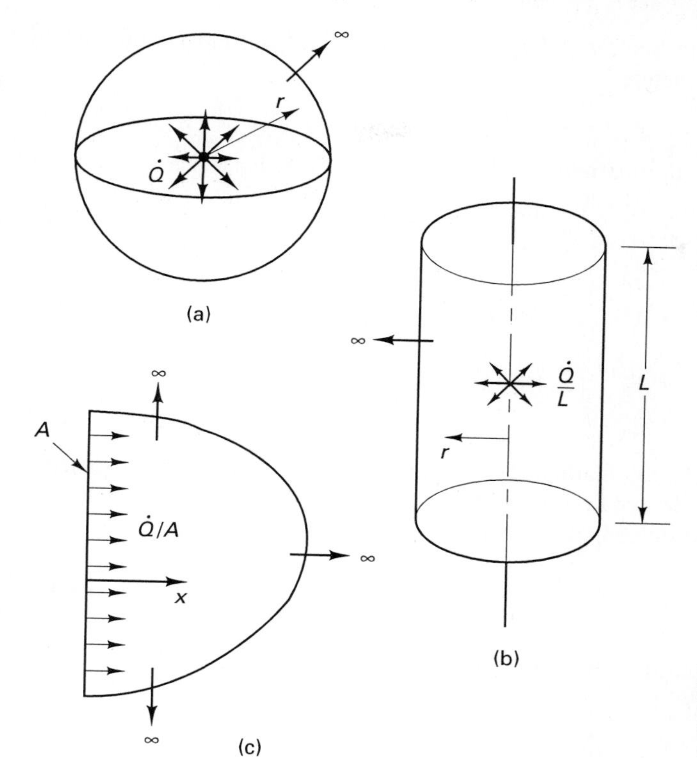


**Figure 7.1** Sources of constant heat production rate: (a) point source; (b) line source; (c) plane source.

## Point Heat Source of Constant Heat Production Rate

Consider a body of large extent possessing a very small region (point) that produces heat continuously, at a constant rate $\dot{Q}$ (Watts). Such is the case, for example, of a canister of nuclear waste buried deep inside the earth. If the initial temperature of the body, which is also the temperature of the body far from the heat-generating region, is denoted by $T_\infty$, the heat conduction process in the body is described by the following model (spherical coordinates are chosen since a large region of the body containing the heat source can be viewed as a sphere with the point source at its center):

$$\frac{1}{\alpha}\frac{\partial T}{\partial t} = \frac{1}{r^2}\frac{\partial}{\partial r}\left(r^2\frac{\partial T}{\partial r}\right) \tag{7.1}$$

$$r \to 0: \qquad -4\pi r^2 k\frac{\partial T}{\partial r} = \dot{Q} \tag{7.2}$$

$$r \to \infty: \qquad T \to T_\infty \tag{7.3}$$

$$t = 0: \qquad T = T_\infty \tag{7.4}$$

Following the discussion in Section 5.3, pertaining to radial transient conduction in spherical coordinates, we define

$$u = r(T - T_\infty) \tag{7.5}$$

In terms of the variable $u$ the mathematical model (7.1)–(7.4) becomes

$$\frac{1}{\alpha}\frac{\partial u}{\partial t} = \frac{\partial^2 u}{\partial r^2} \tag{7.6}$$

$$r \to 0: \qquad -4\pi k\left(r\frac{\partial u}{\partial r} - u\right) = \dot{Q} \tag{7.7}$$

$$r \to \infty: \qquad u \to 0 \tag{7.8}$$

$$t = 0: \qquad u = 0 \tag{7.9}$$

As discussed in Chapter 5, the mathematical model for $u$ describes heat conduction in Cartesian coordinates. To obtain the solution for the variable $u$, we choose to utilize the similarity method described in detail in Section 5.6. Introducing

$$\eta = \frac{r}{2\sqrt{\alpha t}} \tag{7.10}$$

we cast eqs. (7.6)–(7.9) in the following form:

$$\frac{d^2 u}{d\eta^2} + 2\eta\frac{du}{d\eta} = 0 \tag{7.11}$$

$$\eta \to 0: \qquad \dot{Q} = -4\pi k\left(\eta\frac{du}{d\eta} - u\right) \tag{7.12}$$

$$\eta \to \infty: \qquad u \to 0 \tag{7.13}$$

According to the discussion in Section 5.6, the solution to eq. (7.11) is

$$u - u(\eta = 0) = C\frac{\sqrt{\pi}}{2}\operatorname{erf}(\eta) \tag{7.14}$$

Applying boundary conditions (7.12) and (7.13), we obtain the two constants of integration

$$C = -\frac{\dot{Q}}{2k\pi^{3/2}} \tag{7.15}$$

$$u(\eta = 0) = \frac{\dot{Q}}{4\pi k} \tag{7.16}$$

Combining eqs. (7.5), (7.10), and (7.14)–(7.16) yields the following expression for the temperature distribution in the body:

$$T = T_\infty + \frac{\dot{Q}}{4\pi k r}\left[1 - \operatorname{erf}\left(\frac{r}{2\sqrt{\alpha t}}\right)\right] \tag{7.17}$$

At large times ($t \rightarrow \infty$) expression (7.17) reduces to

$$T = T_\infty + \frac{\dot{Q}}{4\pi kr} \tag{7.18}$$

The second term on the right-hand side of eq. (7.18) represents the steady-state temperature rise at any location ($r$) caused by the source.

## Line Heat Source of Constant Heat Production Rate

Consider a large body with a line heat source "buried" within it. Examples of such heat sources are electrical wires or power cables buried within thick layers of insulation or under the surface of the earth. It is assumed that the line source is producing heat at a constant rate per unit length $\dot{Q}/L$ (W/m), where $L$ is the length of the source. Since the extent of the body within which the source is buried is very large, we can assume that the conduction heat transfer occurs within a cylinder of very large radius having the source as its axis. The initial temperature of the body before the source starts generating heat is uniform and is denoted by $T_\infty$. The mathematical model for the above-described problem in cylindrical coordinates is

$$\frac{1}{\alpha}\frac{\partial \Theta}{\partial t} = \frac{1}{r}\frac{\partial}{\partial r}\left(r\frac{\partial \Theta}{\partial r}\right) \tag{7.19}$$

$$r \rightarrow 0: \qquad \frac{\dot{Q}}{L} = -2\pi rk\frac{\partial \Theta}{\partial r} \tag{7.20}$$

$$r \rightarrow \infty: \qquad \Theta \rightarrow 0 \tag{7.21}$$

$$t = 0: \qquad \Theta = 0 \tag{7.22}$$

where, as before, $\Theta = T - T_\infty$.

To solve this model, we again choose to use the similarity method. The similarity variable is defined as

$$\eta = \frac{r^2}{4\alpha t} \tag{7.23}$$

In terms of the similarity variable the mathematical model reads

$$\eta\frac{d^2\Theta}{d\eta^2} + (1 + \eta)\frac{d\Theta}{d\eta} = 0 \tag{7.24}$$

$$\eta \rightarrow 0: \qquad \frac{\dot{Q}}{L} = -4\pi k\eta\frac{d\Theta}{d\eta} \tag{7.25}$$

$$\eta \rightarrow \infty: \qquad \Theta \rightarrow 0 \tag{7.26}$$

The general solution to eq. (7.24) can be written as (Appendix B)

$$\Theta(\eta \rightarrow \infty) - \Theta = C\int_\eta^\infty \frac{e^{-\lambda}}{\lambda}\,d\lambda \tag{7.27}$$

This solution can easily be obtained if one reduces the order of eq. (7.24) by introducing, say, $p = d\Theta/d\eta$ and solves the resulting separable first-order equation. Boundary condition (7.26) yields the fact that $\Theta(\eta \rightarrow \infty) = 0$ and boundary condition (7.25) that $C = -\dot{Q}/4\pi kL$. In addition, if we recall the definition of the exponential integral function (more information on this function is included in Appendix E),

$$-Ei(-x) = Ei(x) = \int_x^\infty \frac{e^{-\lambda}}{\lambda}\, d\lambda \tag{7.28}$$

we can cast eq. (7.27) in the following final form:

$$T = T_\infty + \frac{\dot{Q}}{4\pi kL} Ei\left(\frac{r^2}{4\alpha t}\right) \tag{7.29}$$

The second term on the right-hand side of expression (7.29) represents the temperature rise in the body caused by the heat source. Note that as $t \rightarrow \infty$ the exponential function approaches infinity in eq. (7.29), indicating that no steady state exists in the case of the line source, unlike in the case of the point source.

### Plane Heat Source of Constant Heat Production Rate

If the heat source is a plane of area $A$ producing heat at a rate $\dot{Q}$, the problem becomes one of unidirectional heat conduction in a semi-infinite body, initially isothermal at $T_\infty$, with constant heat transfer rate at its surface. An example of this type is a thick slab with a flat electric heater mounted on its surface. Proceeding as in Chapter 5, we easily realize that the mathematical model for the problem of interest in Cartesian coordinates is

$$\frac{1}{\alpha}\frac{\partial \Theta}{\partial t} = \frac{\partial^2 \Theta}{\partial x^2} \tag{7.30}$$

$$x = 0: \qquad -\dot{Q} = kA\frac{\partial \Theta}{\partial x} \tag{7.31}$$

$$x \rightarrow \infty: \qquad \Theta = 0 \tag{7.32}$$

$$t = 0: \qquad \Theta = 0 \tag{7.33}$$

where, again, $\Theta = T - T_\infty$. The mathematical model above "as is" does not accept a similarity solution. To facilitate the utilization of the similarity method, aided by boundary condition (7.31), we introduce the following variable:

$$S = \frac{\partial \Theta}{\partial x} + \frac{\dot{Q}}{kA} \tag{7.34}$$

Next, we need to prove that the variable $S$ indeed satisfies the conduction equation, that is, that

$$\frac{1}{\alpha}\frac{\partial S}{\partial t} - \frac{\partial^2 S}{\partial x^2} = 0 \tag{7.35}$$

Utilizing eq. (7.34), we cast eq. (7.35) in the following form:

$$\frac{\partial}{\partial x}\left(\frac{1}{\alpha}\frac{\partial \Theta}{\partial t} - \frac{\partial^2 \Theta}{\partial x^2}\right) = 0 \tag{7.36}$$

Equation (7.36) is indeed valid with the help of eq. (7.30). Therefore, the variable $S$ obeys the energy equation. Next, the boundary conditions for $S$ are recovered by using boundary conditions (7.31)–(7.33) together with the definition of $S$, eq. (7.34):

$$x = 0: \qquad S = 0 \tag{7.37}$$

$$x \rightarrow \infty: \qquad S = \frac{\dot{Q}}{kA} \tag{7.38}$$

$$t = 0: \qquad S = \frac{\dot{Q}}{kA} \tag{7.39}$$

The solution for the conduction model [eqs. (7.35) and (7.37)–(7.39)] has already been obtained in Section 5.6. The final result for $S$ is recovered directly from eq. (5.210):

$$S = \frac{\dot{Q}}{kA}\operatorname{erf}(\eta) \tag{7.40}$$

where the similarity variable $\eta$ is given by eq. (5.211).

To obtain the temperature distribution in the body of interest, we first combine eqs. (7.34) and (7.40):

$$\frac{d\Theta}{d\eta} = -\frac{\dot{Q}}{kA}2\sqrt{\alpha t}\operatorname{erfc}(\eta) \tag{7.41}$$

Expression (7.41) can be integrated directly in $\eta$ to yield $\Theta$. Omitting the details for brevity, we report the final expression for the temperature distribution:

$$T = T_\infty + \frac{\dot{Q}x}{kA}\left[\frac{2\sqrt{\alpha t}}{x}\frac{1}{\sqrt{\pi}}e^{-x^2/4\alpha t} - \operatorname{erfc}\left(\frac{x}{2\sqrt{\alpha t}}\right)\right] \tag{7.42}$$

The term added to $T_\infty$ on the right-hand side of eq. (7.42) is the temperature excess caused by the source. By examining the behavior of the temperature at large times ($t \rightarrow \infty$) we realize that in this limit the temperature increases as $t^{1/2}$ and no steady state is reached.

## 7.2 INSTANTANEOUS HEAT SOURCES (THERMAL EXPLOSIONS)

When the heat release by the source is instantaneous (say, at $t = 0$ a single explosion [4] takes place at a small region inside the body and the energy released by the explosion is diffused throughout the body thereafter), energy conservation requires that the total energy increase in the body at all times is constant and it equals the energy released by the source. This fact will be used in the solution process in place of the flux boundary conditions in the neighborhood of the source utilized in Section 7.1 dealing with continuous sources.

### Instantaneous Point Heat Source (Thermal Explosion at a Point Region)

The description of this problem is identical to that discussed in Section 7.1, with the difference that instead of having a source producing heat continuously, at $t = 0$ (only) an amount of energy denoted by $E_0$ is released by the source. This amount of energy raises the temperature of the body, is independent of time, and is related to the temperature of the body by the following expression:

$$\int_0^\infty 4\pi r^2 \rho C_p (T - T_\infty)\, dr = E_0 \tag{7.43}$$

This statement replaces eq. (7.2) in the heat transfer model for the continuous point source of Section 7.1 and the resulting model [eqs. (7.1), (7.3), (7.4), and (7.43)] describes the heat conduction process caused by a point explosion in a large body of initial temperature $T_\infty$. Proceeding as in Section 7.1, we can easily show that in terms of the function $u$ [eq. (7.5)] this heat conduction model becomes

$$\frac{1}{\alpha}\frac{\partial u}{\partial t} = \frac{\partial^2 u}{\partial r^2} \tag{7.44}$$

$$r \to \infty: \qquad u \to 0 \tag{7.45}$$

$$t = 0: \qquad u = 0 \tag{7.46}$$

$$\int_0^\infty ur\, dr = \frac{E_0}{4\pi\rho C_p} \tag{7.47}$$

This mathematical model does not accept a similarity solution, because of the form of the integral in eq. (7.47). To proceed, we notice that the general solution of eq. (7.44) according to the discussion in Sections 5.6 and 7.1 is

$$u = C_1 \operatorname{erf}\left(\frac{r}{2\sqrt{\alpha t}}\right) + C_2 \tag{7.48}$$

where $C_1$ and $C_2$ are constants. It can easily be shown that any order derivative of eq. (7.48) with respect to $r$ is also a solution to eq. (7.44). Taking the first and the second derivatives of eq. (7.48) with respect to $r$ yields, respectively, the following two solutions that satisfy eq. (7.44):

$$u = \frac{C}{\sqrt{\pi\alpha t}} \exp\left(-\frac{r^2}{4\alpha t}\right) \tag{7.49}$$

$$u = -\frac{Cr}{2\sqrt{\pi}(\alpha/t)^{3/2}} \exp\left(-\frac{r^2}{4\alpha t}\right) \tag{7.50}$$

All three expressions (7.48)–(7.50) satisfy the conduction equation (7.44) as well as conditions (7.45) and (7.46). However, only expression (7.50), if substituted in the integral (7.47), makes this integral a constant (independent of both $r$ and $t$). Note that this integral needs to be a constant since it equals a constant [the right-hand side of eq. (7.47)]. Based on the above, we postulate expression (7.50) to be the solution of the conduction model (7.44)–(7.47) and we obtain the constant by substituting eq. (7.50) into eq. (7.47). This last operation yields

$$-\frac{4C}{\sqrt{\pi}}\int_0^\infty \eta^2 e^{-\eta^2}\,d\eta = \frac{E_0}{4\pi\rho C_p} \tag{7.51}$$

Since the integral in eq. (7.51) equals $\frac{1}{4}\pi^{1/2}$, we recover

$$C = -\frac{E_0}{4\pi\rho C_p} \tag{7.52}$$

Therefore,

$$u = \frac{E_0 r}{8\rho C_p(\pi\alpha t)^{3/2}}\exp\left(-\frac{r^2}{4\alpha t}\right) \tag{7.53}$$

and

$$T = T_\infty + \frac{E_0}{8\rho C_p(\pi\alpha t)^{3/2}}\exp\left(-\frac{r^2}{4\alpha t}\right) \tag{7.54}$$

### Instantaneous Line Heat Source (Thermal Explosion at a Line Region)

If the explosion takes place at a line region of length $L$, the integral (7.43) in the earlier discussion needs to be replaced by

$$\int_0^\infty 2\pi r L\rho C_p(T - T_\infty)\,dr = E_0 \tag{7.55}$$

The mathematical model for this problem is identical to the earlier discussion for the continuous line source, with the energy conservation integral (7.55) replacing the boundary condition (7.20). Following that discussion, we postulate as the solution to the problem the first derivative of eq. (7.27) with respect to time, $t$. Note that it is this derivative that upon substitution will render the left-hand side of eq. (7.55) constant:

$$T - T_\infty = \frac{C}{t}\exp\left(-\frac{r^2}{4\alpha t}\right) \tag{7.56}$$

Combining eqs. (7.55) and (7.56) to obtain the constant $C$, we find that

$$4\alpha\pi\rho C_p LC\int_0^\infty e^{-\xi}\,d\xi = E_0 \tag{7.57}$$

Hence

$$C = \frac{E_0}{4\pi\alpha\rho C_p L} \tag{7.58}$$

Therefore, the temperature distribution for the line explosion case satisfying the conduction model consisting of eqs. (7.19), (7.21), (7.22), and (7.55) is given by

$$T = T_\infty + \frac{E_0/t}{4\pi\alpha\rho C_p L} \exp\left(-\frac{r^2}{4\alpha t}\right) \tag{7.59}$$

### Instantaneous Plane Heat Source (Thermal Explosion at a Plane Region)

In this case the instantaneous heat release $E_0$ takes place at a plane region of area $A$. The mathematical model for the heat conduction phenomenon is identical to that of the continuous plane source of Section 7.1 with the energy conservation integral

$$\int_0^\infty A\rho C_p(T - T_\infty)\, dx = E_0 \tag{7.60}$$

replacing the boundary condition (7.31). Proceeding as before, we postulate as the solution to the problem the first derivative of the general solution of eq. (7.30) with respect to $x$, that is,

$$T - T_\infty = \frac{C}{\sqrt{\pi\alpha t}} \exp\left(-\frac{x^2}{4\alpha t}\right) \tag{7.61}$$

This expression satisfies the conduction equation as well as conditions (7.32) and (7.33). In addition, it has the appropriate functional form to make the integral in eq. (7.60) independent of both $r$ and $t$. Combining eqs. (7.60) and (7.61), we obtain

$$C = \frac{E_0}{\rho C_p A} \tag{7.62}$$

Hence the temperature distribution in the body resulting from the plane explosion is

$$T = T_\infty + \frac{E_0}{\rho C_p A(\pi\alpha t)^{1/2}} \exp\left(-\frac{x^2}{4\alpha t}\right) \tag{7.63}$$

Note that while all the point, line, and plane heat sources exhibit the same exponential dependence of temperature on the similarity variable $r^2/4\alpha t$, they exhibit markedly different dependence of temperature on time. At a specific location ($r$ or $x$) and for the same time lapsed after the explosion, the temperature for the case of the point explosion will be the lowest of the three cases examined [$T \sim t^{-3/2}$, eq. (7.54)] and the temperature for the case of the plane explosion will be the highest [$T \sim t^{-1/2}$, eq. (7.63)]. The reason for this fact can be explained if one realizes that the spherical geometry provides the best thermal communication between the cold and the warm regions in the body, the plane geometry the worst, and the cylindrical geometry is in between.

## 7.3 MOVING SOURCES OF CONTINUOUS HEAT PRODUCTION RATE: LARGE-TIME (STEADY-STATE) LIMIT

As mentioned earlier, there exist several engineering applications in which heat is produced in the form of a moving source. Such applications are exemplified by welding, melting, the operation of cutting tools and lasers on workpieces in manufacturing, and various cases of tribology where moving surfaces are in contact [1–3]. In this section, following Sections 7.1 and 7.2, three basic moving heat sources of constant heat production rate are considered sequentially: a point source, a line source, and a plane source. In addition, for simplicity, only the *steady-state* temperature distributions are sought. The complete transient solution is discussed in the problems. Finally, it is assumed that the heat diffusion process is of the *boundary layer* type (i.e., heat diffusion in the direction of motion of the source will be neglected).

### Moving Point Source of Continuous Heat Production Rate

Consider the situation shown graphically in Fig. 7.2a. A continuous point source producing heat at a rate $\dot{Q}$ moves in the negative $x$-direction with velocity $U$. This situation can also be envisioned as one of a stationary source in an initially isothermal region ($T_\infty$) moving with velocity $U$ in the positive $x$-direction (Fig. 7.2b). Heat conduction takes place in the medium. It is convenient to consider the latter situation of a moving medium and a stationary source. The mathematical model of the heat conduction process in the thermally affected region is written in the cylindrical coordinates $(r, x)$ of Fig. 7.2 as follows:

$$\frac{1}{\alpha} U \frac{\partial T}{\partial x} = \frac{1}{r} \frac{\partial}{\partial r}\left(r \frac{\partial T}{\partial r}\right) \tag{7.64}$$

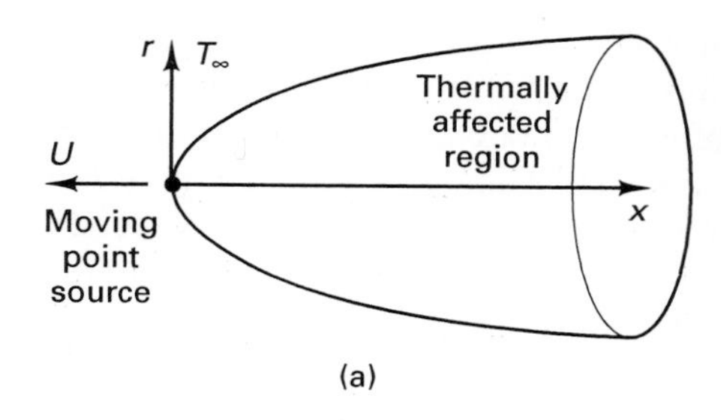


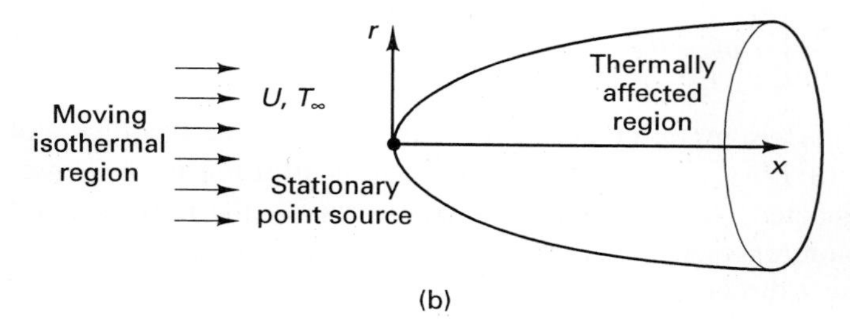


**Figure 7.2** (a) Moving point source in a stationary medium. (b) Stationary point source in a moving medium.

$$r \rightarrow \infty: \qquad T \rightarrow T_\infty \tag{7.65}$$

$$\dot{Q} = \int_0^\infty \rho C_p U(T - T_\infty) 2\pi r \, dr \tag{7.66}$$

Inherent in eq. (7.64) is the assumption that the heat conduction process in the medium is of the boundary layer type; hence the heat diffusion in the direction of motion of the medium is negligible [i.e., $(1/r)(\partial/\partial r)[r(\partial T/\partial r)] \gg \partial^2 T/\partial x^2$]. The left-hand side of eq. (7.64) represents the convection of thermal energy since the medium is moving. This convection effect is balanced by the radial conduction on the right-hand side of eq. (7.64). This equation is a special case of the more general energy equation accounting for the medium motion [5, 6] in the special case of only axial constant velocity and only radial conduction. The derivation of eq. (7.64) was discussed in Problem 2.1. Equation (7.66) states the fact that the heat generation per unit time at the source equals the heat content per unit time of any cross section of the thermally affected region. This condition implies that steady state has been established. To proceed, we note the apparent similarity between this model and the model for the instantaneous line source of Section 7.2 [eqs. (7.19), (7.21), and (7.55)]. The solution for the instantaneous line source [eq. (7.59)] can be used directly to yield the solution for steady-state temperature distribution in the case of the moving point source. After renaming variables as necessary, the final result reads

$$T - T_\infty = \frac{\dot{Q}}{4\alpha\pi\rho C_p x} \exp\left(-U\frac{r^2}{4\alpha x}\right) \tag{7.67}$$

### Moving Line Source of Continuous Heat Production Rate

This problem is shown schematically in Fig. 7.3a. The heat production rate of the line source is denoted by $\dot{Q}/L$ (W/m), where $L$ is the length of the source. Figure 7.3b shows the same problem but from the standpoint of a stationary source and a moving medium. Clearly, the temperature field in the thermally affected region is two-dimensional and it will be sufficient to obtain the temperature field in the arbitrary cross section defined by the $x$–$y$ Cartesian coordinate system of Fig. 7.3b. The mathematical model for the heat conduction process in this cross section is

$$\frac{U}{\alpha}\frac{\partial T}{\partial x} = \frac{\partial^2 T}{\partial y^2} \tag{7.68}$$

$$y \rightarrow \infty: \qquad T \rightarrow T_\infty \tag{7.69}$$

$$\int_{-\infty}^\infty \rho C_p U(T - T_\infty)\, dy = \frac{\dot{Q}}{L} \tag{7.70}$$

Inherent in eq. (7.68) is the assumption that the heat conduction in the medium is of the boundary layer type ($\partial^2 T/\partial y^2 \gg \partial^2 T/\partial x^2$) and the heat diffusion in the direction of motion ($x$) is neglected. As discussed in connection with the moving point source, the energy equation represents a balance between conduction in the $y$-direction and convection in the $x$-direction (see also Problem 2.1). The energy balance in

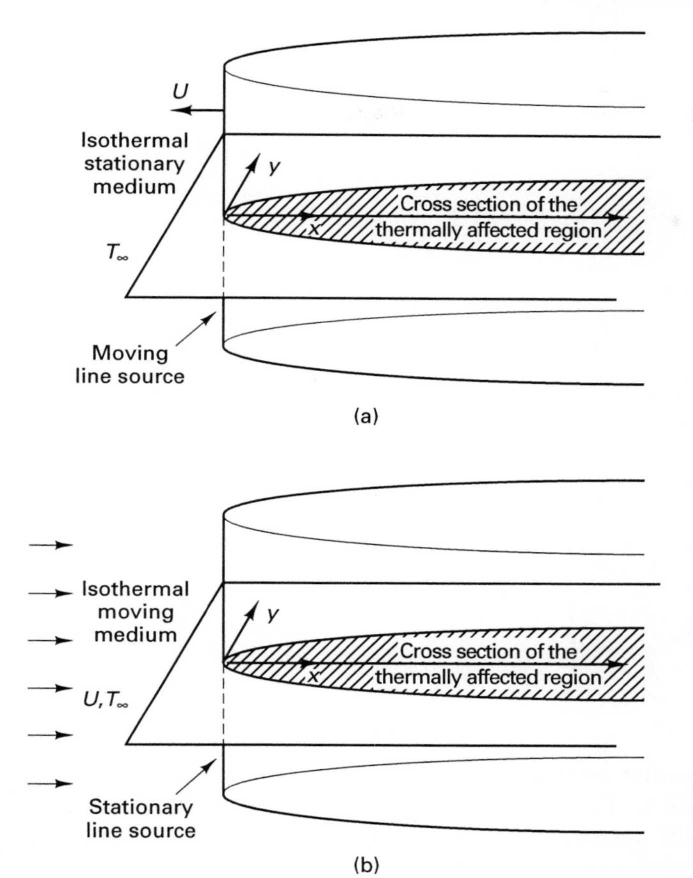


**Figure 7.3** (a) Moving line source in a stationary medium. (b) Stationary line source in a moving medium.

eq. (7.70) states that the energy rate per unit length which is released by the source in the *x–y* plane under consideration equals the energy rate content per unit length of any section of the thermally affected region in the direction perpendicular to the velocity of the medium.

Noting that except for differences in notation the model (7.68)–(7.70) is identical to the model for the instantaneous plane source of Section 7.2, we use eq. (7.63) directly and after renaming variables appropriately, we obtain the temperature distribution in the present case:

$$T = T_\infty + \frac{\dot{Q}/L}{\rho C_p(\pi \alpha U x)^{1/2}} \exp\left(-U\frac{y^2}{4\alpha x}\right) \tag{7.71}$$

### Moving Plane Source of Continuous Heat Production Rate

This case can be visualized as a very large (infinite) plane moving into medium with velocity $U$, while producing heat at a rate $\dot{Q}$. Figure 7.4 shows the alternative way that one can think of this problem. The plane is stationary and the medium is mov-

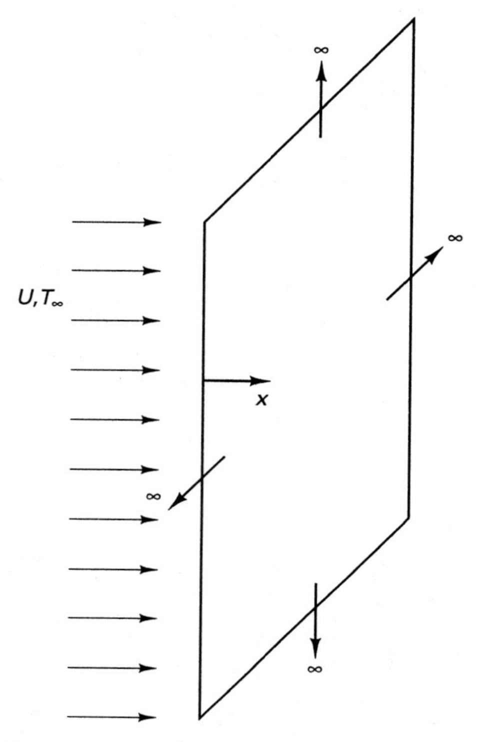


**Figure 7.4** Moving plane source.

ing to the right (in the positive $x$-direction) with velocity $U$. The temperature of the medium prior to the presence of the source is assumed to be uniform ($T_\infty$). The moving plane source can model approximately, for example, the heat conduction during the motion of a flat freezing front or the motion of a wave or flame carrying an exothermic reaction. Based on the discussion earlier in this section, the mathematical model of the heat conduction process in the steady state is

$$\frac{U}{\alpha}\frac{\partial \Theta}{\partial x} = \frac{\partial^2 \Theta}{\partial x^2} \tag{7.72}$$

$$x \to -\infty: \qquad \Theta = 0 \tag{7.73}$$

$$x \geq 0: \qquad \rho C_p A U \Theta = \dot{Q} \tag{7.74}$$

where $\Theta = T - T_\infty$ is the temperature excess in the medium caused by the source. Two distinct regions of conduction can be defined: first, the region $x \geq 0$, which is the part of the body through which the plane source has already traveled; and second, the region $x < 0$, which is the part of the body through which the source has not yet traveled. We consider the two regions separately.

**1.** *The region $x > 0$.* This region is isothermal since the source is an infinite plane. Under steady-state conditions every plane in the region $x \geq 0$ parallel to the source plane, should contain identically the energy rate released by the source plane.

Hence no temperature gradients can be sustained in this region and the temperature will be a constant evaluated directly from eq. (7.74):

$$T = T_\infty + \frac{\dot{Q}}{\rho C_p A U} \tag{7.75}$$

**2.** *The region x < 0.* The solution of eq. (7.72) is

$$\Theta = C\frac{\alpha}{U}e^{(U/\alpha)x} + D \tag{7.76}$$

Equation (7.73) yields $D = 0$. The constant $C$ is obtained by evaluating $\Theta$ at $x = 0$ from eq. (7.76) and by equating this result to that from eq. (7.75) to satisfy temperature continuity at the plane source. The final result for $C$ reads

$$C = \frac{\dot{Q}}{\rho C_p \alpha A} \tag{7.77}$$

From the above,

$$T = T_\infty + \frac{\dot{Q}}{\rho C_p U A}e^{(U/\alpha)x} \tag{7.78}$$

Note that since $x$ is negative, the temperature excess decays exponentially as we move away from the source. The larger the source velocity and the smaller the medium thermal diffusivity, the faster the decay.

## PROBLEMS

**7.1.** Obtain the complete transient solution for the temperature field in an initially isothermal medium at $T_\infty$ if a continuous point source moving with velocity $U$ starts producing heat at a rate $\dot{Q}$ [7, 8].

**7.2.** Obtain the complete transient solution for the temperature field in an initially isothermal medium at $T_\infty$ if a continuous line source moving with velocity $U$ starts producing heat per unit length at a rate $\dot{Q}'$ [7, 8].

**7.3.** A barrel of nuclear waste buried deep under the surface of the earth can be modeled as a point source producing heat continuously at a rate $\dot{Q}$. The temperature of the ground is measured by a temperature sensor placed at a distance $d$ from the barrel. It is desired that the temperature of the ground at this distance does not exceed a threshold value of $T_m$. Can you estimate the time ($t_m$) it will take for the sensor temperature to reach $T_m$? How does this time depend on the properties of the soil? Can you make recommendations based on this dependence such that the barrel is buried in a region that will yield a large $t_m$ value?

**7.4.** Re-solve Problem 7.3, but this time by assuming that a power cable is buried underground producing heat at a rate per unit length $\dot{Q}'$.

**7.5.** During a laser welding process, the seam between two identical thin metal sheets that are being welded together moves with velocity $U$ as it is stricken by the laser beam

(Fig. P7.5). The sheets were initially isothermal at room temperature, $T_\infty$. Assume that a percentage $\eta$ of the power of the laser beam ($\dot{Q}$) that is deposited at the seam is conducted through the sheets. Let $T_m$ denote the melting temperature of the metal. Note also that if the metal is heated above a temperature $T_a (T_m > T_a)$, its microscopic structure changes. Define the region of the metal sheets in which you expect that the structure of the metal will change because of the heating process (this region is termed the *heat-affected region*). Use realistic numerical data of your choice [1, 9] (laser power, $\eta$, material, $T_m$, $T_a$, etc.) and show graphically the heat-affected region corresponding to these data. Comment on the expected accuracy of your solution. What effects did you neglect? How can you account for some of these effects?

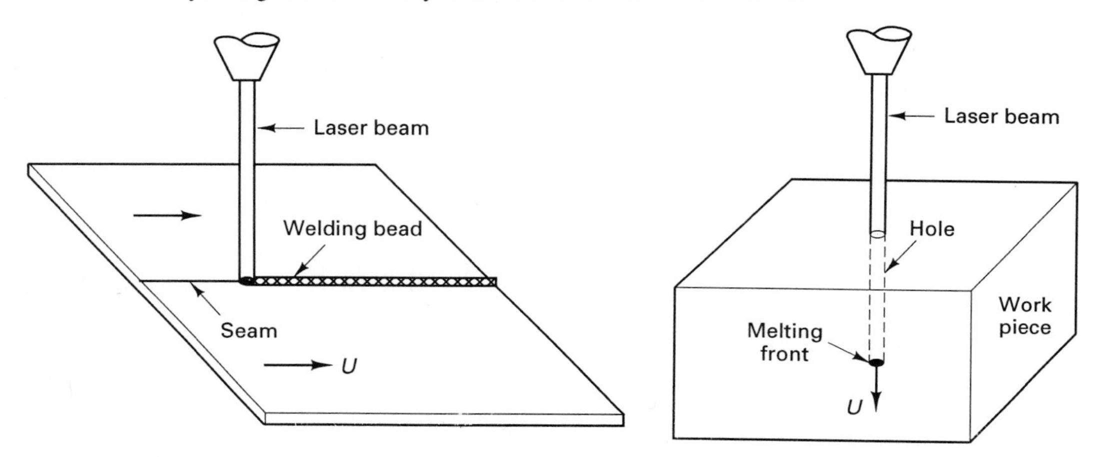


**Figure P7.5** Schematic of the laser welding process.

**Figure P7.7** Schematic of the laser drilling process.

**7.6.** Re-solve Problem 7.5, this time by assuming that the welding occurs at the surface of two *thick* metal plates.

**7.7.** In a laser drilling process, a powerful laser beam strikes a stationary metal workpiece initially at room temperature ($T_\infty$) as shown in Fig. P7.7. As a result of the beam energy ($\dot{Q}$), the metal stricken by the beam melts, evaporates, and eventually, a hole is created. Assume that the "tip" of the laser beam striking the melting and evaporation front advances in the metal body with constant velocity $U$. A large percentage ($\eta$) of the beam energy is conducted through the workpiece. What is the heat-affected region in the workpiece? Note that this region and the associated temperatures were defined earlier in Problem 7.5. Use reasonable data of your choice [1, 9] (much as in Problem 7.5) and show graphically the heat-affected region corresponding to these data. How accurate do you expect your solution to be? List some of the approximations inherent in your model and describe how you can account for some of these approximations to improve the accuracy of your solution.

**7.8.** In an experimental large combustion chamber containing a mixture of fuel and oxidizer of temperature $T_i$, ignition takes place at one end such that the flame propagates through the chamber as a plane front with velocity $U$ generating energy at a rate $\dot{Q}$. What is the temperature in the chamber at a location $L$ from the initial ignition point? If the combustion chamber is modeled as a cylinder of base area $A$ and height $H$, can

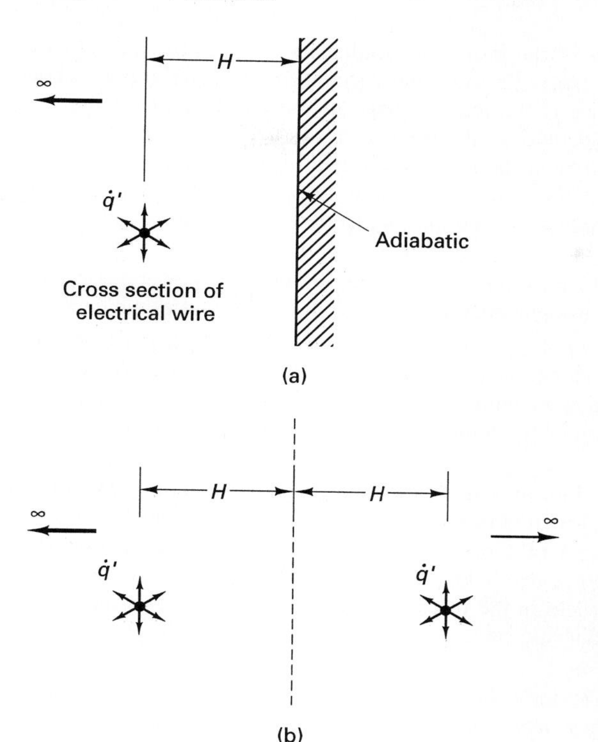


**Figure P7.11** (a) Cross section of heat-generating electrical wire near an adiabatic surface. (b) Cross section of heat-generating electrical wire and its mirror image about an insulated surface.

you estimate the energy of the exhaust gases at the end of the combustion process? After solving the problem for an infinitely thin flame front, correct your solution by accounting for the fact that the flame front has a thickness denoted by $d$.

**7.9.** In a strength of materials experiment, an amount of explosive is inserted deep inside a large block of a composite material of temperature $T_i$ that is being tested. The volume of the explosive is very small compared to the volume of the composite material under testing. Upon detonation, explosion takes place and an amount of energy $E_0$ is generated almost instantaneously at the location of the explosive. If $T_a$ is the temperature of the composite material that defines the heat-affected region (see Problem 7.5), obtain the extend of this region. Using reasonable numerical data of your choice (for the material properties and the energy generated by the explosive), show graphically the heat-affected region. What is the maximum temperature at any distance $r$ from the explosive?

**7.10.** A very short pulse of high-intensity current is passed through a thin wire buried in a thick fiberglass insulation layer. As a result, the wire generates (almost instantaneously) heat $E_0 = 1$ kJ per meter of its length. Obtain the temperature distribution in the insulation and graph it as a function of distance from the wire for characteristic times. What is the maximum temperature occurring at any distance $r$ from the wire? Will the insulation burn?

**7.11.** A thin electrical wire is buried at a distance $H$ from the surface of a very thick insulation layer, as shown in Fig. P7.11a. The wire is producing heat constantly at a rate

$\dot{q}'$ (W/m). The surface of the insulation is adiabatic. Obtain the temperature distribution in the insulation. [*Hint:* Consider the system of two electrical wires shown in Fig. P7.11b. The right wire is the mirror image of the left wire about the vertical axis (from symmetry considerations this axis is adiabatic). Make use of the fact that due to the linearity of the problem the temperature field in the system of Fig. P7.11b is the linear superposition of the temperature fields caused by the two wires when each is present alone in the system, and obtain the temperature distribution in the insulation of Fig. P7.11a.]

**7.12.** Re-solve Problem 7.11 for the case of a wire that due to a power surge produces heat not continuously but instantaneously (at $t = 0$) in the amount $E_0$ (joules).

**7.13.** Reconsider the problem of a point source of continuous heat production rate discussed in Section 7.1. Assume that the point source is buried at a distance $H$ from an adiabatic surface in a very large medium. Utilize the concept of mirror images discussed in Problem 7.11 and obtain the temperature distribution in the medium containing the source.

**7.14.** Re-solve Problem 7.13 for the case of thermal explosion at a point region (Section 7.2) when the thermal explosion occurs at a distance $H$ from an adiabatic surface.

**7.15.** Reconsider the problem of a plane heat source of constant production rate (Section 7.1). Assume that this source is located at a distance $H$ from an adiabatic surface. Obtain the temperature field in the medium containing the source in the region between the source and the adiabatic surface utilizing the concept of mirror images discussed in Problem 7.11.

**7.16.** Re-solve Problem 7.15 for a thermal explosion at a plane region (Section 7.2) for the case where the plane source is located at a distance $H$ from an adiabatic surface in a semi-infinite medium.

## REFERENCES

1. N. Rykalin, A. Uglov, I. Zuev, and A. Kokora, *Laser and Electron Beam Material Processing Handbook,* MIR Publishers, Moscow, 1988.
2. J. Halling (Ed.), *Principles of Tribology,* Macmillan, London, 1975.
3. E. F. Nippes, Coordinator, *Metals Handbook,* Vol. 6 (Welding, Brazing, and Soldering), American Society for Metals, Metals Park, OH, 1983.
4. K. K. Kuo, *Principles of Combustion,* Wiley, New York, 1986.
5. A. Bejan, *Convection Heat Transfer,* Wiley, New York, 1984.
6. V. S. Arpaci, *Convection Heat Transfer,* Prentice Hall, Englewood Cliffs, NJ, 1984.
7. U. Grigull and H. Sandner, *Heat Conduction,* Hemisphere, New York, 1984.
8. P. J. Schneider, *Conduction Heat Transfer,* Addison-Wesley, Reading, MA, 1955.
9. H. F. Boyer and T. L. Gall, Eds., *Metals Handbook,* desk edition, American Society for Metals, Metals Park, OH, 1985.

CHAPTER 8

# SOLUTION OF TRANSIENT HEAT CONDUCTION PROBLEMS UTILIZING LAPLACE TRANSFORMS

In this chapter, transient heat conduction problems are studied and solved by utilizing Laplace transforms. In particular, unidirectional transient conduction problems (of the type studied in Chapters 5 to 7) will be considered in the three common coordinate systems for engineering applications: Cartesian, cylindrical and spherical.

The Laplace transform of a function (temperature) $T(x, t)$ with respect to time ($t$) is an integral defined as follows [1–3]:

$$\mathcal{L}_t[T(x, t)] = \bar{T}(x, s) = \int_{t=0}^{\infty} e^{-st} T(x, t)\, dt \tag{8.1}$$

The subscript $t$ in eq. (8.1) denotes the fact that the Laplace transform is defined with respect to the time variable. The parameter $s$ is assumed to be real or complex. Clearly, since the Laplace transform is an integral between the limits of zero and infinity, it has to be defined with respect to a variable that spans from zero to infinity. The time, $t$, provides such a variable in transient heat conduction problems. Even though a space variable may span from zero to infinity (when one deals with a thick semi-infinite body, for example), to facilitate the solvability of heat conduction problems one needs to utilize Laplace transforms with respect to **time.** The reason for this choice has to do with the resulting mathematical model, after the Laplace transform is applied to the original heat conduction equation, boundary conditions, and initial conditions, as will become apparent later in this chapter.

A great advantage of utilizing Laplace transforms to solve heat conduction problems is that one does not need to worry about the presence of a homogeneous direction which is required in the method of separation of variables (Chapter 5). In addition, this direction does not have to be of finite extent. Note that for the method of separation of variables to yield a series solution of the type discussed in Chapter 5, for example, the extent (thickness) of the body in this direction has to be finite.

A disadvantage of the Laplace transform method in solving heat conduction problems is that inversion of the solution for the Laplace transform of the temperature is required to yield the solution for the temperature itself. This inversion is usually easy if it can be performed with the help of inversion tables (Appendix F). Otherwise, the algebra is rather involved and requires knowledge of integration techniques in the complex plane.

The foregoing comments on the advantages and disadvantages of the Laplace transform method of solution will become clear later in this chapter when specific examples are solved. Before proceeding with the material of the remainder of this chapter, it is recommended that the reader carefully review Appendix F, which contains the essential properties and definitions of Laplace transforms.

The solution of heat conduction problems utilizing Laplace transforms can be organized in a sequence of four steps:

■ ***Step 1***

Obtain the mathematical model of the heat conduction problem of interest (energy equation and boundary and initial conditions) in terms of the temperature, $T(x, t)$.

■ ***Step 2***

Take the Laplace transform of the energy equation as well as of the initial and boundary conditions. One need *not* be overly concerned with the homogeneity of the boundary conditions. It is not essential for the success of the method. At the end of this step the model (equation and boundary conditions) for the Laplace transform of the temperature, $\bar{T}(x, s)$, is obtained.

■ ***Step 3***

Solve the model described above for $\bar{T}(x, s)$.

■ ***Step 4***

Invert the expression for $\bar{T}(x, s)$ utilizing inversion tables or by employing the inversion theorem of Appendix F and obtain the temperature distribution for the problem of inerest. In this book, the inversion process is performed with the help of inversion tables. It is outside the scope of this book to provide the background necessary for the student to perform the inversion by directly evaluating the appropriate integral in Appendix F [eq. (F.2)]. However, if the student already has such a background, Appendix F [eq. (F.2)] does provide an alternative way to perform the inversion process.

## 8.1 TRANSIENT HEAT CONDUCTION IN CARTESIAN COORDINATES

The use of Laplace transforms in solving transient unidirectional heat conduction problems is illustrated with the help of the following three examples.

### Example 8.1

Re-solve Example 5.1 utilizing the Laplace transform method.

**(a)** Assume that the heat transfer coefficient, $h$, is very large.
**(b)** Obtain the solution for a finite value of $h$.

**Solution**

(a) First, the case of a very large heat transfer coefficient will be considered. In this case, the surface temperature of the sheet equals the bath temperature, $T_\infty$.

■ ***Step 1***

The mathematical model of the problem is given by eqs. (5.4)–(5.7). For clarity, we repeat these equations below and also account for the fact that $h \to \infty$.

$$\frac{1}{\alpha}\frac{\partial T}{\partial t} = \frac{\partial^2 T}{\partial x^2} \tag{8.2}$$

$$x = 0: \qquad \frac{\partial T}{\partial x} = 0 \tag{8.3}$$

$$x = a: \qquad T = T_\infty \tag{8.4}$$

$$t = 0 \qquad T = T_0 \tag{8.5}$$

■ ***Step 2***

First, we take the Laplace transform of both sides of the heat conduction equation (8.2):

$$\mathcal{L}\left[\frac{1}{\alpha}\frac{\partial T}{\partial t}\right] = \mathcal{L}\left[\frac{\partial^2 T}{\partial x^2}\right] \tag{8.6}$$

Note that the subscript $t$ has been dropped from the Laplace transform symbol in eq. (8.6). Since only Laplace transforms in time will be taken in this chapter, this subscript is no longer necessary. Utilizing the properties of Laplace transforms outlined in Appendix F [eqs. (F.3), (F.4), and (F.7)], we obtain

$$\frac{1}{\alpha}[s\bar{T}(x, s) - T(x, 0)] = \frac{d^2\bar{T}(x, s)}{dx^2} \tag{8.7}$$

Taking into account the initial condition, eq. (8.5) yields

$$\frac{d^2\bar{T}(x, s)}{dx^2} - \frac{s}{\alpha}\bar{T}(x, s) = \frac{-T_0}{\alpha} \tag{8.8}$$

To complete the problem formulation, we take the Laplace transform of boundary conditions (8.3) and (8.4). Utilizing eqs. (F.3) and (F.6) and property 1 in Table F.1 in Appendix F yields

$$x = 0: \qquad \frac{d\bar{T}}{dx} = 0 \tag{8.9}$$

$$x = a: \qquad \bar{T} = \frac{T_\infty}{s} \tag{8.10}$$

At this point, the model for the Laplace transform of the temperature ($\bar{T}$) is completed and it consists of eqs. (8.8)–(8.10). Note further that the initial condition of the problem [eq. (8.5)] was already satisfied in deriving eq. (8.8).

### ■ *Step 3*

Equation (8.8) is simply a second-order ordinary differential equation with constant coefficients. Its solution reads (Appendix B)

$$\bar{T}(x, s) = A \sinh\left(x\sqrt{\frac{s}{\alpha}}\right) + B \cosh\left(x\sqrt{\frac{s}{\alpha}}\right) + \frac{T_0}{s} \tag{8.11}$$

Applying boundary conditions (8.9) and (8.10) yields

$$A = 0 \qquad B = \frac{T_\infty - T_0}{s \cosh(a\sqrt{s/\alpha})} \tag{8.12, 8.13}$$

Combining eqs. (8.11)–(8.13) yields

$$\bar{T}(x, s) = (T_\infty - T_0)\frac{\cosh(x\sqrt{s/\alpha})}{s \cosh(a\sqrt{s/\alpha})} + \frac{T_0}{s} \tag{8.14}$$

At this point, we have obtained an expression for the Laplace transform of the temperature. Inverting this expression will result in the temperature field in the body.

### ■ *Step 4*

Taking the inverse Laplace transform of eq. (8.14) yields

$$\begin{aligned} \mathcal{L}^{-1}[\bar{T}(x, s)] &= T(x, t) \\ &= \mathcal{L}^{-1}\left[(T_\infty - T_0)\frac{\cosh(x\sqrt{s/\alpha})}{s \cosh(a\sqrt{s/\alpha})}\right] + \mathcal{L}^{-1}\left[\frac{T_0}{s}\right] \end{aligned} \tag{8.15}$$

The inversion of the two terms on the right-hand side of eq. (8.15) is performed with the help of Table F.1. Utilizing properties 94 and 1 in Table F.1 gives

$$\mathcal{L}^{-1}\left[(T_\infty - T_0)\frac{\cosh(x\sqrt{s/\alpha})}{s\cosh(a\sqrt{s/\alpha})}\right] \tag{8.16}$$
$$= (T_\infty - T_0)\left\{1 + \frac{4}{\pi}\sum_{n=1}^{\infty}\frac{(-1)^n}{2n-1}\exp\left[\frac{-(2n-1)^2\pi^2\alpha t}{4a^2}\right]\cos\frac{(2n-1)\pi x}{2a}\right\}$$

$$\cdot\ \mathcal{L}^{-1}\left[\frac{T_0}{s}\right] = T_0 \tag{8.17}$$

Combining eqs. (8.15)–(8.17), we obtain the final expression for the temperature field:

$$T(x, t) = T_\infty + \frac{4(T_\infty - T_0)}{\pi}\sum_{n=1}^{\infty}\frac{(-1)^n}{2n-1}\exp\left[\frac{-(2n-1)^2\pi^2\alpha t}{4a^2}\right] \cdot \cos\frac{(2n-1)\pi x}{2a} \tag{8.18}$$

It can be easily shown that eq. (5.28) becomes identical to eq. (8.18) in the limit of a very large heat transfer coefficient ($h \to \infty$).

Finally, the reader can show that the algebra involved in the solution process of this problem simplifies somewhat if a transformed temperature $\Theta = T - T_0$ is introduced in step 1 and the conduction model for $\Theta$ is solved in steps 2 to 4.

(b) If the heat transfer coefficient is finite, the problem becomes identical to what was considered in Example 5.1.

■ ***Step 1***

The mathematical model of the problem is given by eqs. (5.4)–(5.7)

■ ***Step 2***

Proceeding as in case (a) above, it is straightforward to show that the mathematical model of the Laplace transform of the temperature field is

$$\frac{d^2\bar{T}}{dx^2} - \frac{s}{\alpha}\bar{T} + \frac{T_0}{\alpha} = 0 \tag{8.19}$$

$$x = 0: \quad \frac{d\bar{T}}{dx} = 0 \tag{8.20}$$

$$x = a: \quad -k\frac{d\bar{T}}{dx} = h\left(\bar{T} - \frac{T_\infty}{s}\right) \tag{8.21}$$

■ ***Step 3***

The solution to eq. (8.19) for $\bar{T}$ is given (again) by eq. (8.11). Applying the boundary conditions (8.20) and (8.21) yields

$$A = 0 \tag{8.22}$$

$$B = \frac{h(T_\infty - T_0)}{hs\cosh(a\sqrt{s/\alpha}) + (k/\sqrt{\alpha})\,s^{3/2}\sinh(a\sqrt{s/\alpha})} \tag{8.23}$$

Combining eqs. (8.11), (8.22), and (8.23), we obtain the expression for the Laplace transform of the temperature field:

$$\bar{T} = \frac{h(T_\infty - T_0)\cosh(x\sqrt{s/\alpha})}{hs\cosh(a\sqrt{s/\alpha}) + (k/\sqrt{\alpha})s^{3/2}\sinh(a\sqrt{s/\alpha})} + \frac{T_0}{s} \tag{8.24}$$

■ ***Step 4***

The expression (8.24) is rather complex and cannot be inverted with a straightforward utilization of the material in Apendix F. This fact illustrates the main difficulty inherent in using the Laplace transform method to solve heat conduction problems. Even for rather simple conduction applications (such as the present problems) it may not be sufficient to utilize standard inversion tables to obtain the temperature field and it may be necessary to evaluate the inversion integral [eq. (F.2), Appendix F]. For completeness, we report the final result of this evaluation:

$$T(x, t) = T_\infty + 2(T_0 - T_\infty)\sum_{n=1}^{\infty}\frac{\sin(\lambda_n a)}{\lambda_n a + \sin(\lambda_n a)\cos(\lambda_n a)}\cos(\lambda_n x)e^{-\alpha\lambda_n^2 t} \tag{8.25}$$

where $\lambda_n$ is obtained from the following algebraic equation:

$$\lambda_n \tan(\lambda_n a) = \frac{h}{k} \tag{8.26}$$

This expression is identical to eq. (5.28) obtained with the method of separation of variables.

**Example 8.2**

Re-solve Example 5.7 utilizing the Laplace transform method.

**Solution**

■ ***Step 1***

The mathematical model of the problem is given by eqs. (5.187)–(5.190).

■ ***Step 2***

Taking the Laplace transform of both sides of eq. (5.187) yields

$$\frac{1}{\alpha}[s\bar{T}(x, s) - T(x, 0)] = \frac{d^2\bar{T}(x, s)}{dx^2} \tag{8.27}$$

After accounting for the initial condition (5.190), eq. (8.27) becomes

$$\frac{d^2\bar{T}(x, s)}{dx^2} - \frac{s}{\alpha}\bar{T}(x, s) + \frac{T_i}{\alpha} = 0 \tag{8.28}$$

The Laplace transform of boundary conditions (5.188) and (5.189) yields

$$\bar{T}(0, s) = \frac{T_0}{s} \tag{8.29}$$

$$\bar{T}(\infty, s) = \frac{T_i}{s} \tag{8.30}$$

The solution to eq. (8.28) is

$$\bar{T}(x, s) = Ae^{x\sqrt{s/\alpha}} + Be^{-x\sqrt{s/\alpha}} + \frac{T_i}{s} \tag{8.31}$$

Combining eq. (8.31) with the boundary conditions eqs. (8.29) and (8.30), we obtain

$$A = 0 \tag{8.32}$$

$$B = \frac{T_0 - T_i}{s} \tag{8.33}$$

Based on eqs. (8.31)–(8.33), the final expression for the Laplace transform of the temperature distribution becomes

$$\bar{T}(x, s) = \frac{T_0 - T_i}{s}e^{-x\sqrt{s/\alpha}} + \frac{T_i}{s} \tag{8.34}$$

■ ***Step 3***

To obtain the temperature distribution in the body, we take the inverse Laplace transform of both sides of eq. (8.34):

$$\mathcal{L}^{-1}[\bar{T}(x, s)] = T(x, t) = \mathcal{L}^{-1}\left[\frac{T_0 - T_i}{s}e^{-x\sqrt{s/\alpha}}\right] + \mathcal{L}^{-1}\left[\frac{T_i}{s}\right] \tag{8.35}$$

Utilizing properties 68 and 1 from Table F.1 of Appendix F, we obtain

$$\mathcal{L}^{-1}\left[\frac{T_0 - T_i}{s}e^{-x\sqrt{s/\alpha}}\right] = (T_0 - T_i)\,\text{erfc}\left[\frac{x}{2\sqrt{\alpha t}}\right] \tag{8.36}$$

$$\mathcal{L}^{-1}\left[\frac{T_i}{s}\right] = T_i \tag{8.37}$$

Substituting eqs. (8.36) and (8.37) into eq. (8.35) results into the sought-after expression for the temperature:

$$T(x, t) = (T_0 - T_i)\,\mathrm{erfc}\left(\frac{x}{2\sqrt{\alpha t}}\right) + T_i \tag{8.38}$$

Clearly, the temperature expression of eq. (8.38) is identical to the temperature result of eq. (5.210) obtained with the similarity method.

**Example 8.3**

Re-solve Example 6.2 utilizing the Laplace transform method.

**Solution**

■ ***Step 1***

To simplify the algebra as well as to facilitate the comparison of the present method to that of Chapter 6 (Duhamel's theorem), we utilize the dimensionless model of the problem reported by eqs. (6.47)–(6.50).

■ ***Step 2***

Taking the Laplace transform of the heat conduction equation (6.47) and boundary conditions (6.49) and (6.50) yields

$$\frac{d^2\overline{\Theta}}{d\hat{x}^2} = \hat{s}\overline{\Theta} \tag{8.39}$$

$$\hat{x} = 0: \qquad \overline{\Theta} = \frac{2\pi}{s^2 + (2\pi)^2} \tag{8.40}$$

$$\hat{x} \to \infty: \qquad \overline{\Theta} = 0 \tag{8.41}$$

In obtaining eqs. (8.39)–(8.41), Table F.1 of Appendix F as well as the initial condition (6.48) were utilized in the manner explained in the earlier examples of this chapter.

■ ***Step 3***

The solution of the model (8.39)–(8.41) is straightforward (Appendix B). The final result for the Laplace transform of the temperature reads

$$\overline{\Theta}(\hat{x}, \hat{s}) = 2\pi\frac{e^{-\hat{x}\sqrt{\hat{s}}}}{\hat{s}^2 + (2\pi)^2} \tag{8.42}$$

■ ***Step 4***

Taking the inverse Laplace transform of both sides of eq. (8.42), we obtain

$$\overline{\Theta}(\hat{x}, \hat{s}) = 2\pi\, \mathcal{L}^{-1}\left[\frac{e^{-\hat{x}\sqrt{\hat{s}}}}{\hat{s}^2 + (2\pi)^2}\right] \tag{8.43}$$

To evaluate the inverse Laplace transform, we utilize Table F.1 as well as the convolution theorem, eq. (F.1) in Appendix F. First, utilizing properties 66 and 6 from Table F.1, we obtain

$$\mathcal{L}^{-1}[e^{-\hat{x}\sqrt{\hat{s}}}] = \frac{\hat{x}}{2\sqrt{\pi \hat{t}^3}}e^{-\hat{x}^2/4\hat{t}} \tag{8.44}$$

$$\mathcal{L}^{-1}\left[\frac{1}{\hat{s}^2 + (2\pi)^2}\right] = \frac{\sin(2\pi\hat{t})}{2\pi} \tag{8.45}$$

Next, by a straightforward application of the convolution theorem [eq. (F.11)],

$$\mathcal{L}^{-1}\left[\frac{e^{-\hat{x}\sqrt{\hat{s}}}}{\hat{s}^2 + (2\pi)^2}\right] = \int_{\tau=0}^{\hat{t}} \frac{\sin(2\pi\tau)}{2\pi}\, \frac{\hat{x}}{2\sqrt{\pi}(\hat{t} - \tau)^{3/2}} e^{-\hat{x}^2/4(\hat{t}-\tau)}\, d\tau \tag{8.46}$$

Finally, substituting eq. (8.46) into eq. (8.43), we obtain the final expression for the temperature field:

$$\Theta(\hat{x}, \hat{t}) = \frac{\hat{x}}{2\sqrt{\pi}} \int_{\tau=0}^{\hat{t}} (\hat{t} - \tau)^{-3/2} e^{-\hat{x}^2/4(\hat{t}-\tau)} \sin(2\pi\tau)\, d\tau \tag{8.47}$$

This result is identical to eq. (6.57a).

## 8.2 TRANSIENT HEAT CONDUCTION IN CYLINDRICAL COORDINATES

The utilization of the Laplace transform method for the solution of transient heat conduction problems in cylindrical coordinates follows the procedure outlined in Section 8.1 for Cartesian coordinates. The differences are mainly algebraic. The following examples illustrate this point.

### Example 8.4

Reconsider the transient heat conduction problem of Example 5.2. Obtain the temperature distribution in the rod utilizing the Laplace transform method for the case of a very large heat transfer coefficient ($h \rightarrow \infty$).

**Solution**

■ ***Step 1***

The mathematical model of the problem is

$$\frac{\partial^2\Theta}{\partial r^2} + \frac{1}{r}\frac{\partial\Theta}{\partial r} = \frac{1}{\alpha}\frac{\partial\Theta}{\partial t} \tag{8.48}$$

$$r = 0: \qquad \frac{\partial\Theta}{\partial r} = 0 \tag{8.49}$$

$$r = R: \qquad \Theta = 0 \tag{8.50}$$

$$t = 0: \qquad \Theta = \Theta_0 \tag{8.51}$$

The mathematical model is essentially identical to that of eqs. (5.32)–(5.35) with the exception that it accounts for the fact that the heat transfer coefficient at the rod surface is very large. For algebraic convenience the temperature $\Theta = T - T_\infty$ was introduced.

■ ***Step 2***

Taking the Laplace transform of both sides of the heat conduction equation (8.48) and utilizing eqs. (F.4), (F.6), and (F.7) yields

$$\frac{d^2\overline{\Theta}(r, s)}{dr^2} + \frac{1}{r}\frac{d\overline{\Theta}(r, s)}{dr} = \frac{1}{\alpha}[s\overline{\Theta}(r, s) - \Theta(r, 0)] \tag{8.52}$$

Recalling the initial condition (8.51), eq. (8.52) becomes

$$\frac{d^2\overline{\Theta}}{dr^2} + \frac{1}{r}\frac{d\overline{\Theta}}{dr} - \frac{s}{\alpha}\overline{\Theta} = -\frac{\overline{\Theta}_0}{\alpha} \tag{8.53}$$

To complete the mathematical model for $\overline{\Theta}$ we take the Laplace transform of the boundary conditions (8.49) and (8.50):

$$r = 0: \qquad \frac{d\overline{\Theta}}{dr} = 0 \tag{8.54}$$

$$r = R: \qquad \overline{\Theta} = 0 \tag{8.55}$$

■ ***Step 3***

The solution to eq. (8.53) is the general solution to the corresponding homogenous equation plus a particular solution of the complete equation. The homogeneous equation corresponding to eq. (8.53) is

$$\frac{d^2\overline{\Theta}_h}{dr^2} + \frac{1}{r}\frac{d\overline{\Theta}_h}{dr} - \frac{s}{\alpha}\overline{\Theta}_h = 0 \tag{8.56}$$

This equation can be rewritten as

$$\left(i\sqrt{\frac{s}{\alpha}}r\right)^2 \frac{d^2\overline{\Theta}_h}{d(i\sqrt{s/\alpha}\, r)^2} + \left(i\sqrt{\frac{s}{\alpha}}r\right)\frac{d\overline{\Theta}_h}{d(i\sqrt{s/\alpha}\, r)} + \left(i\sqrt{\frac{s}{\alpha}}r\right)^2\overline{\Theta}_h = 0 \tag{8.57}$$

The solution to this equation is obtained directly from Appendix A [eqs. (A.1) and (A.2)]:

$$\overline{\Theta}_h = C_1 J_0\left(i\sqrt{\frac{s}{\alpha}}r\right) + C_2 Y_0\left(i\sqrt{\frac{s}{\alpha}}r\right) \tag{8.58}$$

A particular solution to eq. (8.53) is (by inspection)

$$\overline{\Theta}_p = \frac{\Theta_0}{s} \tag{8.59}$$

References [1–3] are recommended for systematic methodologies on obtaining particular solutions of nonhomogeneous equations. The general solution for $\overline{\Theta}$ is constructed by adding side by side eqs. (8.58) and (8.59):

$$\overline{\Theta} = \overline{\Theta}_p + \overline{\Theta}_h = C_1 J_0\left(i\sqrt{\frac{s}{\alpha}}r\right) + C_2 Y_0\left(i\sqrt{\frac{s}{\alpha}}r\right) + \frac{\Theta_0}{s} \tag{8.60}$$

Applying boundary conditions (8.54) and (8.55) yields

$$C_2 = 0 \tag{8.61}$$

$$C_1 = -\frac{\Theta_0}{sJ_0(i\sqrt{s/\alpha}\, R)} \tag{8.62}$$

Therefore, the final expression for the Laplace transform of the temperature distribution is [combining eqs. (8.60)–(8.62)]

$$\overline{\Theta} = \Theta_0\left[\frac{1}{s} - \frac{J_0(i\sqrt{s/\alpha}\, r)}{sJ_0(i\sqrt{s/\alpha}\, R)}\right] \tag{8.63}$$

■ ***Step 4***

Taking the inverse Laplace transform of both sides of eq. (8.63) yields

$$\Theta(x, t) = \Theta_0\left\{\mathcal{L}^{-1}\left[\frac{1}{s}\right] - \mathcal{L}^{-1}\left[\frac{J_0(i\sqrt{s/\alpha}\, r)}{sJ_0(i\sqrt{s/\alpha}\, R)}\right]\right\} \tag{8.64}$$

Making use of Table F.1 (properties 1 and 97), we have

$$\Theta(x, t) = 2\Theta_0 \sum_{n=1}^{\infty} \frac{e^{-\lambda_n^2 \alpha t} J_0(\lambda_n r)}{\lambda_n R J_1(\lambda_n R)} \tag{8.65}$$

where $\lambda_n$ are the positive roots of

$$J_0(\lambda_n R) = 0 \tag{8.66}$$

The reader can easily show that the temperature field defined by eqs. (8.65) and (8.66) is identical to the temperature field defined by eqs. (5.47) and (5.52) in the limit of a large heat transfer coefficient.

**Example 8.5**

Re-solve Example 5.4 utilizing the Laplace transform method if $h$ is very large ($h \rightarrow \infty$).

**Solution**

■ ***Step 1***

The mathematical model of this problem was presented in Chapter 5 [eqs. (5.82), (5.83), and (5.85)]. In the present special case ($h \rightarrow \infty$), condition (5.84) needs to be replaced by

$$r = a: \qquad \Theta = 0 \tag{8.67}$$

■ ***Step 2***

Taking the Laplace transform of both sides of the governing equation and boundary conditions in the manner shown in Example 8.2 yields

$$\frac{d^2\overline{\Theta}}{dr^2} + \frac{1}{r}\frac{d\overline{\Theta}}{dr} - \frac{s}{\alpha}\overline{\Theta} + \frac{\dot{q}'''}{sk} = 0 \tag{8.68}$$

$$r = 0: \qquad \frac{d\overline{\Theta}}{dr} = 0 \tag{8.69}$$

$$r = a: \qquad \overline{\Theta} = 0 \tag{8.70}$$

■ ***Step 3***

The solution to the homogeneous part of eq. (8.68) is identical to the solution given in eq. (8.58). By inspection, a particular solution of eq. (8.68) is

$$\overline{\Theta}_p = \frac{\alpha\dot{q}'''}{s^2k} \tag{8.71}$$

As discussed in Example 8.2, the general solution for the Laplace transform of the temperature field is obtained if we add eqs. (8.58) and (8.70) side by side:

$$\overline{\Theta} = C_1 J_0\left(i\sqrt{\frac{s}{\alpha}}r\right) + C_2 Y_0\left(i\sqrt{\frac{s}{\alpha}}r\right) + \frac{\alpha\dot{q}'''}{s^2k} \tag{8.72}$$

Applying boundary conditions (5.83) and (8.65) yields

$$C_1 = \frac{-\alpha\dot{q}'''/k}{s^2J_0(i\sqrt{s/\alpha}\,a)}, \; C_2 = 0 \tag{8.73, 8.74}$$

Combining eqs. (8.72)–(8.74) yields

$$\overline{\Theta} = -\frac{\alpha \dot{q}'''}{k} \frac{J_0(i\sqrt{s/\alpha}\, r)}{s^2 J_0(i\sqrt{s/\alpha}\, a)} + \frac{\alpha \dot{q}'''}{k} \frac{1}{s^2} \tag{8.75}$$

■ *Step 4*

Taking the inverse Laplace transform of both sides of eq. (8.75) and utilizing properties 2 and 98 from Table F.1 gives us

$$\overline{\Theta}(x, t) = -\frac{\dot{q}'''}{k} \left[ \frac{r^2 - a^2}{4} + 2a^2 \sum_{n=1}^{\infty} \frac{e^{-\lambda_n^2 \alpha t} J_0(\lambda_n r)}{(\lambda_n a)^3 J_1(\lambda_n a)} \right] \tag{8.76}$$

where $(\lambda_n a)$, $n = 1, 2, 3, \ldots$ are the positive roots of

$$J_0(\lambda_n a) = 0 \tag{8.77}$$

As an algebraic exercise, the reader is recommended to show that the temperature field given by eqs. (8.76) and (8.77) is identical to the temperature field obtained earlier in Chapter 5 for this problem [eqs. (5.103) and (5.108)] in the limit of a large heat transfer coefficient ($h \to \infty$).

## 8.3 TRANSIENT HEAT CONDUCTION IN SPHERICAL COORDINATES

Following Sections 8.1 and 8.2, utilization of the Laplace transform method to solve heat conduction problems in spherical coordinates will be discussed with the help of illustrative examples. It is worth noting in advance that the general procedure outlined in Sections 8.1 and 8.2 for conduction in Cartesian and cylindrical coordinates is valid for conduction in spherical coordinates as well.

**Example 8.6**

Reconsider Example 5.3. This time assume that the heat transfer coefficient is very large ($h \to \infty$). Obtain the temperature distribution utilizing the Laplace transform method.

**Solution**

■ *Step 1*

According to the analysis in Example 5.3, after applying the variable transformation $u = r\Theta$, the mathematical model for the problem becomes

$$\frac{\partial u}{\partial t} = \alpha \frac{\partial^2 u}{\partial r^2} \tag{8.78}$$

$$r = 0: \qquad u = 0 \tag{8.79}$$

$$r = R: \qquad u = 0 \tag{8.80}$$

$$t = 0: \qquad u = r\Theta_0 \tag{8.81}$$

Note that this mathematical model is identical to that of eqs. (5.61)–(5.64) in the limit of a very large heat transfer coefficient ($h \rightarrow \infty$).

■ ***Step 2***

Taking the Laplace transform of eq. (8.78), we obtain

$$\frac{d^2\overline{u}}{dr^2} - \frac{s}{\alpha}\overline{u} = -\frac{\Theta_0}{\alpha}r \tag{8.82}$$

The Laplace transform of the boundary conditions (8.79) and (8.80) completes the formulation of the model for $\overline{u}$ as follows:

$$r = 0: \qquad \overline{u} = 0 \tag{8.83}$$

$$r = R: \qquad \overline{u} = 0 \tag{8.84}$$

■ ***Step 3***

As discussed in Section 8.2, the solution of eq. (8.82) consists of the solution of the homogeneous part of this equation plus a particular solution of the complete nonhomogeneous equation. The result is (Appendix B)

$$\overline{u}(r, s) = A\sinh\left(\sqrt{\frac{s}{\alpha}}r\right) + B\cosh\left(\sqrt{\frac{s}{\alpha}}r\right) + \frac{\Theta_0}{s}r \tag{8.85}$$

Taking into account eqs. (8.83) and (8.84), we easily obtain

$$A = \frac{-\Theta_0 R}{s\sinh(\sqrt{s/\alpha}\,R)} \qquad B = 0 \tag{8.86, 8.87}$$

Combining eqs. (8.85)–(8.87) yields

$$\overline{u}(r, s) = -\Theta_0 R\frac{\sinh(\sqrt{s/\alpha}\,r)}{s\sinh(\sqrt{s/\alpha}\,R)} + \frac{\Theta_0}{s}r \tag{8.88}$$

■ ***Step 4***

The inversion of the Laplace transform of the solution [eq. (8.88)] is performed term by term by utilizing Table F.1. Based on line 93 of this table, we have

$$\mathcal{L}^{-1}\left[-\Theta_0 R\frac{\sinh(\sqrt{s/\alpha}\,r)}{s\sinh(\sqrt{s/\alpha}\,R)}\right]$$

$$= -\Theta_0 R\left[\frac{r}{R} + \frac{2}{\pi}\sum_{n=1}^{\infty}\frac{(-1)^n}{n}e^{-n^2\pi^2\alpha t/R^2}\sin\left(\frac{n\pi r}{R}\right)\right] \tag{8.89}$$

Similarly, based on line 1 of Table F.1,

$$\mathcal{L}^{-1}\left[\frac{\Theta_0 r}{s}\right] = \Theta_0 r \tag{8.90}$$

Combining eqs. (8.88)–(8.90) and recalling that $u = r\Theta$ yields

$$\Theta = -\frac{2\Theta_0 R}{\pi r}\sum_{n=1}^{\infty}\frac{(-1)^n}{n}e^{-n^2\pi^2\alpha t/R^2}\sin\left(\frac{n\pi r}{R}\right) \tag{8.91}$$

It can be easily shown that for $h \to \infty$ (or $\text{Bi} = hR/k \to \infty$) eq. (5.69) yields $\lambda_n = n\pi/R$. For this value of $\lambda_n$, eq. (5.73) becomes identical to eq. (8.91).

**Example 8.7**

In an ignition experiment a spherical fuel droplet that is suspended at the end of a thin quartz fiber, is placed in a furnace, and is heated radiatively ($\dot{q}''$) as shown in Fig. 8.1. The radius of the droplet is denoted by $R$ and its initial temperature by $T_0$. Obtain the temperature history of the droplet. How much time does it take for the fuel to reach its ignition temperature, $T_i$? Thermal convection effects are negligible in this experiment.

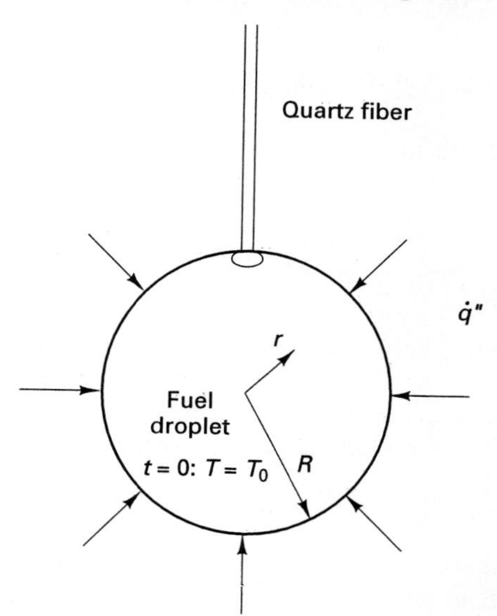


**Figure 8.1** Schematic of a suspended droplet heated radiatively.

**Solution**

■ ***Step 1***

Proceeding as in Example 8.6 it can be shown that the mathematical model of the problem of interest is

$$\frac{1}{\alpha}\frac{\partial u}{\partial t} = \frac{\partial^2 u}{\partial r^2} \tag{8.92}$$

$$r = 0: \qquad u = 0 \tag{8.93}$$

$$r = R: \qquad k\left(\frac{1}{r}\frac{\partial u}{\partial r} - \frac{1}{r^2}u\right) = \dot{q}'' \tag{8.94}$$

$$t = 0: \qquad u = 0 \tag{8.95}$$

■ ***Step 2***

Taking the Laplace transform of both sides of eqs. (8.92)–(8.95), as in Example 8.6, yields

$$\frac{d^2\overline{u}}{dr^2} - \frac{s}{\alpha}\overline{u} = 0 \tag{8.96}$$

$$r = 0: \qquad \overline{u} = 0 \tag{8.97}$$

$$r = R: \qquad \frac{k}{r}\frac{d\overline{u}}{dr} - \frac{k}{r^2}\overline{u} = \frac{\dot{q}''}{s} \tag{8.98}$$

■ ***Step 3***

The solution of eq. (8.96) is (Appendix B)

$$\overline{u}(r, s) = A \sinh\left(\sqrt{\frac{s}{\alpha}}r\right) + B \cosh\left(\sqrt{\frac{s}{\alpha}}r\right) \tag{8.99}$$

Applying boundary conditions (8.97) and (8.98) gives us

$$B = 0 \tag{8.100}$$

$$A = \frac{\dot{q}''R}{(k/\alpha^{1/2})s^{3/2}\cosh(\sqrt{s/\alpha}\,R) - s(k/R)\sinh(\sqrt{s/\alpha}\,R)} \tag{8.101}$$

Combining eqs. (8.99)–(8.101), we obtain the final expression for $\overline{u}$:

$$\overline{u}(r, s) = \frac{\dot{q}''R\sinh(\sqrt{s/\alpha}\,r)}{(k/\alpha^{1/2})s^{3/2}\cosh(\sqrt{s/\alpha}\,R) - s(k/R)\sinh(\sqrt{s/\alpha}\,R)} \tag{8.102}$$

■ ***Step 4***

Expression (8.102) is too complicated to be inverted by utilizing the inversion Table F.1. Therefore, to obtain the temperature field, one needs to evaluate the inversion integral of Appendix F [eq. (F.2)]. No details of this operation are given here [1–3]. The final result, however, reads

$$u(r, t) = \frac{\dot{q}''[(r^3/2) + 3\alpha rt]}{kR} - 2\frac{\dot{q}''}{k}\sum_{n=1}^{\infty}\frac{\sin(\lambda_n r)e^{-\alpha\lambda_n^2 t}}{\lambda_n^2\sin(\lambda_n R)} \tag{8.103}$$

where $\lambda_n$, $n = 1, 2, 3, \ldots$ are the roots of the following equation:

$$\tan(\lambda_n R) = \lambda_n R \tag{8.104}$$

Recalling that $u = r(T - T_0)$, the expression for the temperature field in the droplet is

$$T = T_0 + \frac{\dot{q}''[(r^2/2) + 3\alpha t]}{kR} - 2\frac{\dot{q}''}{kr}\sum_{n=1}^{\infty}\frac{\sin(\lambda_n r)e^{-\alpha\lambda_n^2 t}}{\lambda_n^2 \sin(\lambda_n R)} \tag{8.105}$$

Based on the physics of the problem and since the heating of the droplet occurs at its outer surface, it is reasonable to expect that the ignition temperature of the droplet will be reached at its outer surface (at $r = R$) first. Hence the ignition time ($t_i$) is obtained directly from eq. (8.105) for $T = T_i$ and $r = R$:

$$T_i = T_0 + \frac{\dot{q}''[(R^2/2) + 3\alpha t_i]}{kR} - 2\frac{\dot{q}''}{kR}\sum_{n=1}^{\infty}\frac{e^{-\alpha\lambda_n^2 t_i}}{\lambda_n^2} \tag{8.106}$$

A numerical value for the ignition time can be obtained by solving eq. (8.106) for $t_i$ for any specified fuel properties and droplet radius.

## PROBLEMS

**8.1.** A wall of thickness $L$ of an experimental apparatus is initially at temperature $T_\infty$. An electric current is passed suddenly through the wall generating heat $\dot{q}'''$ (W/m³). The sides of the wall are cooled convectively ($h$, $T_\infty$). Assuming unidirectional heat conduction in the wall, obtain its transient temperature field if $h \to \infty$, utilizing Laplace transforms. Obtain a solution valid at very early times.

**8.2.** Re-solve Example 8.1 for a finite heat transfer coefficient $h$.

**8.3.** A large block of steel of uniform high temperature $T_s$ is suddenly brought into contact with a similar block of ceramic at room temperature $T_b$. Both blocks can be modeled as semi-infinite. Obtain the transient temperature field in the steel block assuming perfect thermal contact by using the Laplace transform method.

**8.4.** A long cylindrical heat-generating electrical wire of radius $R$ has reached a uniform temperature $T_0$. Suddenly, the current is turned off and the wire is cooled convectively ($h$, $T_\infty$). Assuming that the heat transfer coefficient is very large, obtain the transient temperature field in the wire utilizing the Laplace transform method.

**8.5.** Re-solve Problem 8.4 for a finite heat transfer coefficient, $h$.

**8.6.** Obtain an approximate solution for the problem of Example 8.6 valid at very early times of the transient process, utilizing the Laplace transform method.

**8.7.** One side of an isothermal slab (at $T_0$) of thickness $L$ is insulated. The other side is suddenly subjected to a constant radiant heat flux $\dot{q}''$. Utilizing the Laplace transform method, obtain the transient temperature distribution in the slab.

**8.8.** One side of an initialily isothermal slab (at $T_\infty$) of thickness $L$ is suddenly subjected to a uniform heat flux $\dot{q}''$. The other side of the slab is cooled convectively ($h$, $T_\infty$). Assuming that the heat transfer coefficient is very large, obtain the transient temperature field in the slab with the help of the Laplace transform method.

**8.9.** Solve the problem of the point heat source of constant heat production rate outlined in Section 7.1 utilizing the Laplace transform method.

**8.10.** Solve the problem of the line heat source of constant heat production rate outlined in Section 7.1 utilizing the Laplace transform method.

**8.11.** Solve the problem of the plane heat source of constant heat production rate outlined in Section 7.1 utilizing the Laplace transform method.

## REFERENCES

1. C. R. WYLIE and L. C. BARRET, *Advanced Engineering Mathematics,* McGraw-Hill, New York, 1982.

2. M. D. GREENBERG, *Foundations of Applied Mathematics,* Prentice Hall, Englewood Cliffs, NJ, 1978.

3. M. R. SPIEGEL, *Laplace Transforms,* Schaum's Outline Series, McGraw-Hill, New York, 1965.

CHAPTER 9

# HEAT CONDUCTION IN THE PRESENCE OF FREEZING OR MELTING CHARACTERIZED BY A SHARP INTERFACE

Freezing and melting occur in numerous environmental, engineering, medical, and other applications. In our environment, freezing occurs in lakes, estuaries, and the sea during the winter in cold climates. The ice formed subsequently melts in the spring or the summer. With reference to engineering applications, melting and freezing occur in energy storage devices, heat exchangers, welding processes, casting and coating processes, and material removal by high-power laser beams to name just a few [1–4]. Medical applications involving freezing and melting are exemplified by the freezing of tissues in cryosurgery and the preservation of human or laboratory animal organs.

In this chapter unidirectional heat conduction in the presence of freezing or melting is studied. It is assumed that the phase transition takes place at a unique temperature and that the two phases are separated by a sharp interface. Such is the case in freezing or melting of pure substances or eutectic alloys, for example. The topic of freezing of alloys where phase change takes place over a temperature range and a mixed phase (solid and liquid) zone of finite extend exists between the pure solid and the pure liquid regions is examined in Chapter 10.

Convection effects (buoyancy driven or forced) in the liquid phase are largely ignored in the treatment of this chapter. These effects are beyond the scope of this book. Convection with phase change constitutes the subject of an entire subdomain of the convective heat transfer area [5–7].

We proceed by focusing on the new feature offered by melting or freezing problems: namely, the melting or freezing front. Figure 9.1a shows such a front separating a solid from a liquid region. The horizontal dashed lines in Fig. 9.1a represent an infinitely thin control volume surrounding this front.

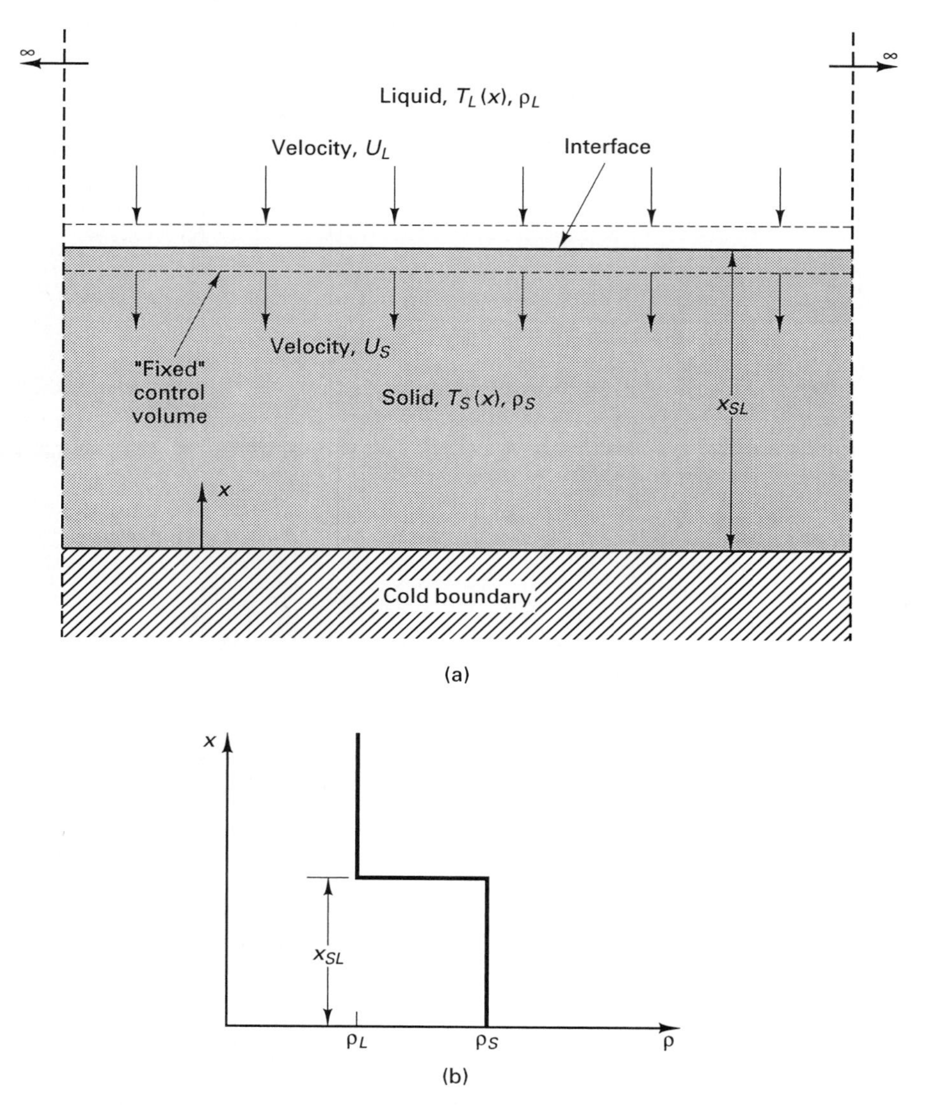


**Figure 9.1** (a) Solidification from a plane cold surface. (b) Density jump across a freezing interface.

It is assumed that *freezing* takes place in the liquid because of the presence of a cold boundary (Fig. 9.1a). The study of a melting process is analogous. As the freezing progresses, the interface moves upward (in the positive $x$-direction) in the stationary medium. It is, we feel, easier to model the heat transfer in the foregoing freezing process if, instead, we assume that the interface and the control volume surrounding it are stationary and that the medium moves downward (in the negative $x$-direction) through the interface. To this end, liquid of density $\rho_L$ and velocity $U_L$ enters the upper side of the control volume and solid of density $\rho_S$ and velocity $U_S$ exits the lower side of the control volume. Note that a density jump occurs across the interface. This jump is shown diagrammatically in Fig. 9.1b, in which the assumption that $\rho_S$ and $\rho_L$ are constant (independent of temperature) with $\rho_S > \rho_L$ has been made. Mass conservation requires that

$$\rho_L U_L A = \rho_S U_S A \tag{9.1}$$

where $A$ is the large area of each side of the control volume.

An energy balance on the control volume yields

$$\dot{Q}_{\text{in cond.}} + \dot{Q}_{\text{in conv.}} - \dot{Q}_{\text{out cond.}} - \dot{Q}_{\text{out conv.}} = 0 \tag{9.2}$$

where $\dot{Q}_{\text{in cond.}}$ is the energy (heat) transferred into the control volume by conduction, $\dot{Q}_{\text{in conv.}}$ is the energy transferred into the control volume by convection (since the medium is moving through the control volume), and $\dot{Q}_{\text{out cond.}}$ and $\dot{Q}_{\text{out conv.}}$ are the analogous quantities exiting the control volume. Fourier's law accounting for the fact that heat diffuses in the negative $x$-direction states that

$$\dot{Q}_{\text{in cond.}} = k_L A\left(\frac{\partial T_L}{\partial x}\right)_{x=x_{SL}} \tag{9.3}$$

$$\dot{Q}_{\text{out cond.}} = k_S A\left(\frac{\partial T_S}{\partial x}\right)_{x=x_{SL}} \tag{9.4}$$

In addition, the energy convected in and out of the control volume is

$$\dot{Q}_{\text{in conv.}} = \rho_L U_L A h_L \tag{9.5}$$

$$\dot{Q}_{\text{out conv.}} = \rho_S U_S A h_S \tag{9.6}$$

where $h_L$ and $h_S$ are the specific enthalpies at the liquid and solid sides of the control volume. Combining eqs. (9.2)–(9.6) yields

$$k_L\left(\frac{\partial T_L}{\partial x}\right)_{x=x_{SL}} + \rho_L U_L h_L - k_S\left(\frac{\partial T_S}{\partial x}\right)_{x=x_{SL}} - \rho_S U_S h_S = 0 \tag{9.7}$$

Rearranging gives

$$k_L\left(\frac{\partial T_L}{\partial x}\right)_{x=x_{SL}} - k_S\left(\frac{\partial T_S}{\partial x}\right)_{x=x_{SL}} + \rho_L U_L h_L - \rho_S U_S h_S = 0 \tag{9.8}$$

Recognizing that the latent heat of fusion is given by

$$h_{SL} = h_L - h_S \tag{9.9}$$

and making use of the mass conservation principle [eq. (9.1)], eq. (9.8) yields

$$k_L\left(\frac{\partial T_L}{\partial x}\right)_{x=x_{SL}} - k_S\left(\frac{\partial T_S}{\partial x}\right)_{x=x_{SL}} + \rho_S U_S h_{SL} = 0 \tag{9.10}$$

Recognizing that the interface velocity can be written as

$$U_S = \frac{dx_{SL}}{dt} \tag{9.11}$$

eq. (9.10) becomes

$$k_L\left(\frac{\partial T_L}{\partial x}\right)_{x=x_{SL}} - k_S\left(\frac{\partial T_S}{\partial x}\right)_{x=x_{SL}} + \rho_S h_{SL}\frac{dx_{SL}}{dt} = 0 \tag{9.12}$$

Equation (9.12) is a direct result of an energy balance at the solidification front and provides a needed matching condition for the solution of the temperature field in the system. In addition, the reader can show that this equation holds for the melting arrangement of Fig. 9.2, where the melting front is also moving upward (in the positive $x$-direction). In the special case where the solid is isothermal at the fusion temperature ($T_F$), eq. (9.12) reduces to

$$k_L\left(\frac{\partial T_L}{\partial x}\right)_{x=x_{SL}} + \rho_S h_{SL}\frac{dx_{SL}}{dt} = 0 \tag{9.13}$$

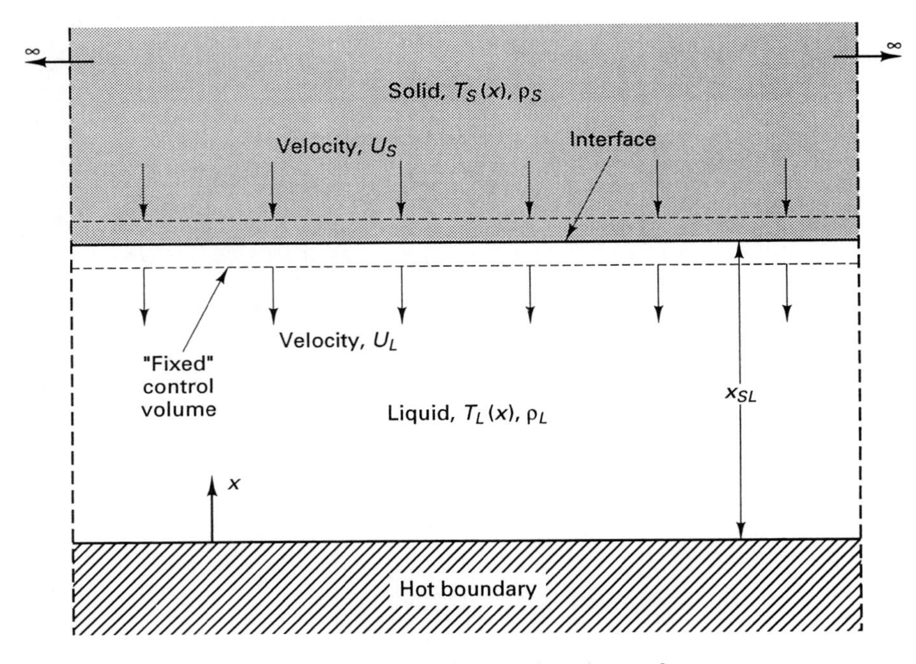


**Figure 9.2** Melting from a plane hot surface.

Similarly, if the liquid is isothermal at the melting temperature ($T_M = T_F$), eq. (9.12) becomes

$$-k_S\left(\frac{\partial T_S}{\partial x}\right)_{x=x_{SL}} + \rho_S h_{SL}\frac{dx_{SL}}{dt} = 0 \tag{9.14}$$

Finally, if convection exists in the melt it is sometimes convenient to use the definition of the convective heat transfer coefficient,

$$h = \frac{-k_L(\partial T_L/\partial x)_{x=x_{SL}}}{T_M - T_\infty} \tag{9.15}$$

where $T_\infty$ is the bulk temperature of the liquid, in place of the temperature gradient on the liquid side of the interface in eq. (9.12):

$$-h(T_M - T_\infty) - k_S\left(\frac{\partial T_S}{\partial x}\right)_{x=x_{SL}} + \rho_S h_{SL}\frac{dx_{SL}}{dt} = 0 \tag{9.16}$$

The reason for this choice is that available information for the exact or approximate value of $h$ may exist in the literature. It is worth noting that eqs. (9.12), (9.13), (9.14), and (9.16) hold for the case of radial conduction in cylindrical or spherical coordinates if the Cartesian coordinate, $x$, is simply replaced by the radial coordinate, $r$, of a cylindrical or spherical system of coordinates.

In what follows, three examples are presented of one-dimensional transient freezing in cartesian, cylindrical, and spherical coordinates, respectively.

**Example 9.1**

Immediately after the pouring of molten plastic of temperature $T_i$ in a rectangular mold (Fig. 9.3), the bottom wall of the mold is chilled to a temperature, $T_C$, well below the fusion temperature of the plastic material, $T_F$. As a result, solidification takes place first near the bottom wall. As time progresses the phase change front moves upward. For simplicity, assume that the solidification as well as the conduction heat transfer phenomena are unidirectional (in the positive $x$-direction of Fig. 9.3). In addition, the mold is assumed to be tall enough such that for a long time period the top wall of the mold does not affect the solidification process. Obtain the temperature history in the mold during this time period. Also, obtain the heat transfer rate at the bottom wall and comment on the effectiveness of the cooling process at this wall. (Note that the

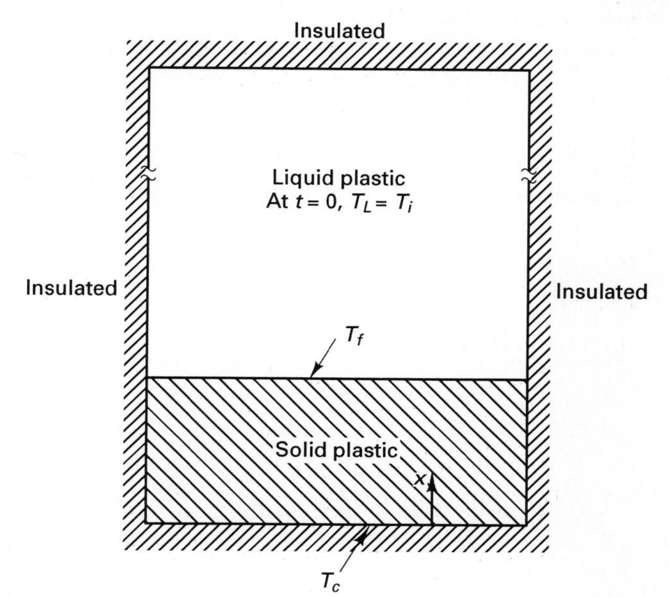


**Figure 9.3** Freezing of plastic in a mold.

mathematical model for this problem is the pioneering Stefan–Neumann model of solidification, published by Neumann in 1912 [8]. Stefan [9] in 1891 published the solution to the same problem for the special case when the initial temperature of the body coincided with the fusion temperature.)

**Solution**

Two distinct regions exist in the system: the solid region ($S$) and the liquid region ($L$). The heat conduction energy equation for these two regions is

$$\frac{1}{\alpha_S}\frac{\partial T_S}{\partial t} = \frac{\partial^2 T_S}{\partial x^2} \qquad 0 < x < x_{SL} \tag{9.17}$$

$$\frac{1}{\alpha_L}\frac{\partial T_L}{\partial t} = \frac{\partial^2 T_L}{\partial x^2} \qquad x_{SL} < x < \infty \tag{9.18}$$

To complete the mathematical model, appropriate initial conditions, boundary conditions, and matching conditions at the solidification front are necessary. These conditions are:

$$x = 0: \qquad T_S = T_C \tag{9.19}$$

$$x = x_{SL}: \qquad T_S = T_L = T_F, \qquad k_L\left(\frac{\partial T_L}{\partial x}\right) - k_S\frac{\partial T_S}{\partial x} + \rho_S h_{SL}\frac{dx_{SL}}{dt} = 0 \tag{9.20, 9.21}$$

$$x \to \infty: \qquad T_L \to T_i \tag{9.22}$$

Note that the first matching condition [eq. (9.20)] states that the temperature at the solidification front is continuous, whereas the second matching condition is the interface energy balance discussed earlier [eq. (9.12)]. To solve the model above, we utilize the similarity method. According to the discussion in Section 5.6 defining

$$\eta = \frac{x}{2\sqrt{\alpha_S t}} \tag{9.23}$$

we cast the governing equations and boundary conditions in terms of the similarity variable:

$$\frac{d^2\Theta_S}{d\eta^2} + 2\eta\frac{d\Theta_S}{d\eta} = 0 \qquad 0 < \eta < \eta_{SL} \tag{9.24}$$

$$\frac{\alpha_L}{\alpha_S}\frac{d^2\Theta_L}{d\eta^2} + 2\eta\frac{d\Theta_L}{d\eta} = 0 \qquad \eta_{SL} < \eta < \infty \tag{9.25}$$

$$\eta = 0: \qquad \Theta_S = T_C - T_i = \Theta_{SC} \tag{9.26}$$

$$\eta = \eta_{SL}: \qquad \Theta_S = \Theta_L = \Theta_F, \qquad k_L\frac{d\Theta_L}{d\eta} - k_S\frac{d\Theta_S}{d\eta} + 2\alpha_S\rho_S h_{SL}\eta_{SL} = 0 \tag{9.27, 9.28}$$

$$\eta \to \infty: \qquad \Theta_L = 0 \tag{9.29}$$

where $\Theta_S = T_S - T_i$, $\Theta_L = T_L - T_i$ and $\Theta_F = T_F - T_i$. Following Section 5.6, the solutions of equations (9.24) and (9.25) are

$$\Theta_S\Big|_0^\eta = A\int_0^\eta e^{-\eta^2}\,d\eta \qquad 0 < \eta < \eta_{SL} \tag{9.30}$$

$$\Theta_L\Big|_{\eta_{SL}}^\eta = B\int_{\eta_{SL}}^\eta e^{-(\alpha_S/\alpha_L)\eta^2}\,d\eta \qquad \eta_{SL} < \eta < \infty \tag{9.31}$$

Using boundary conditions (9.26), (9.27), and (9.29), we obtain

$$\Theta_S(\eta = 0) = \Theta_{SC} \tag{9.32}$$

$$\Theta_L(\eta = \eta_{SL}) = \Theta_F \tag{9.33}$$

$$A = \frac{\Theta_F - \Theta_{SC}}{\int_0^{\eta_{SL}} e^{-\eta^2}\,d\eta} \tag{9.34}$$

$$B = \frac{-\Theta_F}{\int_{\eta_{SL}}^\infty e^{-(\alpha_S/\alpha_L)\eta^2}\,d\eta} \tag{9.35}$$

Combining eqs. (9.30)–(9.35) yields

$$\Theta_S - \Theta_{SC} = (\Theta_F - \Theta_{SC})\frac{\int_0^\eta e^{-\eta^2}\,d\eta}{\int_0^{\eta_{SL}} e^{-\eta^2}\,d\eta} \qquad 0 < \eta < \eta_{SL} \tag{9.36}$$

$$\Theta_L - \Theta_F = -\Theta_F\frac{\int_{\eta_{SL}}^\eta e^{-(\alpha_S/\alpha_L)\eta^2}\,d\eta}{\int_{\eta_{SL}}^\infty e^{-(\alpha_S/\alpha_L)\eta^2}\,d\eta} \qquad \eta_{SL} < \eta < \infty \tag{9.37}$$

Recalling the definition of the error function (Appendix E), the equations above can be written in the following form:

$$\frac{T_S - T_{SC}}{T_F - T_{SC}} = \frac{\text{erf}(\eta)}{\text{erf}(\eta_{SL})} \qquad 0 < \eta < \eta_{SL} \tag{9.38}$$

$$\frac{T_L - T_i}{T_F - T_i} = \frac{\text{erfc}(\sqrt{\alpha_S/\alpha_L}\ \eta)}{\text{erfc}(\sqrt{\alpha_S/\alpha_L}\ \eta_{SL})} \qquad \eta_{SL} < \eta < \infty \tag{9.39}$$

Note that the value of $\eta_{SL}$ is still unknown. To evaluate it, we utilize the matching condition (9.28). Substituting expressions (9.38) and (9.39) into this condition yields

$$\frac{T_F - T_i}{T_F - T_{SC}}\frac{k_L}{k_S}\sqrt{\frac{\alpha_S}{\alpha_L}}\frac{\exp[-(\alpha_S/\alpha_L)\eta_{SL}^2]}{\text{erfc}(\sqrt{\alpha_S/\alpha_L}\eta_{SL})} + \frac{\exp(-\eta_{SL}^2)}{\text{erf}(\eta_{SL})} - \text{Ste}_S\sqrt{\pi}\ \eta_{SL} = 0 \tag{9.40}$$

where

$$\text{Ste}_S = \frac{h_{SL}}{C_{pS}(T_F - T_C)} \tag{9.41}$$

is the Stefan number for the solid region. Expression (9.40) is a nonlinear algebraic equation for $\eta_{SL}$. For a given set of parameters (initial temperatures and properties of the two phases) this equation can be solved numerically to yield $\eta_{SL}$. With $\eta_{SL}$ known, eqs. (9.38) and (9.39) yield the temperature at any point in the system at any time. The heat flux at the bottom wall is obtained from Fourier's law:

$$\dot{q}''_w = -k_S \left( \frac{\partial T_S}{\partial x} \right)_{x=0} \tag{9.42}$$

Combining expressions (9.38) and (9.42) yields

$$\dot{q}''_w = -\frac{k_S(T_F - T_{SC})}{\sqrt{\alpha_S \pi}\, \mathrm{erf}(\eta_{SL})} t^{-1/2} \tag{9.43}$$

Clearly, as time increases, the wall heat flux decreases as $t^{-1/2}$. This is the result of the continuously increasing thermal resistance presented by the growing solid region separating the liquid region from the bottom cold wall.

Finally, it is worth noting that in the special case of freezing of a melt initially at the fusion temperature ($T_i = T_F$), the temperature of the solid region is still given by eq. (9.38) while the liquid remains at the fusion temperature, $T_L = T_F$. Similarly, in the case where the bottom wall temperature equals the fusion temperature ($T_C = T_F$), the solid is isothermal at the fusion temperature ($T_S = T_F$) while the temperature of the liquid region is given by eq. (9.39). An analytical expression for $\eta_{SL}$ can be obtained for small values of $\eta_{SL}(\eta_{SL} \to 0)$. This is the case in slow freezing processes, for example, Replacing the exponentials and the error functions in eq. (9.40) by their Taylor series expansions around zero (Appendix E) one can obtain the following expression (accurate to the first order):

$$\eta_{SL} = \left[ 1 + \frac{k_L}{k_S} \left( \frac{\alpha_S}{\alpha_L} \right)^{1/2} \frac{T_F - T_i}{T_F - T_{SC}} \right]^{1/2} (2\mathrm{Ste}_S)^{-1/2} \tag{9.44}$$

**Example 9.2**

In a cryosurgery experiment a capillary tube of length $L$ is inserted into the muscle of a laboratory animal (Fig. 9.4). A coolant is circulated through the capillary pipe such that freezing is initiated in the vicinity of the pipe. The freezing front propagates outward through the muscle. The muscle size is large enough so that during the freezing process only part of the muscle is affected thermally by the presence of the capillary pipe. For simplicity, it is assumed that the heat transfer rate per unit length absorbed by the pipe is constant, $\dot{Q}' = \dot{Q}/L$ (W/m). The initial temperature of the muscle is denoted by $T_i$. Assuming that the heat transport process is described adequately by transient radial conduction accounting for the presence of the freezing front, obtain the temperature history of the muscle during the above-described treatment. The density of the frozen region is denoted by $\rho_F$.

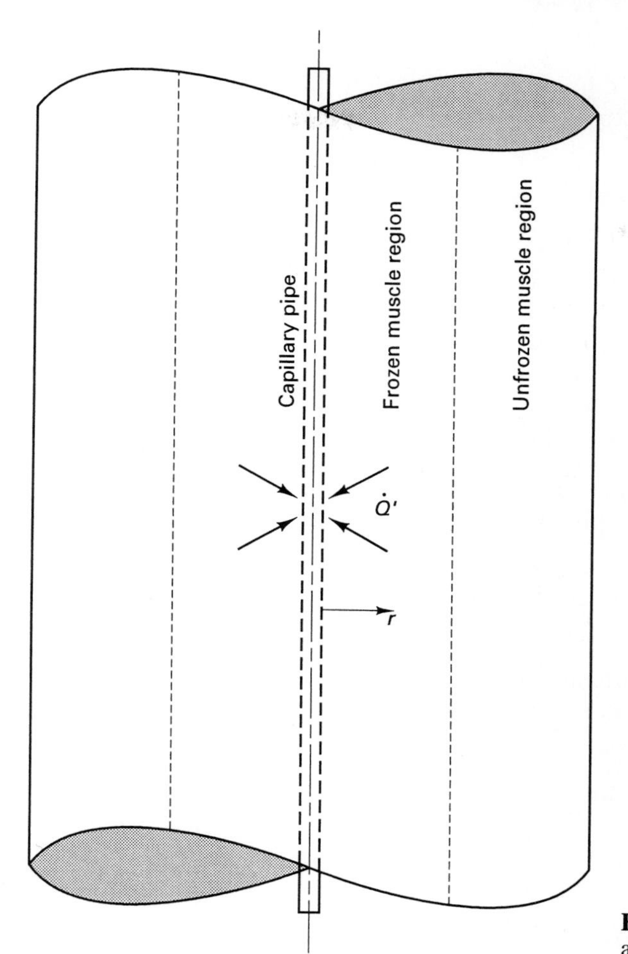


**Figure 9.4** Freezing of tissue around a capillary pipe.

## Solution

This problem resembles the one in Example 9.1 except for the fact that it needs to be modeled in cylindrical coordinates (Fig. 9.4). Due to its capillary size, the pipe can be modeled as a line heat sink. The energy equation in the frozen and the unfrozen regions is

$$\frac{1}{\alpha_F}\frac{\partial \Theta_F}{\partial t} = \frac{1}{r}\frac{\partial}{\partial r}\left(r\frac{\partial \Theta_F}{\partial r}\right) \qquad 0 < r < r_{FU} \tag{9.45}$$

$$\frac{1}{\alpha_U}\frac{\partial \Theta_U}{\partial t} = \frac{1}{r}\frac{\partial}{\partial r}\left(r\frac{\partial \Theta_U}{\partial r}\right) \qquad r_{FU} < r < \infty \tag{9.46}$$

The corresponding boundary conditions in the neighborhood of the pipe and far away from the pipe as well as the matching conditions at the freezing front are

$$r = 0: \qquad \lim_{r\to 0}\left(2\pi r k_F \frac{\partial \Theta_F}{\partial r}\right) = \dot{Q}' \tag{9.47}$$

$$r = r_{FU}: \qquad \Theta_F = \Theta_U = \Theta_M \qquad k_U \frac{\partial \Theta_U}{\partial r} - k_F \frac{\partial \Theta_F}{\partial r} + \rho_F h_{FU} \frac{dr_{FU}}{dt} = 0 \tag{9.48, 9.49}$$

$$r \to \infty: \qquad \Theta_U = 0 \tag{9.50}$$

The initial condition is

$$t = 0: \qquad \Theta_U = 0 \tag{9.51}$$

In the equations above, $\Theta_F = T_F - T_i$, $\Theta_U = T_U - T_i$, $\Theta_M = T_M - T_i$. The subscripts $F$, $U$, $M$, $FU$ stand for frozen region, unfrozen region, melting (or freezing) temperature, and frozen/unfrozen interface, respectively.

We select to solve the mathematical model above with the similarity method. Defining

$$\xi = \frac{r^2}{4\alpha_F t} \tag{9.52}$$

we write the governing equations and boundary conditions, in terms of the similarity variable (a similar procedure was followed in Section 7.1).

$$\frac{d^2\Theta_F}{d\xi^2} + \frac{1 + \xi}{\xi}\frac{d\Theta_F}{d\xi} = 0 \qquad 0 < \xi < \xi_{FU} \tag{9.53}$$

$$\frac{d^2\Theta_U}{d\xi^2} + \frac{1 + \dfrac{\alpha_F}{\alpha_U}\xi}{\xi}\frac{d\Theta_U}{d\xi} = 0 \qquad \xi_{FU} < \xi < \infty \tag{9.54}$$

$$\xi \to 0: \qquad \lim_{\xi\to 0}\left(4\pi k_F \xi \frac{d\Theta_F}{d\xi}\right) = \dot{Q}' \tag{9.55}$$

$$\xi = \xi_{FU}: \qquad \Theta_F = \Theta_U = \Theta_M \qquad k_U\frac{d\Theta_U}{d\xi} - k_F\frac{d\Theta_F}{d\xi} + \rho_F h_{FU}\alpha_F = 0 \tag{9.56, 9.57}$$

$$\xi \to \infty: \qquad \Theta_U \to 0 \tag{9.58}$$

As discussed in Section 7.1, the solution to eqs. (9.53) and (9.54) can be written as

$$\Theta_F = A\int_0^{\xi}\frac{e^{-\lambda}}{\lambda}\,d\lambda + B \qquad 0 < \xi < \xi_{FU} \tag{9.59}$$

$$\Theta_U = C\int_{\xi_{FU}}^{\xi}\frac{e^{-\lambda(\alpha_F/\alpha_U)}}{\lambda}\,d\lambda + D \qquad \xi_{FU} < \xi < \infty \tag{9.60}$$

Applying conditions (9.55), (9.56), and (9.58) yields

$$A = \frac{\dot{Q}'}{4\pi k_F} \tag{9.61}$$

$$B = \Theta_M - \frac{\dot{Q}'}{4\pi k_F} \int_0^{\xi_{FU}} \frac{e^{-\lambda}}{\lambda}\, d\lambda \tag{9.62}$$

$$D = \Theta_M \tag{9.63}$$

$$C = -\frac{\Theta_M}{\displaystyle\int_{\xi_{FU}}^{\infty} \frac{e^{-\lambda(\alpha_F/\alpha_U)}}{\lambda}\, d\lambda} \tag{9.64}$$

Combining eqs. (9.59)–(9.64) gives us

$$\Theta_F = \frac{\dot{Q}}{4\pi k_F}\left(\int_0^{\xi} \frac{e^{-\lambda}}{\lambda}\, d\lambda - \int_0^{\xi_{FU}} \frac{e^{-\lambda}}{\lambda}\, d\lambda\right) + \Theta_M \qquad 0 < \xi < \xi_{FU} \tag{9.65}$$

$$\Theta_U = \Theta_M \left[1 - \frac{\displaystyle\int_{\xi_{FU}}^{\xi} \frac{e^{-\lambda(\alpha_F/\alpha_U)}}{\lambda}\, d\lambda}{\displaystyle\int_{\xi_{FU}}^{\infty} \frac{e^{-\lambda(\alpha_F/\alpha_U)}}{\lambda}\, d\lambda}\right] \qquad \xi_{FU} < \xi < \infty \tag{9.66}$$

Recalling the definition of the exponential integral function [eq. (7.28) and Appendix E], eqs. (9.65) and (9.66) are cast in the following form:

$$T_F = \frac{\dot{Q}}{4\pi k_F}[\mathrm{E}i(\xi_{FU}) - \mathrm{E}i(\xi)] + T_M \qquad 0 < \xi < \xi_{FU} \tag{9.67}$$

$$T_U = (T_M - T_i)\frac{\mathrm{E}i[(\alpha_F/\alpha_U)\xi]}{\mathrm{E}i[(\alpha_F/\alpha_U)\xi_{FU}]} + T_i \qquad \xi_{FU} < \xi < \infty \tag{9.68}$$

In the temperature expression above, $\xi_{FU}$ (the location of the freezing front at a given time) is not known. It is obtained by substituting eqs. (9.67) and (9.68) into the interface energy balance (9.57). The result of this substitution reads

$$\frac{\dot{Q}'}{4\pi} e^{-\xi_{FU}} + \frac{k_U(T_M - T_i)}{\mathrm{E}i[\xi_{FU}(\alpha_F/\alpha_U)]} e^{-\xi_{FU}(\alpha_F/\alpha_U)} - \alpha_F \rho_F h_{FU} \xi_{FU} = 0 \tag{9.69}$$

Once the thermophysical properties of the frozen and unfrozen regions are prescribed, the nonlinear algebraic equation (9.69) can be solved numerically to yield $\xi_{FU}$. With $\xi_{FU}$ known, eqs. (9.67) and (9.68) can be used to obtain the temperature at any point and time in the frozen and the unfrozen regions.

### Example 9.3

It is known that in crystal growth processes particles of dirt or other impurities in the crystal melt act as solid nucleation sites initiating solidification at undesirable times and locations in the melt [10]. As a result, imperfections occur in the solid crystal that may have a deterimental effect on its quality. Figure 9.5 shows a crystal melt tank with a small impurity within it. If the initial temperature of the melt pool is denoted by $T_i$, it is assumed that at the location of the impurity (modeled as a "point") heat is absorbed initiating solidification (Fig. 9.5). Obtain the temperature distribution in the solid and liquid regions in the melt tank near the impurity. To facilitate mathematical tractability, assume that the heat absorption rate at the source is proportional to $t^{1/2}$ [i.e., $\dot{Q} = Q_0 t^{1/2}$, where $Q_0$ is a constant of units W/s$^{1/2}$]. Note that this assumption is not physically realistic and is used for illustrative purposes only.

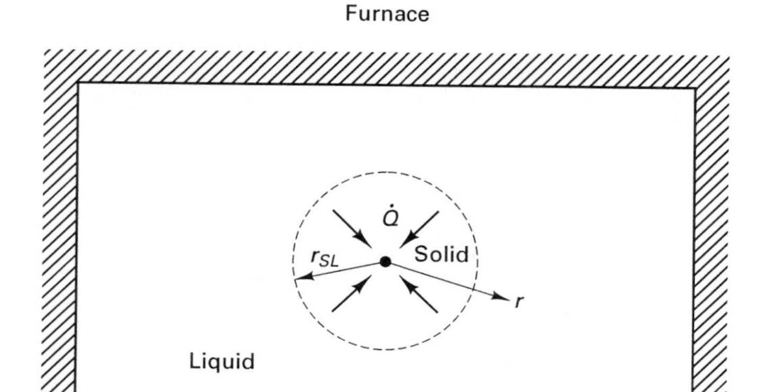


**Figure 9.5** Schematic of crystal growth around an impurity.

### Solution

Assuming that the melt pool is large with respect to the region affected thermally by the impurity, this problem can be modeled as Stefan–Neumann solidification in spherical coordinates around a point heat sink. To this end, proceeding as in Examples 9.1 and 9.2, the mathematical model for the problem reads

$$\frac{1}{\alpha_S}\frac{\partial \Theta_S}{\partial t} = \frac{1}{r^2}\frac{\partial}{\partial r}\left(r^2 \frac{\partial \Theta_S}{\partial r}\right) \qquad 0 < r < r_{SL} \tag{9.70}$$

$$\frac{1}{a_L}\frac{\partial \Theta_L}{\partial t} = \frac{1}{r^2}\frac{\partial}{\partial r}\left(r^2 \frac{\partial \Theta_L}{\partial r}\right) \qquad r_{SL} < r < \infty \tag{9.71}$$

$$r \to 0: \qquad \lim_{r\to 0}\left(4\pi r^2 k_S \frac{\partial \Theta}{\partial r}\right) = \dot{Q} = Q_0 t^{1/2} \tag{9.72}$$

$$r = r_{SL}: \qquad \Theta_S = \Theta_L = \Theta_F$$

$$k_L\left(\frac{\partial \Theta_L}{\partial r}\right)_{r=r_{SL}} - k_S\left(\frac{\partial \Theta_S}{\partial r}\right)_{r=r_{SL}} + \rho_S h_{SL}\frac{dr_{SL}}{dt} = 0 \qquad (9.73, 9.74)$$

$$r \to \infty: \qquad \Theta_L = 0 \tag{9.75}$$

$$t = 0: \qquad \Theta_L = 0 \tag{9.76}$$

Next, we define the similarity variable:

$$\eta = \frac{r^2}{4\alpha_S t} \tag{9.77}$$

In terms of the similarity variable the mathematical model reads

$$\frac{d^2\Theta_S}{d\eta^2} + \left(1 + \frac{3}{2\eta}\right)\frac{d\Theta_S}{d\eta} = 0 \tag{9.78}$$

$$\frac{\alpha_L}{\alpha_S}\frac{d^2\Theta_L}{d\eta^2} + \left(1 + \frac{3}{2\eta}\frac{\alpha_L}{\alpha_S}\right)\frac{d\Theta_L}{d\eta} = 0 \tag{9.79}$$

$$r \to 0: \qquad 16\pi\alpha_S^{1/2}k_S\eta^{3/2}\frac{d\Theta_S}{d\eta} = Q_0 \tag{9.80}$$

$$r = r_{SL}: \qquad \Theta_S = \Theta_L = \Theta_F \qquad k_L\frac{d\Theta_L}{d\eta} - k_S\frac{d\Theta_S}{d\eta} + \rho_S\alpha_S h_{SL} = 0 \tag{9.81, 9.82}$$

$$r \to \infty: \qquad \Theta_L = 0 \tag{9.83}$$

It is now clear that the assumption $\dot{Q} = Q_0 t^{1/2}$ was necessary in order to cast condition (9.80) in terms of $\eta$ only. To obtain the solution of eq. (9.78), we first reduce its order by defining

$$P = \frac{d\Theta_S}{d\eta} \tag{9.84}$$

Hence eq. (9.78) becomes

$$\frac{dP}{d\eta} + \left(1 + \frac{3}{2\eta}\right)P = 0 \tag{9.85}$$

This equation is separable (Appendix B):

$$\frac{dP}{P} = -\left(1 + \frac{3}{2\eta}\right)d\eta \tag{9.86}$$

Integrating both sides and rearranging, we eventually obtain

$$P = C\frac{e^{-\eta}}{\eta^{3/2}} \tag{9.87}$$

To facilitate the remainder of the solution, we define the variable

$$\chi = \eta^{1/2} \tag{9.88}$$

Combining eqs. (9.84), (9.87), and (9.88) gives

$$\frac{d\Theta}{d\chi} = 2C\frac{e^{-\chi^2}}{\chi^2} \tag{9.89}$$

Next, we integrate both sides of eq. (9.89) from $\chi = 0$ to an arbitrary value of $\chi$ in the region $0 < \chi < \chi_{SL}$:

$$\Theta_S = \frac{Q_0}{8\pi\alpha_S^{1/2}k_S}\int_0^{\chi} \frac{e^{-\mu^2}}{\mu^2}\,d\mu + D \tag{9.90}$$

where $\mu$ is a dummy variable. Recognizing that $1/\mu^2 = (d/d\mu)(-1/\mu)$ we apply integration by parts to the integral of eq. (9.90) and obtain

$$\Theta_S = \frac{Q_0}{8\pi\alpha_S^{1/2}k_S}\left(-\left[\frac{e^{-\mu^2}}{\mu}\right]_0^{\chi} - 2\int_0^{\chi} e^{-\mu^2}\,d\mu\right) + D \tag{9.91}$$

Utilizing boundary condition (9.81) yields

$$D = \Theta_F - \frac{Q_0}{8\pi\alpha_S^{1/2}k_S}\left(-\left[\frac{e^{-\mu^2}}{\mu}\right]_0^{\chi_{SL}} - 2\int_0^{\chi_{SL}} e^{-\mu^2}\,d\mu\right) \tag{9.92}$$

Following this, we combine equations (9.91) and (9.92) to obtain

$$\Theta_S = \Theta_F + \frac{Q_0}{8\pi\alpha_S^{1/2}k_S}\left(-\left[\frac{e^{-\mu^2}}{\mu}\right]_0^{\chi_{SL}} + 2\int_0^{\chi_{SL}} e^{-\mu^2}\,d\mu - 2\int_0^{\chi} e^{-\mu^2}\,d\mu\right) \tag{9.93}$$

Recalling the definition of the error function (Appendix E), the final result for the temperature distribution in the solid phase region is

$$T_S = T_F + \frac{Q_0}{8\pi\alpha_S^{1/2}k_S}\left\{\frac{e^{-\chi_{SL}^2}}{\chi_{SL}} - \frac{e^{\chi^2}}{\chi} + \sqrt{\pi}\left[\operatorname{erf}(\chi_{SL}) - \operatorname{erf}(\chi)\right]\right\} \tag{9.94}$$

where $\chi$ was as defined in eq. (9.88), $\chi = r/2\sqrt{\alpha_S t}$, $\chi_{SL} = r_{SL}/2\sqrt{\alpha_S t}$. Following the identical procedure, it can be shown that the solution for the temperature in the liquid region is

$$T_L = (T_F - T_i)\left\{\frac{\dfrac{e^{-\chi^2(\alpha_S/\alpha_L)}}{\chi} - \left(\pi\dfrac{\alpha_S}{\alpha_L}\right)^{1/2}\operatorname{erfc}\left[\chi\left(\dfrac{\alpha_S}{\alpha_L}\right)^{1/2}\right]}{\dfrac{e^{-\chi_{SL}^2(\alpha_S/\alpha_L)}}{\chi_{SL}} - \left(\pi\dfrac{\alpha_S}{\alpha_L}\right)^{1/2}\operatorname{erfc}\left[\chi_{SL}\left(\dfrac{\alpha_S}{\alpha_L}\right)^{1/2}\right]} - 1\right\} + T_F \tag{9.95}$$

Much as in Examples 9.1 and 9.2, $\chi_{SL}$ is obtained by substituting eqs. (9.94) (9.95) into the interface energy balance (9.82). The result of this substitution is

$$\frac{k_L(T_F - T_i)e^{-\chi_{SL}^2(\alpha_S/\alpha_L)}}{2\chi_{SL}^2\left\{e^{-\chi_{SL}^2(\alpha_S/\alpha_L)} - \left(\pi\dfrac{\alpha_S}{\alpha_L}\right)^{1/2}\chi_{SL}\operatorname{erfc}\left[\chi_{SL}\left(\dfrac{\alpha_S}{\alpha_L}\right)^{1/2}\right]\right\}} + \frac{Q_0}{16\pi\alpha_S^{1/2}}\frac{e^{-\chi_{SL}^2}}{\chi_{SL}^3} - \rho_S\alpha_S h_{SL} = 0 \tag{9.96}$$

Solving the above nonlinear algebraic equation numerically yields $\chi_{SL}$. For a material with known thermophysical properties, after $\chi_{SL}$ is obtained, eqs. (9.94) and (9.95) determine the temperature at any point in the solid or liquid regions.

## PROBLEMS

**9.1.** Utilize the results of Example 9.1 and obtain a graph showing the temperature history in the solid and liquid phases during the freezing of water, initially at 10°C, in contact with a cold boundary kept at −30°C. Also show graphically the dependence of the heat flux at the cold boundary on time.

**9.2.** Utilize the results of Example 9.2 and obtain a graph showing the temperature history in both the solid and liquid phases during the process of freezing initiated by a long capillary pipe of coolant immersed in a large bath of water. For simplicity, assume that the pipe absorbs heat at approximately a constant rate $\dot{Q}' = 100$ W/m. The initial temperature of the water bath is 5°C.

**9.3.** Utilize the results of Example 9.3 and obtain a graph that can be used to estimate the temperature distribution in the solid and liquid phases in the following experiment. A specially designed miniature heater that is embedded at the center of a large sphere of ice of temperature −10°C is suddenly turned on. The current to the heater is regulated such that the power it generates is given by $\dot{Q} = Q_0 t^{1/2}$, where $Q_0 = 100$ W/s$^{1/2}$ and $t$ is time (s). If the diameter of the sphere is 0.5 m, can you *estimate* the time it will take for the sphere to melt?

**9.4.** A molten material of temperature $T_i$ is poured into a large chill with a very thick bottom wall. The bottom wall of the chill has been precooled to a low temperature $T_C$, considerably lower than the freezing temperature of the molten material $T_F$. As a result, solidification takes place as shown in Fig. P9.4. Obtain the temperature history of the solidifying material and of the chill wall. What is the heat flux at the chill wall?

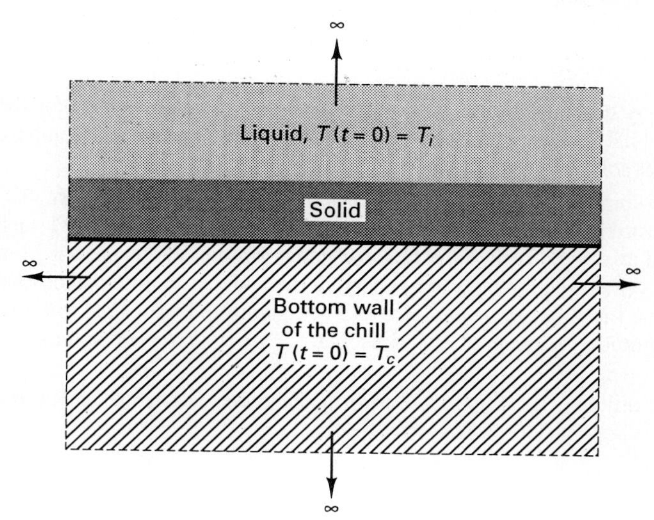


**Figure P9.4**

**9.5.** A conductive, thin-walled tank of a liquid refrigerant with low boiling point ($T_B$) is placed into contact with a similar water tank as shown in Fig. P9.5. The freezing temperature of water is denoted by $T_F$. The initial temperature of the water tank is denoted by $T_{iW}$, where $T_{iW} > T_B$. The initial temperature of the refrigerant tank is denoted by $T_{iR}$, where $T_{iR} < T_B$. As a result of the above-mentioned experiment, the refrigerant is heated by the water and starts vaporizing. In doing so, it absorbs heat from the water and the water freezes. The freezing and vaporization processes occur in the idealized manner shown in Fig. P9.5. Find the temperature history of the water and the refrigerant. List all the assumptions you make in the solution process.

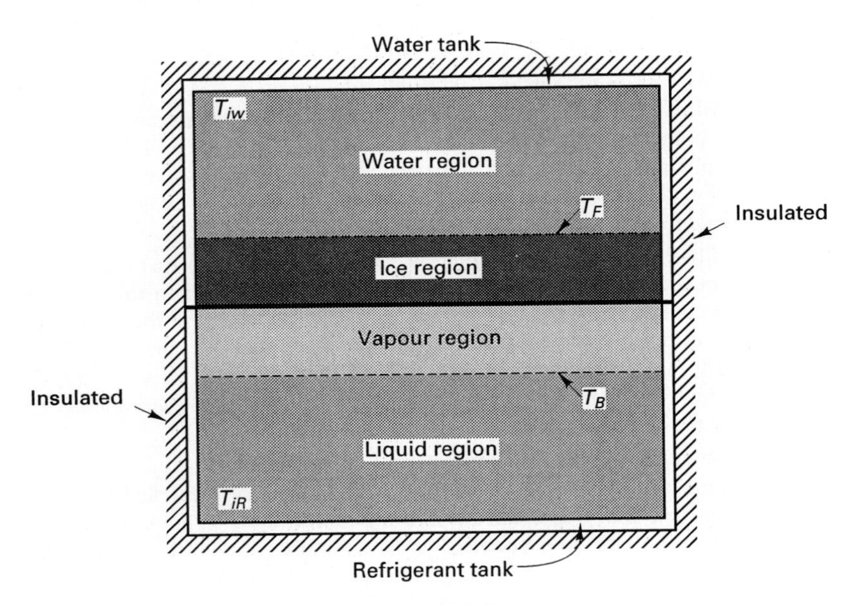


**Figure P9.5**

**9.6.** Re-solve Example 9.1 by utilizing the integral method (described in Chapter 5) for the case where $T_C = T_F$ [11]. Use a quadratic and a cubic temperature profile and compare your results to the exact solution of Example 9.1.

**9.7.** A thick solid slab of a composite material is heated up to its melting temperature, $T_m$. Suddenly, the temperature of one side of the slab (at $x = 0$) is raised to $T_w > T_m$ such that melting is initiated at $x = 0$. Utilize the Laplace transform method as described in Ku and Chan [12] and obtain the temperature history in the slab, for a time period that is short enough such that the slab can be modeled as a semi-infinite body. Can you propose an order-of-magnitude estimate of the time period for which your solution is valid?

**9.8.** Re-solve Example 9.1 utilizing the Laplace transform technique described in Ku and Chan [12].

## REFERENCES

1. N. Rykalin, A. Uglov, I. Zuev, and A. Kokora, *Laser and Electron Beam Material Processing Handbook,* MIR Publishers, Moscow, 1988.
2. E. F. Nippes, Coordinator, *Metals Handbook,* Vol. 6 (Welding Brazing and Soldering), American Society for Metals, Metals Park, OH, 1983.
3. H. E. Boyer and T. L Gall, Eds., *Metals Handbook,* desk ed., American Society for Metals, Metals Park, OH, 1985.
4. L. E. Murr, Ed., *Solar Materials Science,* Academic Press, New York, 1980.
5. R. Viskanta, Phase change heat transfer, in *Solar Heat Storage: Latent Heat Materials,* G. A. Lane, Ed., CRC Press, Boca Raton, FL, pp. 153–222, 1983.
6. R. Viskanta, Natural convection in melting and solidification, in *Natural Convection: Fundamentals and Applications,* S. Kakac, W. Aung, and R. Viskanta, Eds., Hemisphere, New York, pp. 845–877, 1985.
7. R. Viskanta, Heat transfer during melting and solidification of metals, *J. Heat Transfer,* Vol. 110, pp. 1205–1219, 1988.
8. F. Neumann, *Die partiellen Differentialgleichunger der mathematischen Physik,* Vol. 2, p. 121, 1912.
9. J. Stefan, Uber die Theorie der Eisbildung, Insbesondere Ueber die Eisbildung im Polarmeere, *Ann. Phys. Chem.,* Vol. 42, pp. 269–286, 1991.
10. F. Rosenberger, *Fundamentals of Crystal Growth,* Springer-Verlag, Berlin, 1979.
11. M. N. Özisik, *Heat Conduction,* Wiley, New York, pp. 416–419, 1980.
12. J. Y. Ku and S. H. Chan, A generalized Laplace transform technique for phase-change problems, *J. Heat Transfer,* Vol. 112, pp. 495–497, 1990.

CHAPTER **10**

# HEAT CONDUCTION DURING THE FREEZING OF BINARY ALLOYS

Alloy solidification has a plethora of industrial and environmental applications, ranging from casting and coating processes to crystal growth and ice formation in the sea. The modeling of the heat transfer and mass transfer phenomena occurring during the solidification of alloys is complex. The development of reliable universal models describing the above-mentioned transport phenomena in the presence of convective flow is a current research area [1–9].

The complexity of the problem of alloy freezing is better appreciated if, for a moment, we reconsider the solidification of pure substances discussed in Chapter 9. When the solid is created from a pure melt cooled by conduction, the two phases (solid and liquid) are separated by a distinct interface. This interface moves as the solid grows.

In solidification of binary alloys, the phase change takes place in a manner that is markedly different from the description above. The liquid and solid phases are often separated by a mixed-phase region, consisting of an intricate mixture of liquid and dendrites commonly termed the *mushy zone*. This region is of finite and sometimes dominant extent and the heat transfer process within it greatly affects the solidification phenomenon as well as the growth and the properties of the resulting solid. The mushy zone diminishes in the case of eutectic freezing (discussed later in this section).

Figures 10.1 and 10.2 are photographs of solidifying binary alloys obtained in the author's laboratory. Figure 10.1 shows a solidifying mixture of water and ammonium chloride initially at room temperature and concentration 5 percent by weight. The solidification occurs on the underside of a horizontal surface in contact with the

**Figure 10.1** Photograph of solidification of a binary mixture of water and $NH_4Cl$ of initial concentration 5% by weight on the underside of a cold surface, showing the solid, mixed-phase, and liquid regions.

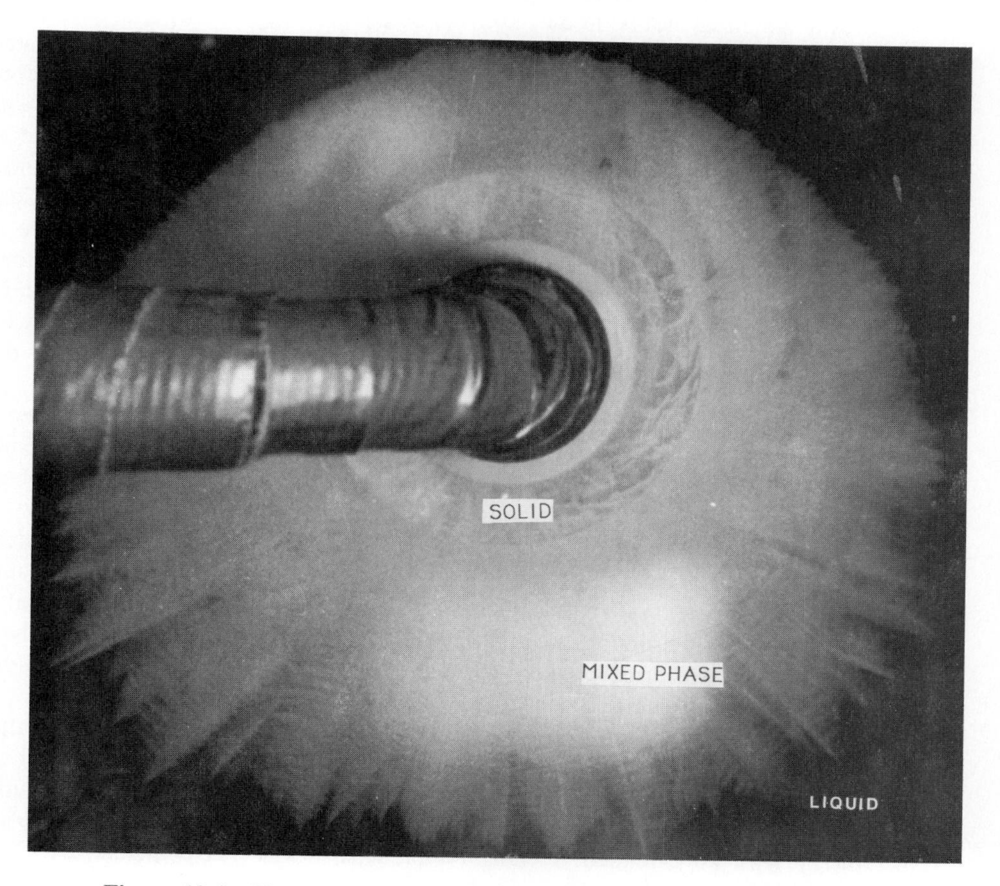


**Figure 10.2** Photograph of solidification of a binary mixture of water and $NH_4Cl$ of initial concentration 25% by weight around a horizontal cold pipe showing the solid, mixed-phase, and liquid regions.

alloy. This surface was suddenly cooled to about −20°C. The three regions (solid, mushy, and liquid) are clearly distinct. The size of the mushy (mixed phase) region is noteworthy. Figure 10.2 shows a solidifying alloy of water and ammonium chloride around a horizontal pipe. The alloy had an initial concentration of 25 percent by weight and an initial temperature of about 20°C. A coolant was suddenly circulated through the pipe and its surface temperature was lowered quickly to about −20°C. Again three distinct regions existed in the freezing process: the solid region, the (sizable) mushy region, and the liquid region. Note that due to the presence of double diffusive convection in the liquid region, the solidification process is not axisymmetric.

In the modeling of heat conduction during solid growth from the melt of an alloy, accounting for the mushy zone is of primary importance. This fact complicates the modeling significantly. Note that the heat and mass transfer processes inside the mushy zone may not be in thermodynamic equilibrium. To this end, several attempts have been made to study theoretically the problem of conduction-dominated alloy solidification. For example, O'Callaghan et al. [10] used a model that accounts for the mushy region to study heat and mass transfer during solidification of a eutectic binary solution. The wall temperature was assumed to be lower than the eutectic temperature of the mixture. The speed of propagation of the solid–mush and mush–liquid interfaces was taken to be identical. Fang et al. [11] studied theoretically and experimentally the selective freezing of a dilute salt solution on a cold ice surface. They neglected the mass transfer effect on the growth of the solid. Therefore, their analysis is accurate for mixtures in which the solidification process is driven by thermal diffusion alone. Comparisons between theory and experiment showed good agreement. A similar study (experimental and analytical) was performed by Webb and Viskanta [12]. These authors used paraffins as phase-change materials, and they focused on solidification from a horizontal cold boundary. They also neglected species diffusion in both the solid and the liquid regions. Their observations showed that the dendrite tips in the mush–liquid interface advanced in a horizontal plane, suggesting the absence of instabilities.

Alexiadis et al. [13] proposed a macroscopic mathematical model describing the evolution of the phases of a binary mixture undergoing solidification under the simultaneous action of heat and mass diffusion. Their formulation is global (i.e., it holds in the whole region occupied by the alloy). A theoretical model for the prediction of the characteristics of a dendritic interface during the controlled solidification of binary alloys was also developed by Trivedi [14].

Worster [15] solved theoretically the problem of solidification of an alloy from a horizontal cold flat plate by using a model that accounts for the simultaneous action of heat and species diffusion. He reported interesting results on the growth of the mushy region in aqueous solutions. An exact solution for freezing in cylindrical symmetry with an extended freezing temperature range was published by Ozisik and Uzzell [16].

Poulikakos [17] and Poulikakos and Cao [18] reported theoretical studies on diffusion dominated binary alloy solidification from a cold capillary pipe or wire. In the first study [17] it was assumed that the solidification is dominated by heat diffusion. In the second study [18] the simultaneous presence of heat and species diffusion was taken into account. Hayashi and Komori [19] conducted experiments on the freezing of salt solutions in cells. They also published a simple heat conduction model that predicted reasonably well the growth rates of the solid–mush and mush–liquid interfaces. Cao and Poulikakos [20] extended the model of Hayashi and Komori to predict the solidification process in a bed of beads saturated with a binary alloy. When compared with experimental results in [20] this model performed acceptably well for cases where the buoyancy-driven flow in the system was weak. The discussion above is intended to give the reader a "flavor" of the fast-growing area of transport phenomena in alloy solidification and is by no means inclusive of all published studies. Only examples of works that focus on conduction-dominated solidification that is pertinent to this book were discussed. It is worth noting that significant research has been reported in the literature of convection effects on alloy

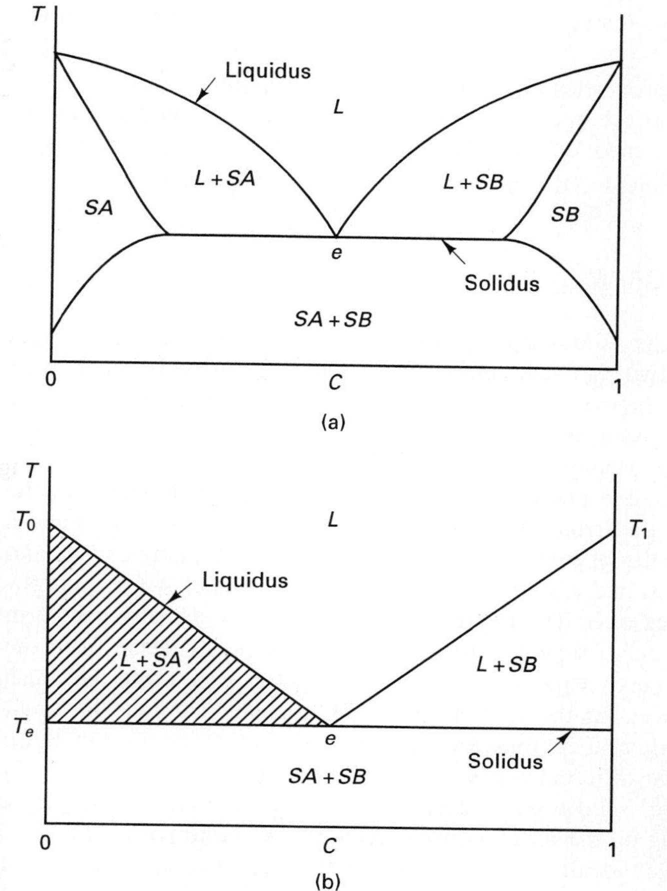


**Figure 10.3** (a) Typical equilibrium-phase diagram of a binary mixture. (b) Simplified equilibrium-phase diagram of a water and salt mixture.

solidification (Refs. [1–9] and references therein, for example). This topic falls outside the scope of this book.

The purpose of this chapter is to present a classroom-level treatment of basic problems in alloy solidification in the *absence of convection*. In addition, the following assumptions are introduced in the conduction model presented in this chapter.

1. Mass diffusion does not affect the solidification process, which is governed by heat diffusion. This assumption is justified for alloys that have a large thermal diffusivity compared to the species thermal diffusivity.
2. Equilibrium conditions exist locally in the mixed-phase region, so that the temperature–concentration relationship is adequately represented by an equilibrium-phase diagram (Fig. 10.3), which will be discussed shortly.
3. The solid fraction ($\chi$) and the amount of internal heat generation corresponding to the latent heat of fusion released in the mushy zone are functions of temperature only. Furthermore, an average constant value for the rate of change of the solid fraction with temperature in the mushy zone will be utilized in the model.
4. The thermophysical properties of the mushy zone are assumed to be the volume fraction weighted averages of the properties of the individual phases. In addition, they are assumed to be constant at an average value in the model.
5. The densities of the liquid and solid phases do not vary appreciably.

## 10.1 EQUILIBRIUM-PHASE DIAGRAM

Of central importance to the solidification process is the equilibrium-phase diagram for a system consisting of two chemical components, $A$ and $B$. Figure 10.3a shows a typical equilibrium-phase diagram for a binary eutectic alloy. Given a mixture of concentration $C$ (weight percent of component $B$) with uniform temperature $T$, the coordinates $C$ and $T$ define a point on the phase diagram. The region in which this point is located determines the phase present in the system at equilibrium. In the phase diagram, $L$ denotes the liquid, $S$ the solid, $L + SA$ a mixture of liquid and a solid in which solid molecules of $B$ are incorporated into the crystal lattice of $A$, and $L + SB$ a mixture of liquid and a solid in which solid molecules of $A$ are incorporated into the crystal lattice of $B$. The liquidus and solidus curves are also shown in Fig. 10.3a. The liquidus curve separates the region of pure liquid phase from the mushy zone. The solidus curve separates the region of a solid phase from the mush or from another solid phase. At the eutectic point ($e$), there is a unique melting–freezing temperature and there is no mushy zone. Note that in the equilibrium phase diagram of Fig. 10.3a two different types of mushy zones ($L + SA$ and $L + SB$) and three different types of solid regions ($SA$, $SB$, $SA + SB$) are possible during a freezing process depending on the initial concentration of the liquid phase ($L$).

Figure 10.3b shows an idealized phase equilibrium diagram commonly used in modeling. In this diagram the solidus curve coincides with the ordinate. Pure solid of type *SA* or *SB* cannot exist alone. This equilibrium diagram is valid for many binary systems. For example, in almost all aqueous solutions of salts used in laboratory experiments in which molecules of the diluted substance (salt) do not easily fit into the crystalline lattice of the solid (ice), and vice versa. In these cases, as shown in Fig. 10.3b, the concentration of the solid dendrites in the mushy zone for all initial concentrations below $C_e$ (the eutectic concentration) is $C = 0$ (pure ice in an aqueous salt solution, for example). Similarly, the concentration of the solid in the mushy zone for initial concentrations greater than $C_e$ is $C = 1$ (pure crystallized salt in an aqueous salt solution).

Another interesting feature of the equilibrium-phase diagram of Fig. 10.3b is that both the segments of the liquidus line (the segment for $C < C_e$ and for $C > C_e$) are straight lines. This approximation is rather good for all aqueous salt solutions and aids the mathematical modeling. The equation of the liquidus line in the equilibrium diagram of Fig. 10.3b is

$$T = \begin{cases} \dfrac{T_e - T_0}{C_e} C + T_0 & 0 \le C \le C_e \\[2ex] \dfrac{T_1 - T_e}{1 - C_e} C + \dfrac{T_e - T_1 C_e}{1 - C_e} & C_e \le C \le 1 \end{cases} \tag{10.1}$$

Recall that since it was assumed that the equilibrium-phase diagram of Fig. 10.3b is adequately valid during heat conduction in the mushy zone, eq. (10.1) can be used to relate the temperature to concentration when needed in a heat conduction model. To proceed, two simple case studies are presented to illustrate the modeling of conduction during solidification of binary alloys.

## 10.2 SOLIDIFICATION OF A BINARY ALLOY FROM A FLAT COLD SURFACE

A binary alloy of initial concentration ($C_i$), considerably smaller than the eutectic concentration, is described adequately by the approximate equilibrium phase diagram of Fig. 10.3b. This alloy is contained in a large rectangular mold (Fig. 10.4a) and has an initial temperature, $T_i$, higher than the liquidus temperature, $T_{eq}$, corresponding to the initial concentration, $C_i$ (Fig. 10.4b). Suddenly, the bottom wall temperature is cooled down to $T_c$, which is significantly lower than the eutectic temperature, $T_e$ (Fig. 10.4b). The remaining walls of the mold are adiabatic. As a consequence of the above-described cooling process, solidification takes place. It is initiated near the bottom cold wall of the mold and proceeds upward. The solid and the liquid region are separated by a mixed-phase (mushy) region. The mold is assumed to be "tall enough" such that for a long time period the top wall of the mold does not affect the solidification process. We need to obtain the temperature history of the binary alloy in the mold during this time period as well as the growth rates of the solid–mush and mush–liquid interfaces. Finally, the determination of the heat flux at the bottom wall of the mold is desired.

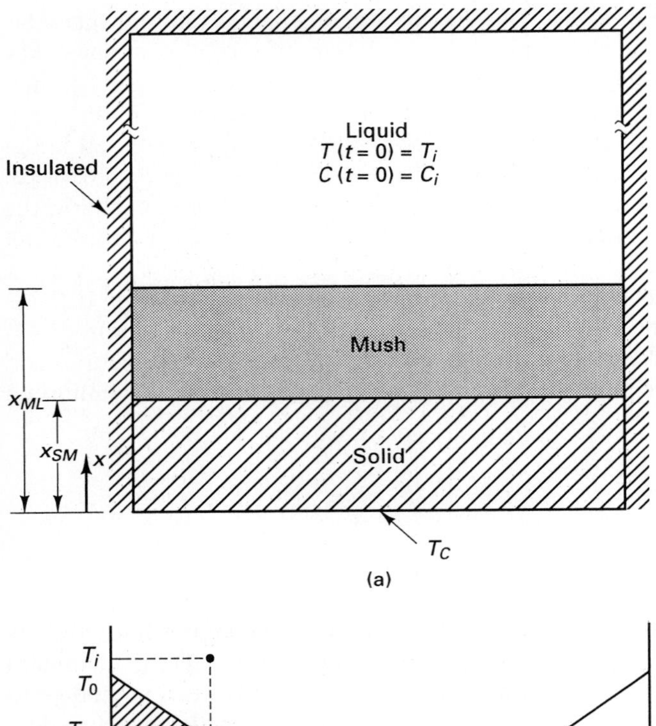


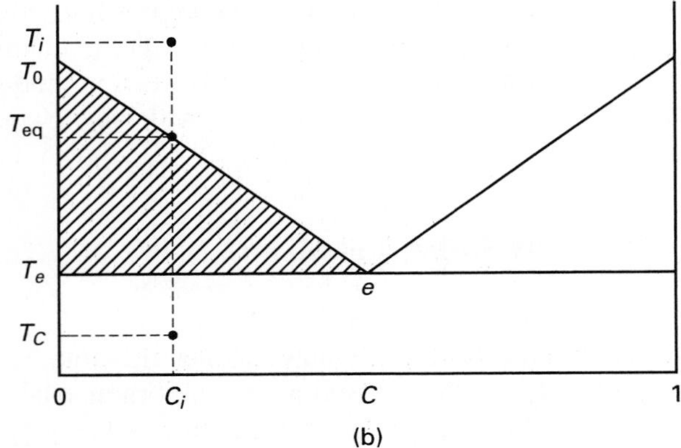


**Figure 10.4** (a) Solidification of a binary mixture from a cold surface. (b) Equilibrium-phase diagram.

**Modeling and solution.** Based on the description of the problem it is reasonable to assume that the solidification phenomenon depends only on the $x$ Cartesian coordinate of Fig. 10.4a. Three distinct regions exist in the system: The solid region (S), the mixed-phase of mushy region (M), and the liquid region (L). The heat conduction equation for these three regions is

$$\frac{1}{\alpha_S}\frac{\partial T_S}{\partial t} = \frac{\partial^2 T_S}{\partial x^2} \qquad 0 < x < x_{SM} \tag{10.2}$$

$$\frac{1}{\alpha_M}\frac{\partial T_M}{\partial t} = \frac{\partial^2 T_M}{\partial x^2} + \frac{\rho L}{k_M}\frac{\partial \chi}{\partial t} \qquad x_{SM} < x < x_{ML} \tag{10.3}$$

$$\frac{1}{\alpha_L}\frac{\partial T_L}{\partial t} = \frac{\partial^2 T_L}{\partial x^2} \qquad x_{ML} < x < \infty \tag{10.4}$$

Subscripts *SM* and *ML* stand for the solid–mush and the mush–liquid interfaces, respectively. Note that the term $\rho L(\partial\chi/\partial t)$ on the right-hand side of eq. (10.3) accounts for the heat generated in the mushy region by the creation of the solid dendrites or other crystalline structures constituting the solid matrix in the mushy zone. The solid fraction in the mushy zone is denoted by $\chi$ and it represents the percentage of volume at any point in the mush occupied by solid. Clearly, then, $\partial\chi/\partial t$ is the time rate of change of the solid volume fraction which when multiplied by $\rho L$ yields the volumetric heat generation (W/m$^3$) caused by the partial solidification inside the mushy zone. In addition, noticing that

$$\frac{\partial\chi}{\partial t} = \frac{\partial\chi}{\partial T_M}\frac{\partial T_M}{\partial t} \tag{10.5}$$

and combining eqs. (10.3) and (10.5) yields

$$\left(\frac{1}{\alpha_M} - \frac{\rho L}{k_M}\frac{\partial\chi}{\partial T_M}\right)\frac{\partial T_M}{\partial t} = \frac{\partial^2 T_M}{\partial x^2} \qquad x_{SM} < x < x_{ML} \tag{10.6}$$

As mentioned earlier in this chapter, for the purposes of this analysis $\partial\chi/\partial T_M$ will be assumed constant at an average value which, if needed for a numerical solution, can be estimated from the equilibrium phase diagram of the alloy of interest. The boundary and matching conditions necessary to complete the formulation of the conduction model are:

$x = 0$: $\quad T_S = T_c \qquad (10.7)$

$x = x_{SM}$: $\quad T_S = T_M = T_e \qquad (10.8)$

$$k_S\left(\frac{\partial T_S}{\partial x}\right)_{x=x_{SM}} - k_M\left(\frac{\partial T_M}{\partial x}\right)_{x=x_{SM}} = \rho L(1 - \chi_{SM})\frac{dx_{SM}}{dt} \tag{10.9}$$

$x = x_{ML}$: $\quad T_M = T_L = T_{\text{eq}} \qquad k_M\left(\frac{\partial T_M}{\partial x}\right)_{x=x_{ML}} - k_L\left(\frac{\partial T_L}{\partial x}\right)_{x=x_{ML}} = \rho L\chi_{ML}\frac{dx_{ML}}{dt}$

(10.10, 10.11)

$x \to \infty$: $\quad T_L = T_i \qquad (10.12)$

It is worth noting that the matching conditions (10.9) and (10.11) can be derived by integrating the governing equation (10.3), across infinitesimally thin control volumes surrounding the solid–mush interface [for condition (10.9)] and the mush–liquid interface [for condition (10.11)]. This derivation is the topic of Problem 10.1. The model described above accepts a similarity solution. The similarity variable utilized is the usual one:

$$\eta = \frac{x}{2\sqrt{\alpha_S t}} \tag{10.13}$$

In terms of the similarity variable, the governing conduction equations (10.2), (10.4), and (10.6) and boundary and matching conditions (10.7)–(10.12) read

$$\frac{d^2\Theta_S}{d\eta^2} + 2\eta\frac{d\Theta_S}{d\eta} = 0 \qquad 0 < \eta < \eta_{SM} \tag{10.14}$$

$$\frac{d^2\Theta_M}{d\eta^2} + 2A\eta\frac{d\Theta_M}{d\eta} = 0 \qquad \eta_{SM} < \eta < \eta_{ML} \tag{10.15}$$

$$\frac{d^2\Theta_L}{d\eta^2} + 2B\eta\frac{d\Theta_L}{d\eta} = 0 \qquad \eta_{ML} < \eta < \infty \tag{10.16}$$

$$\eta = 0: \qquad \Theta_S = 0 \tag{10.17}$$

$$\eta = \eta_{SM}: \qquad \Theta_S = 1, \quad \Theta_M = 0$$

$$\Theta_e\left(\frac{d\Theta_S}{d\eta}\right)_{\eta=\eta_{SM}} - R_M\left(\frac{d\Theta_M}{d\eta}\right)_{\eta=\eta_{SM}} = 2\eta_{SM}\frac{1-\chi_{SM}}{\text{Ste}}$$

(10.18, 10.19, 10.20)

$$\eta = \eta_{ML}: \qquad \Theta_M = \Theta_{\text{eq}}, \quad \Theta_L = 1$$

$$R_M\left(\frac{d\Theta_M}{d\eta}\right)_{\eta=\eta_{ML}} - R_L\Theta_i\left(\frac{d\Theta_L}{d\eta}\right)_{\eta=\eta_{ML}} = 2\eta_{ML}\frac{\chi_{ML}}{\text{Ste}}$$

(10.21, 10.22, 10.23)

$$\eta \to \infty: \qquad \Theta_L = 0 \tag{10.24}$$

In the equations above, the temperatures in the solid, mushy, and liquid regions were nondimensionalized according to the following definitions:

$$\Theta_S = \frac{T_S - T_c}{T_e - T_c} \qquad \Theta_M = \frac{T_M - T_e}{T_0 - T_e} \qquad \Theta_L = \frac{T_L - T_i}{T_{\text{eq}} - T_i} \tag{10.25}$$

In addition, the nondimensionalization process resulted in the appearance of the following groups:

$$A = \frac{C_{pM} + L(\partial\chi/\partial T_M)}{C_{pS}\,(k_M/k_S)} \tag{10.26}$$

$$B = \frac{C_{pL}}{C_{pS}}\frac{k_S}{k_L} \tag{10.27}$$

$$R_M = \frac{k_M}{k_S} \tag{10.28}$$

$$R_L = \frac{k_L}{k_S} \tag{10.29}$$

$$\Theta_i = \frac{T_{\text{eq}} - T_i}{T_0 - T_e} \tag{10.30}$$

$$\Theta_e = \frac{T_e - T_c}{T_0 - T_e} \tag{10.31}$$

$$\Theta_{\text{eq}} = \frac{T_{\text{eq}} - T_e}{T_0 - T_e} \tag{10.32}$$

$$\text{Ste} = \frac{C_{pS}(T_0 - T_e)}{L} \tag{10.33}$$

The dimensionless groups above represent various ratios of properties and of characteristic temperature differences in the three regions of interest (solid, mush, and liquid). The last group [eq. 10.33] is the Stefan number.

The solution of the similarity model of eqs. (10.14)–(10.24) can be obtained in terms of the error function. The procedure is similar to that discussed in Example 9.1. Omitting the algebraic details for brevity and because they are the topic of Problem 10.3, we report the final results for the temperature distribution.

*Solid region* $0 < \eta < \eta_{SM}$:

$$T_S(\eta) = T_c + (T_e - T_c)\frac{\text{erf}(\eta)}{\text{erf}(\eta_{SM})} \tag{10.34}$$

*Mushy region* $\eta_{SM} < \eta < \eta_{ML}$:

$$T_M = \frac{T_e \,\text{erf}(A^{1/2}\eta_{ML}) - T_{\text{eq}}\,\text{erf}(A^{1/2}\eta_{SM}) + (T_{\text{eq}} - T_e)\,\text{erf}(A^{1/2}\eta)}{\text{erf}(A^{1/2}\eta_{ML}) - \text{erf}(A^{1/2}\eta_{SM})} \tag{10.35}$$

*Liquid region* $\eta_{ML} < \eta < \infty$:

$$T_L = T_i + (T_{\text{eq}} - T_i)\,\frac{\text{erfc}(B^{1/2}\eta)}{\text{erfc}(B^{1/2}\eta_{ML})} \tag{10.36}$$

At this point of the solution process, all matching and boundary conditions have been utilized, except for conditions (10.20) and (10.23). To obtain the values of the similarity variable at the solid–mush and mush–liquid interfaces ($\eta_{SM}$ and $\eta_{ML}$) expressions (10.34)–(10.36) are substituted into conditions (10.20) and (10.23). The result of this operation reads

$$\frac{\Theta_e e^{-\eta_{SM}^2}}{\text{erf}(\eta_{SM})} - \frac{A^{1/2}R_M\Theta_{\text{eq}}e^{-A\eta_{SM}^2}}{\text{erf}(A^{1/2}\eta_{ML}) - \text{erf}(A^{1/2}\eta_{SM})} = \frac{\sqrt{\pi}\,\eta_{SM}(1 - \chi_{SM})}{\text{Ste}} \tag{10.37}$$

$$\frac{A^{1/2}R_M\Theta_{\text{eq}}e^{-A\eta_{ML}^2}}{\text{erf}(A^{1/2}\eta_{ML}) - \text{erf}(A^{1/2}\eta_{SM})} + \frac{B^{1/2}R_L\Theta_i e^{-B\eta_{ML}^2}}{\text{erfc}(B^{1/2}\eta_{ML})} = \frac{\sqrt{\pi}\,\eta_{ML}\,\chi_{ML}}{\text{Ste}} \tag{10.38}$$

Expressions (10.37) and (10.38) are two nonlinear algebraic equations for $\eta_{SM}$ and $\eta_{ML}$. Once the values of the various parameters are prescribed, these equations can be solved simultaneously numerically to yield the values of $\eta_{SM}$ and $\eta_{ML}$. Note that a plethora of numerical methods exists in the literature [21, 22] for the solution of nonlinear algebraic equations. With $\eta_{SM}$ and $\eta_{ML}$ known, eqs. (10.34)–(10.36) yield the temperature at any point and time in the system.

The growth rates of the solid–mush and mush–liquid interfaces are given by

$$V_{SM} = \frac{dx_{SM}}{dt} \qquad V_{ML} = \frac{dx_{ML}}{dt} \tag{10.39, 10.40}$$

Recalling that $\eta_{SM} = x_{SM}/2\sqrt{\alpha_S t}$, $\eta_{ML} = x_{ML}/2\sqrt{\alpha_S t}$, the equations above become

$$V_{SM} = \eta_{SM}\frac{\alpha_S^{1/2}}{t^{1/2}} \qquad V_{ML} = \eta_{ML}\frac{\alpha_S^{1/2}}{t^{1/2}} \tag{10.41, 10.42}$$

The growth rates of both interfaces are inversely proportional to the square root of time. Numerical values for $\eta_{SM}$ and $\eta_{ML}$ in eqs. (10.41) and (10.42) are obtained in the manner explained earlier.

The heat flux at the bottom wall of the mold is calculated by direct application of Fourier's law:

$$\dot{q}'' = -k_S\left(\frac{\partial T_S}{\partial x}\right)_{x=0} = \frac{-k_S}{2\sqrt{\alpha_S t}}\left(\frac{dT_S}{d\eta}\right)_{\eta=0} \tag{10.43}$$

Combining eqs. (11.43) and (11.34) yields

$$\dot{q}'' = -\frac{k_S(T_e - T_c)}{\sqrt{\pi\alpha_S t}}\frac{1}{\operatorname{erf}(\eta_{SM})} \tag{10.44}$$

## 10.3 SOLIDIFICATION OF A BINARY ALLOY FROM A COLD CAPILLARY PIPE

A long capillary pipe is immersed in a large container of a binary alloy of temperature $T_i$ and concentration $C_i$ described by the equilibrium-phase diagram of Fig. 10.4b. Suddenly, a coolant is circulated through the pipe. The flow rate and the temperature of the coolant are externally controlled such that the pipe absorbs heat at a constant rate $\dot{Q}'$ (W/m). After some time solidification takes place and three distinct regions surround the pipe: a solid region, a mixed-phase region (mushy zone), and a liquid region (Fig. 10.5). It is of interest to obtain the temperature distribution in these three regions and the growth rates of the solid–mush and the mush–liquid interfaces.

**Modeling and solution.** Following the problem description and according to the approach adopted in Section 10.2 for solidification from a flat surface, it is reasonable to assume that heat conduction occurs in the radial direction ($r$) only. In addition, since the pipe is very thin (capillary), it can conveniently be modeled as a line heat sink absorbing heat per unit length at a constant rate $\dot{Q}'$. The heat conduction equations in the three regions (solid, mushy, and liquid) present during the freezing process (Fig.10.5) are

$$\frac{1}{\alpha_S}\frac{\partial T_S}{\partial t} = \frac{1}{r}\frac{\partial}{\partial r}\left(r\frac{\partial T_S}{\partial r}\right) \qquad 0 < r < r_{SM} \tag{10.45}$$

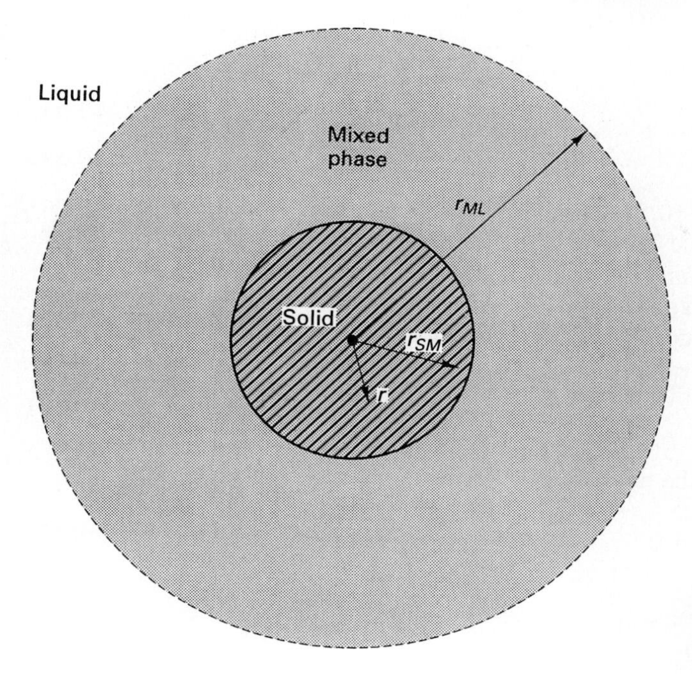


**Figure 10.5** Solidification of a binary mixture around a capillary pipe.

$$\frac{1}{\alpha_M}\frac{\partial T_M}{\partial t} = \frac{1}{r}\frac{\partial}{\partial r}\left(r\frac{\partial T_M}{\partial r}\right) + \frac{\rho_M}{k_M}L\frac{\partial \chi}{\partial t} \qquad r_{SM} < r < r_{ML} \tag{10.46}$$

$$\frac{1}{\alpha_L}\frac{\partial T_L}{\partial t} = \frac{1}{r}\frac{\partial}{\partial r}\left(r\frac{\partial T_L}{\partial r}\right) \qquad r_{ML} < r < \infty \tag{10.47}$$

The notation and the subscripts are identical to what was defined in Section 10.2. In addition, taking eq. (10.5) into account, the conduction equation for the mushy zone (eq. 10.46) becomes

$$\left(\frac{1}{\alpha_M} - \frac{\rho_M L}{k_M}\frac{\partial \chi}{\partial T_M}\right)\frac{\partial T_M}{\partial t} = \frac{1}{r}\frac{\partial}{\partial r}\left(r\frac{\partial T_M}{\partial r}\right) \qquad r_{SM} < r < r_{ML} \tag{10.48}$$

As in Section 10.2, an average value for $\partial\chi/\partial T_M$ will be utilized for the purposes of the analytical solution. This average value can be estimated from the equilibrium phase diagram of the alloy of interest.

The boundary and matching conditions necessary to complete the mathematical model are

$$r \to 0: \qquad \dot{Q}' = k_S 2\pi r\frac{\partial T_S}{\partial r} \tag{10.49}$$

$$r = r_{SM}: \qquad T_S = T_M = T_e \qquad k_S\frac{\partial T_S}{\partial r} - k_M\frac{\partial T_M}{\partial r} = \rho_M L(1 - \chi_{SM})\frac{dr_{SM}}{dt} \qquad (10.50,\ 10.51)$$

$$r = r_{ML}: \qquad T_M = T_L = T_{eq} \qquad k_M \frac{\partial T_M}{\partial r} - k_L \frac{\partial T_L}{\partial r} = \rho_L L \chi_{ML} \frac{dr_{ML}}{dt} \tag{10.52, 10.53}$$

$$r \to \infty: \qquad T_L = T_i \tag{10.54}$$

The matching conditions (10.51) and (10.53) can be derived by integrating the conduction equation for the mushy zone across infinitesimally thin control volumes surrounding the solid–mush and the mush–liquid interfaces, respectively. This derivation is the object of Problem 10.2.

The mathematical model described earlier accepts a similarity solution. The similarity variable is defined as

$$\eta = \frac{r^2}{4\alpha_S t} \tag{10.55}$$

In terms of this variable the model for the freezing process reads

$$\eta \frac{d^2\Theta_S}{d\eta^2} + (1 + \eta) \frac{d\Theta_S}{d\eta} = 0 \qquad 0 < \eta < \eta_{SM} \tag{10.56}$$

$$\eta \frac{d^2\Theta_M}{d\eta^2} + (A\eta + 1) \frac{d\Theta_M}{d\eta} = 0 \qquad \eta_{SM} < \eta < \eta_{ML} \tag{10.57}$$

$$\eta \frac{d^2\Theta_L}{d\eta^2} + (B\eta + 1) \frac{d\Theta_L}{d\eta} = 0 \qquad \eta_{ML} < \eta < \infty \tag{10.58}$$

$$\eta \to 0: \qquad \eta \frac{d\Theta_S}{d\eta} = \frac{1}{4\pi} \tag{10.59}$$

$$\eta = \eta_{SM}: \qquad \Theta_S = 0, \quad \Theta_M = 0$$

$$\frac{d\Theta_S}{d\eta} - D \frac{d\Theta_M}{d\eta} = (1 - \chi_{SM}) \frac{D}{\mathrm{Ste} R_M} \tag{10.60, 10.61, 10.62}$$

$$\eta = \eta_{ML}: \qquad \Theta_M = \Theta_{eq}, \quad \Theta_L = 1$$

$$R_M \frac{d\Theta_M}{d\eta} - R_L \Theta_i \frac{d\Theta_L}{d\eta} = \frac{\chi_{ML}}{\mathrm{Ste}} \tag{10.63, 10.64, 10.65}$$

$$\eta \to \infty: \qquad \Theta_L = 0 \tag{10.66}$$

In the equations above, the nondimensionalization was carried out according to the following definitions:

$$\Theta_S = \frac{T_S - T_e}{\dot{Q}'/k_S} \qquad \Theta_M = \frac{T_M - T_e}{T_0 - T_e} \qquad \Theta_L = \frac{T_L - T_e}{T_{eq} - T_i} \tag{10.67}$$

In addition, the following dimensionless groups appeared in the dimensionless similarity model:

$$A = \frac{C_{pM} + L(\partial\chi/\partial T_M)}{C_{pS}(k_M/k_S)} \tag{10.68}$$

$$B = \frac{C_{pL} k_S}{C_{pS} k_L} \tag{10.69}$$

$$D = \frac{k_M (T_0 - T_e)}{\dot{Q}'} \tag{10.70}$$

$$R_M = \frac{k_M}{k_S} \tag{10.71}$$

$$R_L = \frac{k_L}{k_S} \tag{10.72}$$

$$\Theta_{\text{eq}} = \frac{T_{\text{eq}} - T_e}{T_0 - T_e} \tag{10.73}$$

$$\Theta_i = \frac{T_{\text{eq}} - T_i}{T_0 - T_e} \tag{10.74}$$

$$\text{Ste} = \frac{C_{pS}(T_0 - T_e)}{L} \tag{10.75}$$

These dimensionless groups represent ratios of properties and of characteristic temperature differences in the solid, mushy, and liquid regions. The last group is the Stefan number.

The solution of the similarity model can be obtained in terms of the exponential integral function (Problem 10.4). Omitting the details for brevity, we present only the final expressions for the temperature.

*Solid region* $0 < \eta < \eta_{SM}$:

$$T_S = T_e + \frac{\dot{Q}'}{4\pi k_S}[\text{Ei}(\eta_{SM}) - \text{Ei}(\eta)] \tag{10.76}$$

*Mushy zone* $\eta_{SM} < \eta < \eta_{ML}$:

$$T_M = T_e + (T_{\text{eq}} - T_e)\frac{\text{Ei}(A\eta_{SM}) - \text{Ei}(A\eta)}{\text{Ei}(A\eta_{SM}) - \text{Ei}(A\eta_{ML})} \tag{10.77}$$

*Liquid region* $\eta_{ML} < \eta < \infty$:

$$T_L = T_i + (T_{\text{eq}} - T_i)\frac{\text{Ei}(B\eta)}{\text{Ei}(B\eta_{ML})} \tag{10.78}$$

To obtain $\eta_{SM}$ and $\eta_{ML}$ (which are still unknown), eqs. (10.76)–(10.78) are substituted into conditions (10.62) and (10.65) to yield

$$\frac{1}{4\pi}e^{-\eta_{SM}} - D\Theta_{\text{eq}}\frac{e^{-A\eta_{SM}}}{\text{Ei}(A\eta_{SM}) - \text{Ei}(A\eta_{ML})} = \eta_{SM}(1 - \chi_{SM})\frac{D}{\text{Ste}R_M} \tag{10.79}$$

$$R_M\Theta_{\text{eq}}\frac{e^{-A\eta_{ML}}}{\text{Ei}(A\eta_{SM}) - \text{Ei}(A\eta_{ML})} + R_L\Theta_i\frac{e^{-B\eta_{ML}}}{\text{Ei}(\eta_{ML})} = \eta_{ML}\frac{\chi_{ML}}{\text{Ste}} \tag{10.80}$$

Expressions (10.79) and (10.80) are two nonlinear algebraic equations for $\eta_{SM}$ and $\eta_{ML}$. Once the values of the various parameters are prescribed, these equations can be solved simultaneously numerically to yield the values of $\eta_{SM}$ and $\eta_{ML}$. Note that a plethora of numerical methods exists in the literature [21, 22] for the solution of nonlinear algebraic equations. With $\eta_{SM}$ and $\eta_{ML}$ known, eqs. (10.76)–(10.78) yield the temperature at any point and time in the system.

The growth rates of the solid–mush and mush–liquid interfaces are given by

$$V_{SM} = \frac{dr_{SM}}{dt} \qquad V_{ML} = \frac{dr_{ML}}{dt} \tag{10.81, 10.82}$$

Recalling that $\eta_{SM} = r_{SM}^2/4\alpha_s t$ and $\eta_{ML} = r_{ML}^2/4\alpha_s t$, the equations above become

$$V_{SM} = \eta_{SM}^{1/2} \frac{\alpha_S^{1/2}}{t^{1/2}} \qquad V_{ML} = \eta_{ML}^{1/2} \frac{\alpha_S^{1/2}}{t^{1/2}} \tag{10.83, 10.84}$$

The growth rates of both interfaces are inversely proportional to the square root of time. Numerical values for $\eta_{SM}$ and $\eta_{ML}$ in eqs. (10.83) and (10.84) are obtained in the manner explained earlier.

The procedure described in this example can be applied identically to model approximately conduction-dominated unidirectional alloy solidification in spherical coordinates. This is the topic of Problem 10.6.

## PROBLEMS

**10.1.** Derive the matching conditions (10.9) and (10.11). (*Hint:* You may begin by integrating the conduction equation for the mushy zone across infinitesimally thin control volumes surrounding the solid–mush and mush–liquid interfaces.)

**10.2.** Derive the matching conditions (10.51) and (10.53). (*Hint:* You may begin by integrating the conduction equation for the mushy zone across infinitesimally thin control volumes surrounding the solid–mush and the mush–liquid interfaces.)

**10.3.** Prove that the solution of the similarity model (10.14)–(10.23) is given by eqs. (10.34)–(10.38).

**10.4.** Prove that the solution to the similarity model (10.56)–(10.66) is given by eqs. (10.76)–(10.80).

**10.5.** A binary alloy is saturating a large packed bed of beads. Utilizing the procedure of Example 10.1 as well as the analysis contained in Ref. [20], obtain the temperature distribution in the solid, mushy, and liquid regions during the freezing of this alloy. Assume that the initial temperature of the alloy is $T_i$. Suddenly, the temperature of bottom wall of the packed bed is lowered down to $T_C$, much lower than the eutectic temperature of the alloy. Assume also that the binary alloy is adequately described by the equilibrium phase diagram of Fig. 10.2b. What is the heat flux at the cold wall? What is the growth rate of the solid–mush and the mush–liquid interface?

**10.6.** Assume that solidification is initiated by an impurity in the melt of a binary alloy of temperature $T_i$ and concentration $C_i$, described by the equilibrium phase diagram of Fig. 10.3b. The impurity can be modeled (for illustration purposes) as a point sink absorbing heat at a rate $Q = Q_0 t^{1/2}$, where $Q_0$ is a constant and where $t$ is time. Obtain the temperature distribution in the alloy and the growth rate of the solid-mush and the mush–liquid interfaces.

**10.7.** Reconsider the problem of freezing of a binary alloy around a cold capillary pipe (Section 10.3) for the case where the binary alloy is saturating a solid porous matrix. Obtain the temperature field in the alloy. Consult Ref. [20] and Problem 10.5. Such a solidification process may occur in food freezing or the freezing of tissue.

**10.8.** Re-solve Problem 10.6 for the case where the alloy is saturating a porous solid matrix. Follow the procedure outlined in Ref. [20].

## REFERENCES


1. W. D. BENNON and F. P. INCROPERA, A continuum model for momentum, heat and species transport in binary solid–liquid phase change systems. I. Model formulation, *Int. J. Heat Mass Transfer,* Vol. 30, pp. 2161–2170, 1987.
2. P. J. PRESCOTT, F. P. INCROPERA and W. D. BENNON, Modelling of dendritic solidification systems: reassessment of the continuum momentum equation, *Int. J. Heat Mass Transfer,* Vol. 34, pp. 2351–2359, 1991.
3. V. R. VOLLER, A. D. BRENT, and C. PRAKASH, The modelling of heat, mass and solute transport in solidification systems, *Int. J. Heat Transfer,* Vol. 32, pp. 1718–1731, 1989.
4. R. VISKANTA, Mathematical modeling of transport processes during solidification of binary systems, *J. SME Int. Ser. II,* Vol. 33, pp. 409–423, 1990.
5. J. NI and C. BECKERMANN, A two-phase model for mass, momentum, heat, and species transport during solidification, *AIAA/ASME Joint Thermophysics and Heat Transfer Conference,* Seattle, WA, June 18–20, 1990.
6. C. BECKERMANN, Melting and solidification of binary mixtures with double-diffusive convection in the melt, *Ph.D. thesis,* Purdue University, West Lafayette, IN, 1987.
7. C. BECKERMANN and R. VISKANTA, Double-diffusive convection during dendritic solidification of binary mixture, *PhysicoChem. Hydrodynam.,* Vol. 10, pp. 195–213, 1988.
8. S. GANESAN and D. R. POIRIER, Conservation of mass and momentum for the flow of interdendritic liquid during solidification, *Metall. Trans. B,* Vol. 21B, pp. 173–181, 1990.
9. I. KECECIOGLU and B. RUBINSKY, A continuum model for the propagation of discrete phase-change fronts in porous media in the presence of coupled heat flow, fluid flow and species transport processes, *Int. J. Heat Mass Transfer,* Vol. 32, pp. 1111–1130, 1989.
10. M. G. O'CALLAGHAN, E. G. GRAVALHO, and C. E. HUGGINS, An analysis of heat and solute transport during solidification of an aqueous binary solution, *Int. J. Heat Mass Transfer,* Vol. 25, pp. 553–573, 1982.
11. L. J. FANG, F. B. CHEUNG, J. H. LINEHAN, and D. R. PEDERSON, Selective freezing of a dilute salt solution on a cold ice surface, *J. Heat Transfer,* Vol. 106, pp. 385–393, 1984.
12. B. W. WEBB and R. VISKANTA, An experimental and analytical study of solidification of a binary mixture, in *Heat Transfer 1986: Proceedings of the 8th International Heat*

*Transfer Conference,* San Francisco, August 1986, Vol. 4, C. L. Tien, V. P. Carey, and J. K. Ferrell, Eds. Hemisphere, New York, pp. 1739–1744, 1986.

13. V. Alexiadis, D. G. Wilson, and A. D. Solomon, Macroscopic global modelling of binary alloy solidification processes, *Q. Appl. Math.,* Vol. 43, pp. 143–158, 1985.
14. R. Trivedi, Theory of dendritic growth during the directional solidification of binary alloys, *J. Cryst. Growth,* Vol. 49, pp. 219–232, 1980.
15. M. G. Worster, Solidification of an alloy from a cooled boundary, *J. Fluid Mech.,* Vol. 167, pp. 481–501, 1986.
16. M. N. Ozisik and J. C. Uzzell, Jr., Exact solution for freezing in cylindrical symmetry with extended freezing temperature range, *J. Heat Transfer,* Vol. 101, pp. 331–334, 1979.
17. D. Poulikakos, On the growth of a solid from a line heat sink in a binary alloy, *Numer. Heat Transfer,* Vol. 14, pp. 113–126, 1988.
18. D. Poulikakos and W.-Z. Cao, Solidification of a binary alloy from a cold wire or pipe: modelling of the mixed phase region, *Numer. Heat Transfer Part A,* Vol. 15, pp. 197–219, 1989.
19. Y. Hayashi and T. Komori, Investigation of freezing of salt solutions in cells, *J. Heat Transfer,* Vol. 101, pp. 459–464, 1979.
20. W.-Z. Cao and D. Poulikakos, Freezing of a binary alloy saturating a packed bed of spheres, *AIAA J. Thermophys. Heat Transfer,* Vol. 5, pp. 46–53, 1991.
21. Y. Jaluria, *Computer Methods for Engineering,* Allyn and Bacon, Boston, 1988.
22. B. Carnahan, H. A. Luther, and J. O. Wilkes, *Applied Numerical Methods,* Wiley, New York, 1969.

CHAPTER **11**

# NUMERICAL SOLUTION OF HEAT CONDUCTION PROBLEMS

In this chapter we discuss the numerical solution of heat conduction problems. Often in practical applications, the heat conduction models describing the phenomena of interest are complex. As a result, the analytical methods presented so far in this book cannot be used to determine the temperature field. These methods can always be used to solve simplified models of complex phenomena. However, when simplified models are not adequate, the utilization of numerical methods is necessary. Complexities that make heat conduction models not anemeable to accurate analytical solutions are exemplified by the temperature dependence of the thermophysical properties of a material, complicated boundary conditions, the presence of heat sources or sinks, complicated geometries (other than simple rectangular, cylindrical, and spherical), and multidimensional freezing or melting.

Two major methodologies are usually employed to obtain numerical solutions of heat conduction problems: the finite difference methodology and the finite element methodology. This chapter focuses on the former (finite difference). Information on the finite element methodology can be found in numerous books dedicated to the topic of numerical methods in engineering in general and in thermal sciences in particular [1–4]. Owing to space limitations as well as to the fact that this is a heat conduction textbook and not a textbook on numerical methods, the classroom-level discussion of only one of the two methodologies was possible. We selected to present the finite difference method in this chapter because of its simplicity as well as its great popularity in teaching, research, and engineering practice in the area of thermal sciences [1,4–7].

## 11.1 GENERAL PRINCIPLES FOR THE FINITE DIFFERENCING OF HEAT CONDUCTION MODELS

To obtain the solution of a heat conduction problem numerically, the mathematical model of the problem needs to be cast in a form that allows for the processing of information for the temperature by a computer at *discrete* points inside the body of interest. Since all the mathematical models discussed in this book deal with heat conduction in a continuum, a process must be introduced by which a continuum heat conduction model can be converted to a *discretized* heat conduction model representing a number of discrete points of the system of interest. If the number of these points is large enough, an accurate picture of the temperature distribution and the heat conduction phenomenon in the body of interest can be pieced together. The above-mentioned process is termed *discretization.*

A sequence of steps in obtaining the numerical solution of heat conduction problems is structured as follows:

■ ***Step 1***

Obtain the mathematical model of the problem of interest. This model consists of the conduction equation and the corresponding boundary and initial conditions, all written for a medium (continuum) within which heat conduction takes place.

■ ***Step 2***

Apply the discretization process. This step involves, first, the identification of a sufficiently large group of points inside the continuum. Knowledge of the temperature at these points should adequately describe the temperature field in the body. Next, discretization of the heat conduction model of step 1 with respect to the previously identified group of points needs to take place. At the end of step 2 we are left with a set of algebraic equations (in place of the differential equations of the continuum model) describing the temperature of the body at a group of discrete points.

■ ***Step 3***

Solve the set of algebraic equations of step 2 and obtain the temperature of the body at a number of discrete points. As mentioned earlier, if the number of these points is large enough, knowledge of their temperature can be used to piece together an approximate, yet acceptably accurate, picture of the temperature field.

The first step in the sequence above does not differ from what has been discussed so far in this book. It is the second and third steps that constitute the focus of this chapter. Clearly, the initial task of the discretization process of step 2 is to define a group of discrete points in the body of interest the temperature of which is to be determined. In the finite difference method, this task is performed by overlaying the region of interest with a *net* or *grid* or *mesh* of intersecting lines (gridlines). The discrete intersection points of these lines are then the *grid points*. Figure 11.1a shows a typical set of grid points in a rectangular region. The gridlines are drawn equidistantly and parallel to the sides of the rectangular region ($m$ vertically and $n$ horizontally) for simplicity. Note that, in general, the distance between two sequential gridlines can vary depending on the location in the region of interest. The number and the distancing of the gridlines affect the number of the resulting grid points and therefore the accuracy of the numerically obtained temperature field. The placement of the grid lines is usually determined from experience, after a trial-and-error

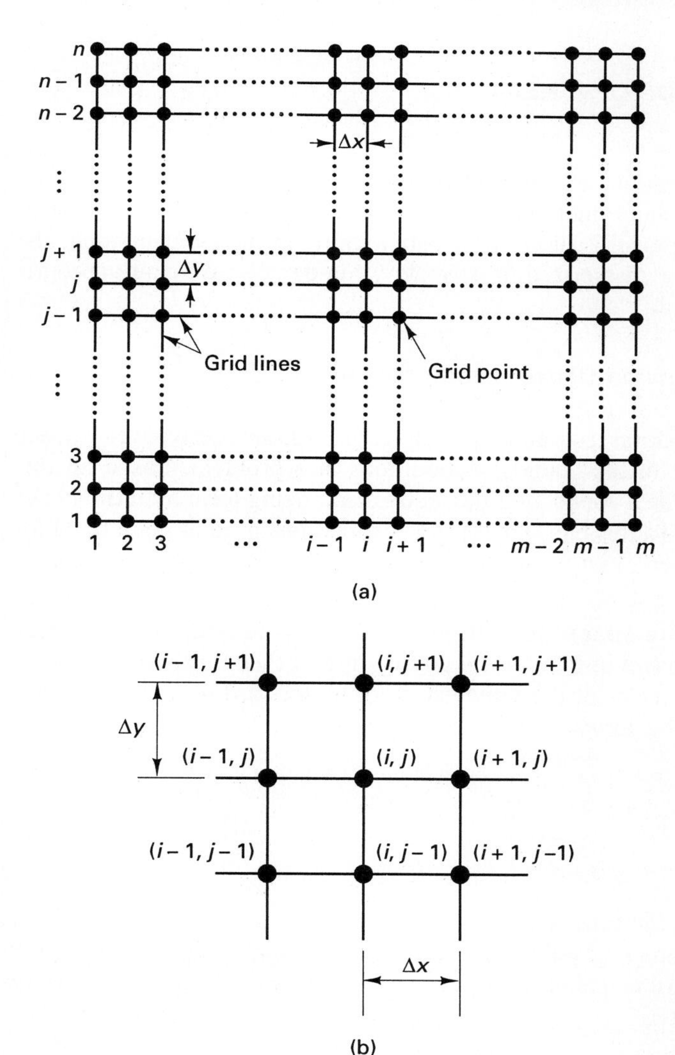


**Figure 11.1** (a) Grid network. (b) Group of grid points surrounding point $(i,j)$.

process, and depends on the individual features of the problem of interest. Once a set of grid points is generated in the region of interest, the mathematical model of the heat conduction problem of interest can be discretized. It is worth noting that the time variable, $t$, in a heat conduction model is discretized in a manner similar to that of the space variables. To this end, it is assumed that the heat conduction process progresses in a sequence of time steps each of which is denoted by $\Delta t$. The choice of magnitude of $\Delta t$ depends on the problem under consideration and the solution method. Its proper determination (sometimes a result of a trial-and-error process) is crucial to the efficiency and the convergence of the solution. Two popular discretization methods [1–7] are discussed next.

## 11.2 DISCRETIZATION METHODS OF A HEAT CONDUCTION MODEL

Two methods of discretization are presented here. The Taylor series expansion method and the control volume method. While in the former method, each derivative in a heat conduction model is discretized individually; in the latter method, the heat conduction equation is discretized as a whole by integration over an appropriately chosen control volume.

### The Taylor Series Expansion Discretization Method

The heat conduction equation contains both time and space derivatives. Space derivatives may also exist in the boundary conditions of a problem. Based on this fact, one may discretize a heat conduction model by discretizing term by term all the derivatives in this model. To this end, appropriate expressions need to be derived for both the space and time derivatives usually appearing in heat conduction models.

**Discretization of the space derivatives of the temperatures.** Consider the group of grid points shown in Fig. 11.1b. At any instant the temperature of point $(i, j)$ can be expressed in terms of the temperature of its forward neighboring point $(i + 1, j)$ with the help of a forward Taylor series expansion:

$$T^P_{i+1,j} = T^P_{i,j} + \left(\frac{\partial T}{\partial x}\right)^P_{i,j} \Delta x + \frac{1}{2!}\left(\frac{\partial^2 T}{\partial x^2}\right)^P_{i,j} (\Delta x)^2 + \frac{1}{3!}\left(\frac{\partial^3 T}{\partial x^3}\right)^P_{i,j} (\Delta x)^3 + (\text{higher-order terms}) \tag{11.1}$$

The superscript $P$ denotes the time step at which eq. (11.1) is applied ($t = P\,\Delta t$ if $\Delta t$ is constant). The temperature of point $(i, j)$ can also be expressed in terms of the temperature of its backward neighboring point $(i - 1, j)$ with the help of a backward Taylor series approximation.

$$T_{i-1,j}^P = T_{i,j}^P - \left(\frac{\partial T}{\partial x}\right)_{i,j}^P \Delta x + \frac{1}{2!}\left(\frac{\partial^2 T}{\partial x^2}\right)_{i,j}^P (\Delta x)^2 - \frac{1}{3!}\left(\frac{\partial^3 T}{\partial x^3}\right)_{i,j}^P (\Delta x)^3 + \text{(higher-order terms)} \tag{11.2}$$

The expressions (11.1) and (11.2) can be used to yield useful approximations for the first and second derivatives of the temperature with respect of $x$ at the $p$th time step. Three such approximations are offered here for the first derivative of the temperature.

**1.** *Forward-difference approximation.* This approximation can be obtained by neglecting all the terms of order $(\Delta x)^2$ or higher in eq. (11.1) and by solving the resulting expression for the temperature derivative:

$$\left(\frac{\partial T}{\partial x}\right)_{i,j}^P = \frac{T_{i+1,j}^P - T_{i,j}^P}{\Delta x} \tag{11.3}$$

Equation (11.3) gives the temperature derivative at point $(i, j)$ as a function of the temperature at that point and at its forward point. The accuracy of eq. (11.3) is $O(\Delta x)$ since all the higher-order terms in the Taylor series (11.1) were neglected.

**2.** *Backward-difference approximation.* Utilizing eq. (11.2) this time and omitting all terms of order $(\Delta x)^2$ and higher yields the following expression for the temperature derivative:

$$\left(\frac{\partial T}{\partial x}\right)_{i,j}^P = \frac{T_{i,j}^P - T_{i-1,j}^P}{\Delta x} \tag{11.4}$$

Equation (11.4) approximates the derivative of the temperature at point $(i, j)$ utilizing the temperature at this point and the temperature at its backward point. The accuracy of eq. (11.4) is $O(\Delta x)$ since it was derived from eq. (11.2) by eliminating all the higher-order terms.

**3.** *Centered-difference approximation.* If eq. (11.2) is subtracted from eq. (11.1), we obtain

$$T_{i+1,j}^P - T_{i-1,j}^P = 2\left(\frac{\partial T}{\partial x}\right)_{i,j}^P \Delta x + \frac{1}{3!}\left(\frac{\partial^3 T}{\partial x^3}\right)_{i,j}^P (\Delta x)^3 + \text{(higher-order terms)} \tag{11.5}$$

Note that terms of order $(\Delta x)^2$ canceled out in the subtraction process. Neglecting the terms of order $(\Delta x)^3$ and higher and solving the resulting expression for the derivative of the temperature yields

$$\left(\frac{\partial T}{\partial x}\right)_{i,j}^P = \frac{T_{i+1,j}^P - T_{i-1,j}^P}{2\Delta x} \tag{11.6}$$

Equation (11.6) is the centered-difference approximation of the temperature derivative at point $(i, j)$ and it expresses this derivative in terms of the temperature of the backward point $(i - 1, j)$ and the forward point $(i + 1, j)$ of point $(i, j)$. Expression (11.6) is accurate to $O(\Delta x)^2$ since all the terms of order higher than 2 in eq. (11.5) were neglected.

A third-order-accurate centered-difference expression for the second derivative of the temperature at point $(i, j)$ can be derived in a manner similar to that used to obtain eq. (11.6). To this end, adding eq. (11.2) to eq. (11.1) yields

$$T_{i+1,j}^{P} + T_{i-1,j}^{P} = 2T_{i,j}^{P} + \left(\frac{\partial^2 T}{\partial x^2}\right)_{i,j}^{P} (\Delta x)^2 + (\text{higher-order terms}) \tag{11.7}$$

Neglecting the higher-order terms in this equation, we obtain

$$\left(\frac{\partial^2 T}{\partial x^2}\right)_{i,j}^{P} = \frac{T_{i+1,j}^{P} + T_{i-1,j}^{P} - 2T_{i,j}^{P}}{(\Delta x)^2} \tag{11.8}$$

Note that the second derivative of the temperature at point $(i, j)$ is expressed in terms of the temperature at this point and the temperature at its forward and backward points.

Expressions similar to eqs. (11.3), (11.4), (11.6), and (11.8) can be derived for the temperature derivatives in any other direction by utilizing Taylor series expansions in that direction. The results of such procedure in the $y$-direction, for example, are

*Forward-difference approximation:*

$$\left(\frac{\partial T}{\partial y}\right)_{i,j}^{P} = \frac{T_{i,j+1}^{P} - T_{i,j}^{P}}{\Delta y} \tag{11.9}$$

*Backward-difference approximation:*

$$\left(\frac{\partial T}{\partial y}\right)_{i,j}^{P} = \frac{T_{i,j}^{P} - T_{i,j-1}^{P}}{\Delta y} \tag{11.10}$$

*Centered-difference approximation:*

$$\left(\frac{\partial T}{\partial y}\right)_{i,j}^{P} = \frac{T_{i,j+1}^{P} - T_{i,j-1}^{P}}{2\Delta y}$$

$$\left(\frac{\partial^2 T}{\partial y^2}\right)_{i,j}^{P} = \frac{T_{i,j+1}^{P} + T_{i,j-1}^{P} - 2T_{i,j}^{P}}{(\Delta y)^2} \qquad (11.11,\ 11.12)$$

Based on our discussion so far, it is reasonable to conclude that the centered-difference approximation should be preferable to both the forward- and backward-difference approximations since it has accuracy $O(\Delta x)^2$, whereas the accuracy of the forward- and backward-difference approximations is $O(\Delta x)$. Although the statement above is generally true, a problem arises when it is desired to discretize a derivative at a grid point located on a boundary. As shown in Fig. 11.2, the group of points $(1, j)$ for $j = 1, \ldots, n$ are located on the left boundary of the region of interest

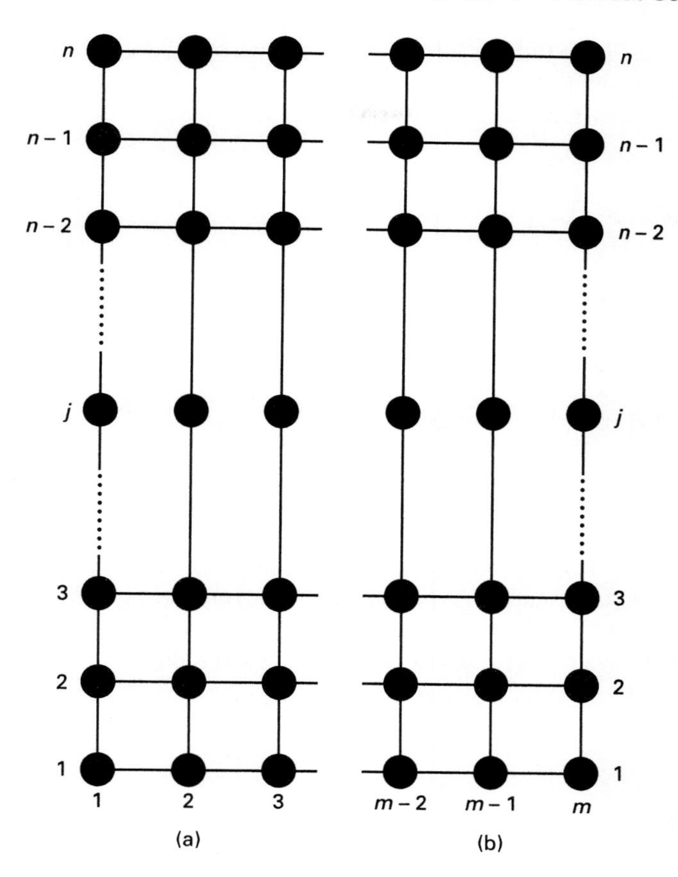


**Figure 11.2** (a) Grid points near left boundary. (b) Grid points near right boundary.

(Fig. 11.2a), while the group of points $(m, j)$ for $j = 1, \ldots, n$ are located on the right boundary of the region of interest. Since the centered-difference approximation involves both the backward and forward points of the groups $(1, j)$ and $(m, j)$ it cannot be used to estimate the temperature derivatives at the boundaries. For example, if eq. (11.6) is used to approximate the temperature derivatives at the left and right boundaries of Fig. 11.2a and b, it will yield

$$\left(\frac{\partial T}{\partial x}\right)_{1,j}^{P} = \frac{T_{2,j}^{P} - T_{0,j}^{P}}{2\Delta x} \tag{11.13}$$

$$\left(\frac{\partial T}{\partial x}\right)_{m,j}^{P} = \frac{T_{m+1,j}^{P} - T_{m-1,j}^{P}}{2\Delta x} \tag{11.14}$$

Clearly, the points $(0, j)$ in eq. (11.13) and $(m + 1, j)$ in eq. (11.14) do not exist in the domain of interest. Therefore, it appears that one may be forced to use the first-order-accurate eqs. (11.3) and (11.4) to discretize derivatives at a boundary location. Based on the earlier discussion, eq. (11.3) would have to be used to discretize $(\partial T/\partial x)_{1,j}^{P}$ and eq. (11.4) to discretize $(\partial T/\partial x)_{m,j}^{P}$.

To avoid the loss of accuracy resulting from the use of first-order-accurate formulas in the evaluation of boundary derivatives, it is desirable to obtain second-order-accurate formulas. To this end, a forward Taylor series expansion of $T^P_{i+2,j}$ about $T^P_{i,j}$ (Fig. 11.3) yields

$$T^P_{i+2,j} = T^P_{i,j} + \left(\frac{\partial T}{\partial x}\right)^P_{i,j} (2\Delta x) + \frac{1}{2!}\left(\frac{\partial^2 T}{\partial x^2}\right)^P_{i,j} (2\Delta x)^2 + \frac{1}{3!}\left(\frac{\partial^3 T}{\partial x^3}\right)^P_{i,j} (2\Delta x)^3 + (\text{higher-order terms}) \tag{11.15}$$

Similarly, a backward Taylor series expansion of $T^P_{i-2,j}$ about $T^P_{i,j}$ yields

$$T^P_{i-2,j} = T^P_{i,j} - \left(\frac{\partial T}{\partial x}\right)^P_{i,j} (2\Delta x) + \frac{1}{2!}\left(\frac{\partial^2 T}{\partial x^2}\right)^P_{i,j} (2\Delta x)^2 - \frac{1}{3!}\left(\frac{\partial^3 T}{\partial x^3}\right)^P_{i,j} (2\Delta x)^3 + (\text{higher-order terms}) \tag{11.16}$$

To derive a second-order-accurate forward-difference approximation for $(\partial T/\partial x)^P_{i,j}$, we multiply eq. (11.1) by 4 and subtract the resulting equation from eq. (11.15), after neglecting the terms of order $(\Delta x)^3$ and higher [note that in the subtraction process the terms of order $(\Delta x)^2$ cancel out]. After rearranging, we obtain

$$\left(\frac{\partial T}{\partial x}\right)^P_{i,j} = \frac{-3T^P_{i,j} + 4T^P_{i+1,j} - T^P_{i+2,j}}{2\Delta x} \tag{11.17}$$

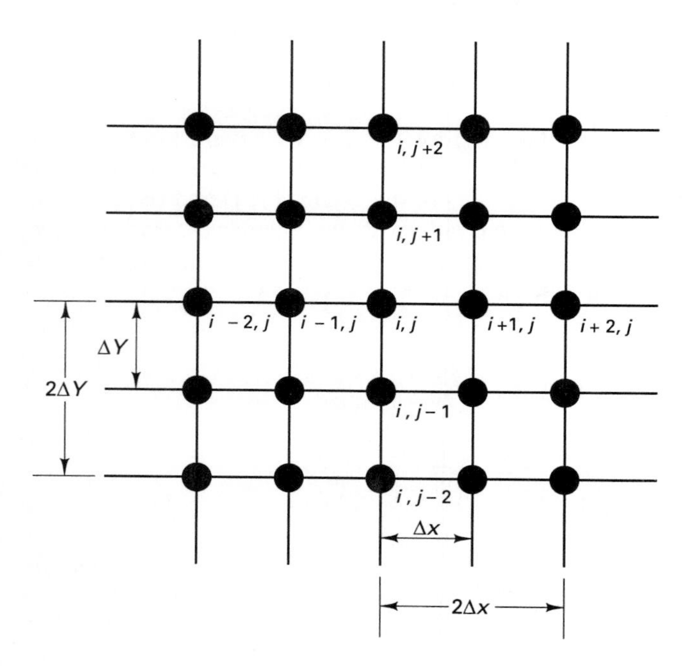


**Figure 11.3**

Working with eqs. (11.2) and (11.16) in a manner similar to that which led us to eq. (11.17) yields the following backward-difference approximation:

$$\left(\frac{\partial T}{\partial x}\right)_{i,j}^{P} = \frac{3T_{i,j}^{P} - 4T_{i-1,j}^{P} + T_{i-2,j}^{P}}{2\Delta x} \tag{11.18}$$

Note that eq. (11.18) is second-order accurate, like eq. (11.16). Note further that eqs. (11.16) and (11.17) can be conveniently used to discretize a temperature derivative at the boundary shown in Fig. 11.2a and b, respectively.

Second-order-accurate discretization formulas for the second derivative of the temperature at the boundaries can be derived in a similar manner. For example, multiplying eq. (11.1) by a factor of 2, subtracting the result from eq. (11.15) and neglecting all terms of order higher than 3 yields, after some rearranging, the following three-point forward-difference formula:

$$\left(\frac{\partial^2 T}{\partial x^2}\right)_{i,j}^{P} = \frac{T_{i,j}^{P} - 2\,T_{i+1,j}^{P} + T_{i+2,j}^{P}}{(\Delta x)^2} \tag{11.19}$$

On the other hand, multiplying eq. (11.2) by a factor of 2, subtracting it from eq. (11.16), and neglecting all terms of order greater than 3 yields a three-point backward discretization formula for the second derivative of the temperature:

$$\left(\frac{\partial^2 T}{\partial x^2}\right)_{i,j}^{P} = \frac{T_{i,j}^{P} - 2T_{i-1,j}^{P} + T_{i-2,j}^{P}}{(\Delta x)^2} \tag{11.20}$$

Expressions similar to eqs. (11.17)–(11.20) can be derived in any desired space direction by utilizing the appropriate Taylor series expansions in that direction. The result of such operation in the $y$-direction of Fig. 11.3, for example, is:

*Three-point forward-difference approximations:*

$$\left(\frac{\partial T}{\partial y}\right)_{i,j}^{P} = \frac{-3T_{i,j}^{P} + 4T_{i,j+1}^{P} - T_{i,j+2}^{P}}{2\Delta y} \tag{11.21}$$

$$\left(\frac{\partial^2 T}{\partial y^2}\right)_{i,j}^{P} = \frac{T_{i,j}^{P} - 2T_{i,j+1}^{P} + T_{i,j+2}^{P}}{(\Delta y)^2} \tag{11.22}$$

*Three-point backward-difference approximations:*

$$\left(\frac{\partial T}{\partial y}\right)_{i,j}^{P} = \frac{3T_{i,j}^{P} - 4T_{i,j-1}^{P} + T_{i,j-2}^{P}}{2\Delta y} \tag{11.23}$$

$$\left(\frac{\partial^2 T}{\partial y^2}\right)_{i,j}^{P} = \frac{T_{i,j}^{P} - 2T_{i,j-1}^{P} + T_{i,j-2}^{P}}{(\Delta y)^2} \tag{11.24}$$

**Discretization of the time derivative of temperature.** The first derivative of the temperature with respect to time appears in the heat conduction equation. This derivative can be discretized, in principle, in a manner identical to that of the space derivatives. To this end, a forward Taylor series expansion in time of the temperature $T_{i,j}^P$ yields

$$T_{i,j}^{P+1} = T_{i,j}^P + \left(\frac{\partial T}{\partial t}\right)_{i,j}^P (\Delta t) + \frac{1}{2!}\left(\frac{\partial^2 T}{\partial t^2}\right)_{i,j}^P (\Delta t)^2 + \frac{1}{3!}\left(\frac{\partial^3 T}{\partial t^3}\right)_{i,j}^P (\Delta t)^3 + (\text{higher-order terms}) \tag{11.25}$$

A forward-difference approximation of accuracy $O(\Delta t)$ of the first time derivative can be obtained directly from eq. (11.25) by neglecting the terms of order $(\Delta t)^2$ and higher. The result is

$$\left(\frac{\partial T}{\partial t}\right)_{i,j}^P = \frac{T_{i,j}^{P+1} - T_{i,j}^P}{\Delta t} \tag{11.26}$$

This discretization formula uses information for the temperature at time step $p$ (present) and at time step $p + 1$ (the time step following immediately). When used in the heat conduction equation, eq. (11.26) aids the advancement of the temperature field by $\Delta t$ (from the present time to the immediate future time). Note that since time advances only in the positive direction, information at time levels $p + 2$ and beyond is not available, and three-point forward discretization formulas cannot be used in a straightforward manner. A centered-difference formula analogous to eq. (11.8) can easily be derived:

$$\left(\frac{\partial T}{\partial t}\right)_{i,j}^P = \frac{T_{i,j}^{P+1} - T_{i,j}^{P-1}}{2\Delta t} \tag{11.27}$$

However, if this formula is used to discretize the heat conduction equation, it may yield unstable algorithms [8, 9]. In summary, the first-order accurate forward-difference approximation of eq. (11.26) that is commonly used [1, 4–7] is recommended for the discretization of the time derivative of the heat conduction equation.

At this point, the discretization of both the space derivatives and the time derivative of a heat conduction model with the method of Taylor series expansions has been discussed. In the following examples the utilization of the discretization formulas for these derivatives in obtaining the discretization of a complete heat conduction model is illustrated.

**Example 11.1**

Figure 11.4a shows a metallic bar of square cross section through which an electric current is passing, generating heat at a rate $\dot{q}'''$ (W/m$^3$). The bottom of this bar is resting on an adiabatic electrical insulator. Its remaining three sides are cooled convectively ($h$, $T_\infty$). Figure 11.4b shows the cross section of the bar. Assume that the lengthwise temperature gradients are negligible and that steady-state heat conduction takes place. Obtain the mathematical model for the heat conduction process and discretize it (a) for a very large heat transfer

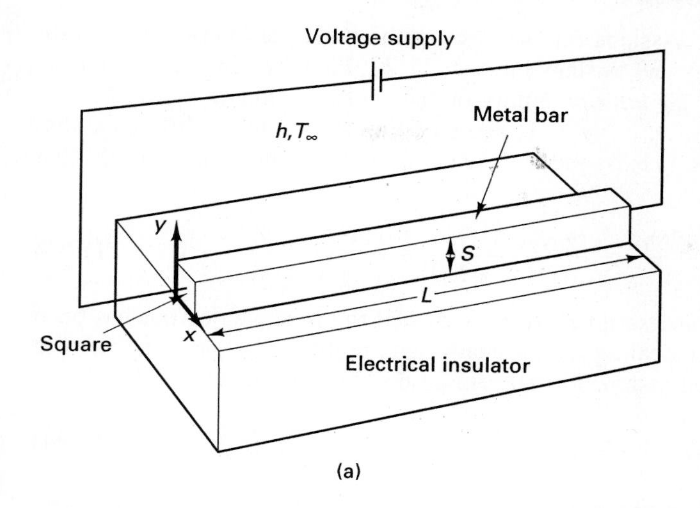


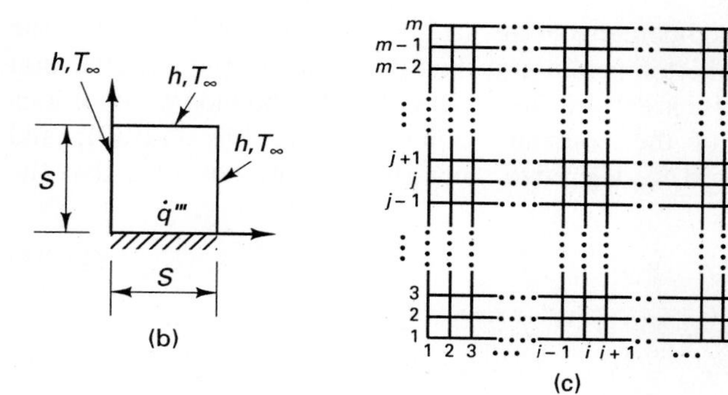


**Figure 11.4** (a) Schematic of a heat-generating metal bar. (b) Bar cross section. (c) Grid network.

coefficient $h \to \infty$ and (b) for a finite heat transfer coefficient. Utilize a uniform grid in both directions ($\Delta x = \Delta y$).

**Solution**

(a) *The case for a large heat transfer coefficient ($h \to \infty$)* The mathematical model of the problem is

$$\frac{\partial^2 \Theta}{\partial x^2} + \frac{\partial^2 \Theta}{\partial y^2} + \frac{\dot{q}'''}{k} = 0 \tag{11.28}$$

$$x = 0: \qquad \Theta = 0 \tag{11.29}$$

$$x = s: \qquad \Theta = 0 \tag{11.30}$$

$$y = 0: \qquad \frac{\partial \Theta}{\partial y} = 0 \tag{11.31}$$

$$y = s: \qquad \Theta = 0 \tag{11.32}$$

where $\Theta = T - T_\infty$. Assume that we overlay the cross section of the bar with a uniform grid ($\Delta x = \Delta y$) as shown in Fig. 11.4c. First, we discretize the conduction equation at all *interior* points of the cross section ($i = 2, 3, \ldots, m - 1, m, j = 2, 3, \ldots, m - 1, m$). Utilizing the centered-difference discretization formulas (11.8) and (11.12), eq. (11.28) becomes (recall that $\Delta x = \Delta y$)

$$\frac{\Theta_{i+1,j} + \Theta_{i-1,j} - 2\Theta_{i,j}}{(\Delta x)^2} + \frac{\Theta_{i,j+1} + \Theta_{i,j-1} - 2\Theta_{i,j}}{(\Delta x)^2} + \frac{\dot{q}'''_{i,j}}{k} = 0 \qquad (11.33)$$

In eq. (11.33) the superscript $P$ which indicated the time step earlier has been dropped since we are dealing with a steady-state problem. In addition, the heat source term has been assigned to all grid points. Solving for $\Theta_{i,j}$ yields

$$\Theta_{i,j} = \frac{1}{4}\left[\Theta_{i+1,j} + \Theta_{i-1,j} + \Theta_{i,j+1} + \Theta_{i,j-1} + \frac{\dot{q}'''_{i,j}(\Delta x)^2}{k}\right] \qquad (11.34)$$

Equation (11.34) gives the temperature of any interior point of the cross section as a function of the temperature of the grid points before and after it in the $x$ and the $y$ directions. The temperature at the boundaries of the cross section of the bar has not been discussed yet. To achieve this, the boundary conditions are discretized. Three of the boundary conditions (11.29), (11.30), and (11.32) are straightforward to discretize since they simply postulate that the temperature is constant:

$$\Theta_{1,j} = 0: \qquad j = 1, \ldots, m \qquad (11.35)$$

$$\Theta_{m,j} = 0: \qquad j = 1, \ldots, m \qquad (11.36)$$

$$\Theta_{i,m} = 0: \qquad i = 1, \ldots, m \qquad (11.37)$$

The last boundary condition [eq. (11.31)] is discretized by invoking the second-order-accurate eq. (11.21):

$$\frac{-3\Theta_{i,1} + 4\Theta_{i,2} - \Theta_{i,3}}{2\Delta y} = 0 \qquad (11.38)$$

Solving for $\Theta_{i,1}$, we obtain

$$\Theta_{i,1} = \frac{4\Theta_{i,2} - \Theta_{i,3}}{3} \qquad i = 1, \ldots, m \qquad (11.39)$$

At this point, the mathematical model of the problem of interest has been discretized. Equations (11.34)–(11.37) and (11.39) constitute the discretized heat conduction model. The solution process of the discretized model that yields the temperature field in the bar is discussed in Section 11.3.

(b) *Case of a finite heat transfer coefficient* The only difference between the model for this case and the model for case (a) is the boundary conditions at the left side, the right side, and the top of the bar. To accommodate the existence of a finite heat transfer coefficient, eqs. (11.29), (11.30), and (11.32) need to be replaced (respectively) by

$$x = 0: \qquad k\frac{\partial \Theta}{\partial x} = h\Theta \tag{11.40}$$

$$x = s: \qquad -k\frac{\partial \Theta}{\partial x} = h\Theta \tag{11.41}$$

$$y = s: \qquad -k\frac{\partial \Theta}{\partial y} = h\Theta \tag{11.42}$$

The conduction equation (11.28) and the boundary condition (11.31) remain the same. Hence here we focus on the discretization of eqs. (11.40)–(11.42). Utilizing the second-order-accurate discretization formula (11.17) in boundary condition (11.40) yields

$$k\frac{-3\Theta_{1,j} + 4\Theta_{2,j} - \Theta_{3,j}}{2\Delta x} = h\Theta_{1,j} \tag{11.43}$$

Solving for $\Theta_{1,j}$ gives us

$$\Theta_{1,j} = \frac{4\Theta_{2,j} - \Theta_{3,j}}{2(h\,\Delta x/k) + 3} \qquad j = 1, \ldots, m \tag{11.44}$$

Similarly, utilizing eqs. (11.18) and (11.23) to discretize boundary conditions (11.41) and (11.42), respectively, we obtain

$$\Theta_{m,j} = \frac{4\Theta_{m-1,j} - \Theta_{m-2,j}}{2(h\,\Delta x/k) + 3} \qquad j = 1, \ldots, m \tag{11.45}$$

$$\Theta_{i,m} = \frac{4\Theta_{i,m-1} - \Theta_{i,m-2}}{2(h\,\Delta y/k) + 3} \qquad i = 1, \ldots, m \tag{11.46}$$

The discretized conduction model for case (b) in which the heat transfer coefficient is finite consists of eqs. (11.34), (11.39), and (11.44)–(11.46). Methods of solution of this model are discussed in Section 11.3.

**Example 11.2**

A large steak is placed in a nonstick frying pan (Fig. 11.5a). The thickness of the steak is denoted by $L$ and its initial temperature by $T_\infty$. Upon placement in the frying pan, the bottom surface of the steak becomes (say, instantaneously) equal to the "hot" temperature of the frying pan, $T_H$. The top of the steak is cooled by natural convection to the environment ($h$, $T_\infty$). The heat transfer through the lateral surface (edge) of the steak is assumed negligible. As a result, unidirectional heat conduction takes place in the steak. Obtain the mathematical model of this problem and discretize it utilizing a uniform grid.

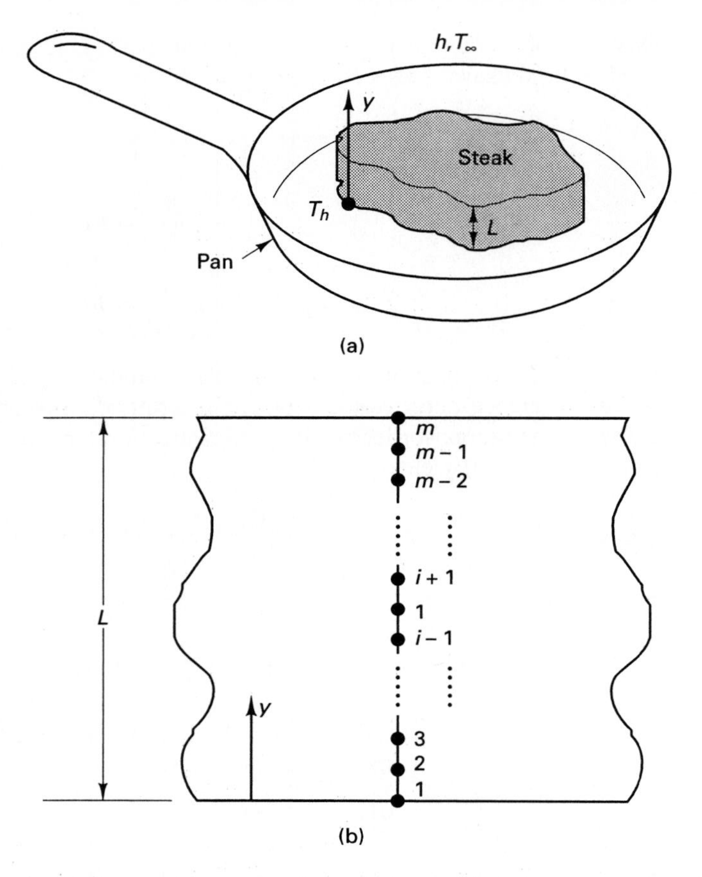


**Figure 11.5** (a) Schematic of steak being fried in a pan. (b) Grid network.

## Solution

The mathematical model for this problem reads

$$\frac{\partial \Theta}{\partial t} = \alpha \frac{\partial^2 \Theta}{\partial y^2} \tag{11.47}$$

$$y = 0: \qquad \Theta = T_H - T_\infty = \Theta_H \tag{11.48}$$

$$y = L: \qquad -k \frac{\partial \Theta}{\partial y} = h\Theta \tag{11.49}$$

$$t = 0: \qquad \Theta = 0 \tag{11.50}$$

where, as usual,

$$\Theta = T - T_\infty \tag{11.51}$$

To discretize this model, we utilize the grid shown in Fig. 11.5b. The governing equation is discretized first at all interior points. We choose to utilize the centered-difference formula (11.8) for the space derivative and the forward-difference formula (11.26) for the time derivative. Substituting these formulas into eq. (11.47), we obtain

$$\frac{\Theta_i^{P+1} - \Theta_i^P}{\Delta t} = \alpha\left[\frac{\Theta_{i+1}^P + \Theta_{i-1}^P - 2\Theta_i^P}{(\Delta y)^2}\right] \tag{11.52}$$

Note that only one subscript ($i$) is necessary since the temperature depends only on one space direction ($y$). Next, we solve eq. (11.52) for $\Theta_i^{P+1}$:

$$\Theta_i^{P+1} = \Theta_i^P + \frac{\alpha\ \Delta t}{(\Delta y)^2}(\Theta_{i+1}^P + \Theta_{i-1}^P - 2\Theta_i^P) \qquad i = 2, \ldots, m - 1 \tag{11.53}$$

The discretization of the boundary conditions (11.48) and (11.49) will provide us with information on the temperature at the boundary points. Equation (11.48) states that the temperature at the first grid point is constant:

$$\Theta_1^P = \Theta_H \qquad p = 1, \ldots, n \tag{11.54}$$

where it was assumed that the solution will be carried through a total of $n$ time steps. The boundary condition (11.49) is discretized with the help of eq. (11.18):

$$-k\frac{-3\Theta_m^P - 4\Theta_{m-1}^P + \Theta_{m-2}^P}{2\Delta y} = h\Theta_m^P \tag{11.55}$$

Solving for the temperature at the upper surface of the steak yields

$$\Theta_m^P = \frac{4\Theta_{m-1}^P + \Theta_{m-2}^P}{2(h\ \Delta y/k) + 3} \qquad p = 1, \ldots, n \tag{11.56}$$

The discretized initial condition of the problem is

$$\Theta_i^1 = 0 \qquad i = 2, \ldots, m - 1 \tag{11.57}$$

At this point, the discretization of the mathematical model of the problem of frying of the steak in Fig. 11.5a is complete.

### The Control Volume Discretization Method

The main premise of this method is to discretize the conduction equation directly by integration over properly defined control volumes. To this end, the calculation domain is subdivided into a number of control volumes. These control volumes should not overlap, and each of them should surround one grid point. The conduction equation is, next, integrated over each control volume. To perform the integration, the distribution of the temperature needs to be assumed for each control volume. The exact form of this distribution (linear, exponential, etc.) depends on the application of interest. Linear temperature profiles are common when exceedingly steep temperature variations are not expected within a control volume. After performing the integration, the discretized conduction equation containing the values of the temperature at each grid point is obtained. The method is presented in great detail in numerical heat transfer articles and textbooks, exemplified by [1, 5, 7, 10–12]. In the following section, the discretization of the two-dimensional steady-state heat conduction equation will be performed.

**Discretization of the two-dimensional steady-state conduction equation.** Consider the typical grid point ($A$) shown in Fig. 11.6. The grid points surrounding point ($A$) are conveniently denoted by ($E$), ($W$), ($N$), and ($S$) (east, west, north, and south of point $A$). The dashed lines in Fig. 11.6 define the four faces of a rectangular control volume surrounding point $A$. The intersections of the dashed lines with the grid lines are denoted by $e$, $w$, $n$, and $s$. For simplicity, we assume that the grid is uniform in the $x$-direction ($\Delta y$ = const), but we will discretize the conduction equation by allowing for $\Delta x \neq \Delta y$. In addition, we assume that the control volume faces are located halfway between two sequential grid points (Fig. 11.6). This is not a necessary assumption, but it simplifies the bookkeeping required in the discretization process. Allowing for the presence of variable thermal conductivity of the material ($k$) and of variable heat generation ($\dot{q}'''$), the two-dimensional steady-

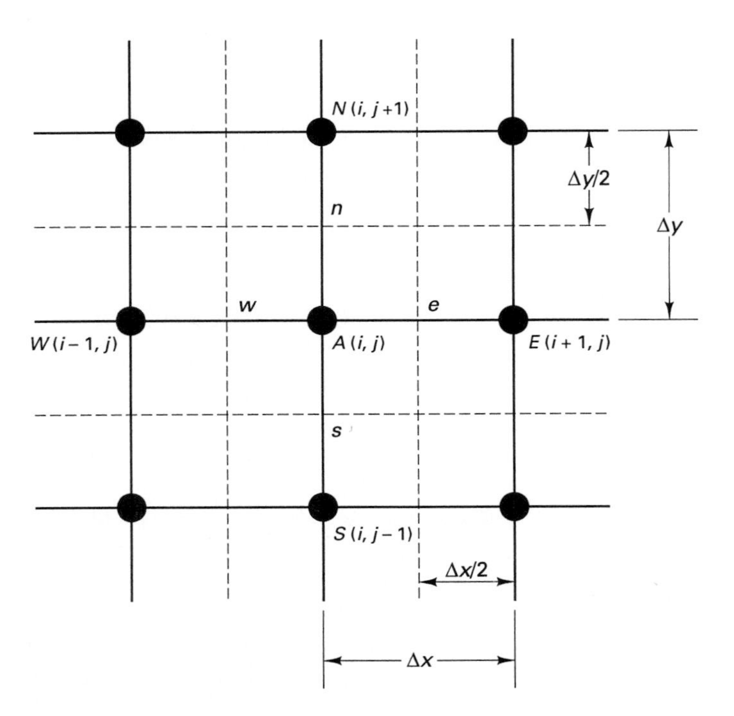


**Figure 11.6** Group of grid points in a control volume formulation.

state heat conduction equation to be discretized reads

$$\frac{\partial}{\partial x}\left(k\frac{\partial T}{\partial x}\right) + \frac{\partial}{\partial y}\left(k\frac{\partial T}{\partial y}\right) + \dot{q}''' = 0 \tag{11.58}$$

Integrating this equation over the control volume of Fig. 11.6 yields

$$\int_w^e \int_s^n \frac{\partial}{\partial x}\left(k\frac{\partial T}{\partial x}\right) dx\, dy + \int_w^e \int_s^n \frac{\partial}{\partial y}\left(k\frac{\partial T}{\partial y}\right) dx\, dy + \int_w^e \int_s^n \dot{q}'''\, dx\, dy = 0 \tag{11.59}$$

Next, we perform the east-west ($x$) integration in the first term of eq. (11.59) and the south-north ($y$) integration in the second term:

$$\int_s^n \left[\left(k\frac{\partial T}{\partial x}\right)_e - \left(k\frac{\partial T}{\partial x}\right)_w\right] dy + \int_w^e \left[\left(k\frac{\partial T}{\partial y}\right)_n - \left(k\frac{\partial T}{\partial y}\right)_s\right] dx + \bar{\dot{q}}'''\, \Delta x\, \Delta y = 0 \tag{11.60}$$

where $\bar{\dot{q}}'''$ is the average value of the heat generation term over the control volume. To proceed, we assume that the heat fluxes $(k(\partial T/\partial x))_e$, $(k(\partial T/\partial x))_w$, $(k(\partial T/\partial y))_n$, $(k(\partial T/\partial y))_s$ prevail over the entire faces of the control volume at $e$, $w$, $n$, and $s$, respectively. Hence eq. (11.60) becomes

$$\left[\left(k\frac{\partial T}{\partial x}\right)_e - \left(k\frac{\partial T}{\partial x}\right)_w\right] \Delta y + \left[\left(k\frac{\partial T}{\partial y}\right)_n - \left(k\frac{\partial T}{\partial y}\right)_s\right] \Delta x + \bar{\dot{q}}'''\, \Delta x\, \Delta y = 0 \tag{11.61}$$

We have now arrived at the point where an assumption on the shape of the temperature variation in the control volume, and therefore between grid points, is necessary. The most commonly used shape is the piecewise-linear shape (linear temperature variation from grid point to grid point). This shape combines easy implementation and good performance [5]. Utilization of the piecewise-linear temperature variation yields

$$\left(\frac{\partial T}{\partial x}\right)_w = \frac{T_A - T_W}{\Delta x} \tag{11.62}$$

$$\left(\frac{\partial T}{\partial x}\right)_e = \frac{T_E - T_A}{\Delta x} \tag{11.63}$$

$$\left(\frac{\partial T}{\partial y}\right)_s = \frac{T_A - T_S}{\Delta y} \tag{11.64}$$

$$\left(\frac{\partial T}{\partial y}\right)_n = \frac{T_N - T_A}{\Delta y} \tag{11.65}$$

Combining eqs. (11.61)–(11.65) yields

$$\left(k_e \frac{T_E - T_A}{\Delta x} - k_w \frac{T_A - T_W}{\Delta x}\right) \Delta y + \left(k_n \frac{T_N - T_A}{\Delta y} - k_s \frac{T_A - T_S}{\Delta y}\right) + \bar{\dot{q}}'''\, \Delta x\, \Delta y = 0 \tag{11.66}$$

Solving for $T_A$ gives us

$$T_A = \frac{1}{(k_e + k_w)\dfrac{\Delta y}{\Delta x} + (k_n + k_s)\dfrac{\Delta x}{\Delta y}}\left(k_e \frac{\Delta y}{\Delta x} T_E + k_w \frac{\Delta y}{\Delta x} T_W + k_n \frac{\Delta x}{\Delta y} T_N + k_s \frac{\Delta x}{\Delta y} T_S + \bar{\dot{q}}''' \, \Delta x \, \Delta y\right) \tag{11.67}$$

Equation (11.67) is the discretized version of the energy equation (11.58). Note that discretizing the three-dimensional heat conduction equation is a straightforward extension of the procedure discussed in this section.

**Discretization of boundary conditions.** When the boundary temperature is prescribed, one should simply assign to the boundary the given value of the temperature as done in Examples 11.1 and 11.2. The difficulty arises when one needs to discretize a boundary condition containing a heat flux (i.e., a temperature gradient). This boundary condition may be of the second or third kind, as discussed in Chapter 2.

To derive a discretization formula for a boundary condition containing a heat flux, consider the boundary point $D$ shown in Fig. 11.7a. Since this point lies on the boundary, only one-half of a control volume in the discretization domain can surround it, as marked by the shading in Fig. 11.7a. Integrating the conduction equation (11.58) over this half control volume, we obtain

$$\int_w^e \int_D^n \frac{\partial}{\partial x}\left(k \frac{\partial T}{\partial x}\right) dx \, dy + \int_w^e \int_D^n \frac{\partial}{\partial y}\left(k \frac{\partial T}{\partial y}\right) dx \, dy + \int_w^e \int_D^n \dot{q}''' \, dx \, dy = 0 \tag{11.68}$$

Proceeding as earlier, eq. (11.68) becomes

$$\left[\left(k \frac{\partial T}{\partial x}\right)_e - \left(k \frac{\partial T}{\partial x}\right)_w\right] \frac{\Delta y}{2} + \left[\left(k \frac{\partial T}{\partial y}\right)_n - \left(k \frac{\partial T}{\partial y}\right)_D\right] \Delta x + \bar{\dot{q}}''' \frac{\Delta x \, \Delta y}{2} = 0 \tag{11.69}$$

Piecewise-linear temperature profiles are assumed to approximate the temperature gradients at points $e$, $w$, and $n$:

$$\left(\frac{\partial T}{\partial x}\right)_e = \frac{T_E - T_D}{\Delta x} \tag{11.70}$$

$$\left(\frac{\partial T}{\partial x}\right)_w = \frac{T_D - T_W}{\Delta x} \tag{11.71}$$

$$\left(\frac{\partial T}{\partial y}\right)_n = \frac{T_N - T_D}{\Delta y} \tag{11.72}$$

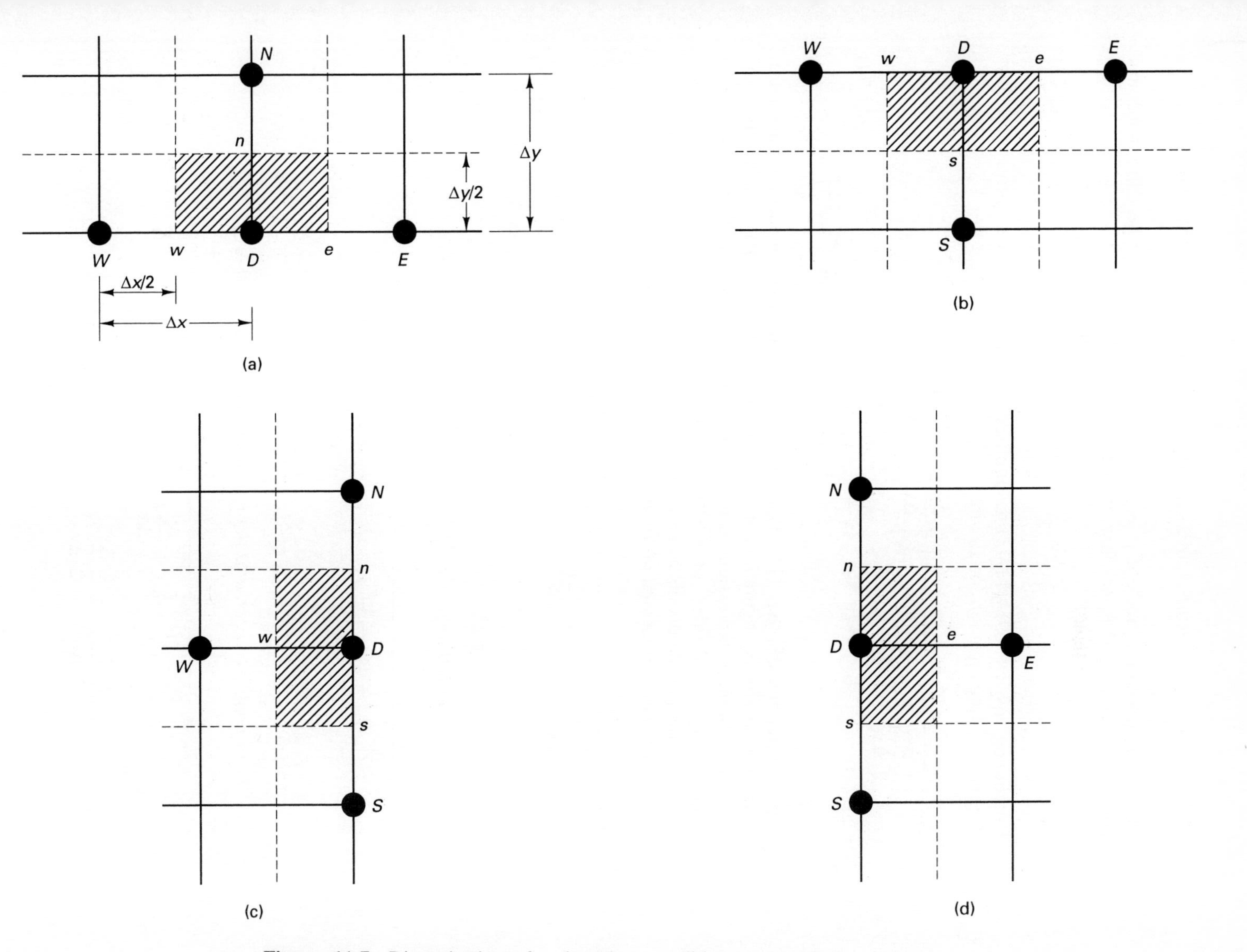


**Figure 11.7** Discretization of a boundary condition at (a) bottom, (b) top, (c) right, and (d) left boundary.

The last temperature gradient in eq. (11.69) is dictated by the heat flux boundary condition. Utilizing Fourier's law, we obtain

$$\dot{q}_D'' = -k_D \left(\frac{\partial T}{\partial y}\right)_D \tag{11.73}$$

Combining eqs. (11.69)–(11.73) to eliminate the temperature derivatives and solving the resulting equation for $T_D$ yields

$$T_D = \frac{1}{\dfrac{1}{2}\dfrac{\Delta y}{\Delta x}(k_e + k_w) + \dfrac{\Delta x}{\Delta y}k_n}\left(\frac{k_e}{2}\frac{\Delta y}{\Delta x}T_E + \frac{k_w}{2}\frac{\Delta y}{\Delta x}T_W + k_n\frac{\Delta x}{\Delta y}T_N + \dot{q}_D''\,\Delta x + \bar{\dot{q}}'''\frac{\Delta x\,\Delta y}{2}\right) \tag{11.74}$$

For a prescribed $\dot{q}_D''$, eq. (11.74) is an expression for the boundary temperature. Note that in the configuration of Fig. 11.7a, point $D$ is located at a "bottom" boundary. Similar expressions can be easily obtained if point $D$ is located on a "top," a "right-hand side," or a "left-hand side" boundary (Fig. 11.7b to d, respectively). Omitting the details as an exercise for the reader, these expressions are:

*Grid point (D) located at the top boundary (Fig. 11.7b):*

$$T_D = \frac{1}{\dfrac{1}{2}\dfrac{\Delta y}{\Delta x}(k_e + k_w) + \dfrac{\Delta x}{\Delta y}k_s}\left(\frac{k_e}{2}\frac{\Delta y}{\Delta x}T_E + \frac{k_w}{2}\frac{\Delta y}{\Delta x}T_W + k_s\frac{\Delta x}{\Delta y}T_S - \dot{q}_D''\,\Delta x + \bar{\dot{q}}'''\frac{\Delta x\,\Delta y}{2}\right) \tag{11.75}$$

*Grid point (D) located at the right-hand-side boundary (Fig. 11.7c):*

$$T_D = \frac{1}{\dfrac{1}{2}\dfrac{\Delta x}{\Delta y}(k_n + k_s) + \dfrac{\Delta y}{\Delta x}k_w} \cdot \left(\frac{k_n}{2}\frac{\Delta x}{\Delta y}T_N + \frac{k_s}{2}\frac{\Delta x}{\Delta y}T_S + k_w\frac{\Delta y}{\Delta x}T_W - \dot{q}_D''\,\Delta y + \bar{\dot{q}}'''\frac{\Delta x\,\Delta y}{2}\right) \tag{11.76}$$

*Grid point (D) located at the left-hand-side boundary (Fig. 11.7d):*

$$T_D = \frac{1}{\dfrac{1}{2}\dfrac{\Delta x}{\Delta y}(k_n + k_s) + \dfrac{\Delta y}{\Delta x}k_e} \cdot \left(\frac{k_n}{2}\frac{\Delta x}{\Delta y}T_N + \frac{k_s}{2}\frac{\Delta x}{\Delta y}T_S + k_e\frac{\Delta y}{\Delta x}T_E + \dot{q}_D''\,\Delta y + \bar{\dot{q}}'''\frac{\Delta x\,\Delta y}{2}\right) \tag{11.77}$$

**Discretization of the two-dimensional unsteady heat conduction equation.** The general form of this equation is

$$\rho c \frac{\partial T}{\partial t} = \frac{\partial}{\partial x}\left(k \frac{\partial T}{\partial x}\right) + \frac{\partial}{\partial y}\left(k \frac{\partial T}{\partial y}\right) + \dot{q}''' \tag{11.78}$$

For simplicity, we assume that the thermal capacity of the material is constant ($\rho c$ = const). To discretize eq. (11.78) the grid module of Fig. 11.6 will be used. The object of the discretization will be to obtain the temperature at a grid point ($A$) at the ($p$ + 1) time step (or at time $t + \Delta t$) when the temperature of all grid points at the previous time step ($p$) (or at time $t$) is known. To proceed, we integrate eq. (11.78) over the control volume of Fig. 11.6 and over the time interval from $t$ to $t + \Delta t$:

$$\begin{aligned} \rho c \int_w^e \int_s^n \int_t^{t+\Delta t} \frac{\partial T}{\partial t}\, dx\, dy\, dt &= \int_w^e \int_s^n \int_t^{t+\Delta t} \frac{\partial}{\partial x}\left(k \frac{\partial T}{\partial x}\right) dx\, dy\, dt \\ &+ \int_w^e \int_s^n \int_t^{t+\Delta t} \frac{\partial}{\partial y}\left(k \frac{\partial T}{\partial y}\right) dx\, dy\, dt \\ &+ \int_w^e \int_s^n \int_t^{t+\Delta t} \dot{q}'''\, dx\, dy\, dt \end{aligned} \tag{11.79}$$

For the space integration of the space derivatives the procedure and the earlier assumptions of this section are followed identically. For the space integration of the time derivative it is assumed that the value $(\partial T/\partial t)_A$ (the value of the derivative at point $A$) prevails over the entire control volume [5]. Hence, after performing the space integration in eq. (11.79), we obtain

$$\begin{aligned} \rho c\, \Delta x\, \Delta y \int_t^{t+\Delta t} \left(\frac{\partial T}{\partial t}\right)_A dt &= \Delta y \int_t^{t+\Delta t} \left(k_e \frac{T_E - T_A}{\Delta x} - k_w \frac{T_A - T_W}{\Delta x}\right) dt \\ &+ \Delta x \int_t^{t+\Delta t} \left(k_n \frac{T_N - T_A}{\Delta y} - k_s \frac{T_A - T_S}{\Delta y}\right) dt \\ &+ \bar{\dot{q}}'''\, \Delta x\, \Delta y\, \Delta t \end{aligned} \tag{11.80}$$

where $\dot{q}'''$ is the average heat generation over the control volume ($\Delta x\, \Delta y$) and over the time step $\Delta t$.

To perform the integration of the time derivative of eq. (11.80) is straightforward:

$$\int_t^{t+\Delta t} \left(\frac{\partial T}{\partial t}\right)_A dt = T_A^{P+1} - T_A^P \tag{11.81}$$

On the other hand, an assumption is necessary to perform the integrations at the right-hand side of eq. (11.80). A popular and rather general choice of assumption

[1, 5, 7] is to postulate that the temperature at each grid point (point $A$, for example) is given by

$$\int_t^{t+\Delta t} T_A \, dt = [\lambda T_A^{P+1} + (1 - \lambda) T_A^P] \, \Delta t \tag{11.82}$$

where $\lambda$ is a weighing factor in the range between zero and unity. When $\lambda = 0$ it is implied that the temperature at the current (known) $P$th time step prevails. When $\lambda = 1$ it is implied that the temperature at the future (unknown) $(P + 1)$ time step prevails. Combining eqs. (11.80)–(11.82), we obtain

$$\begin{aligned} \rho c \, \Delta x \, \Delta y (T_A^{P+1} - T_A^P) = {} & \Delta y \, \Delta t \left[ \lambda \left( k_e \frac{T_E^{P+1} - T_A^{P+1}}{\Delta x} - k_w \frac{T_A^{P+1} - T_W^{P+1}}{\Delta x} \right) \right. \\ & \left. + (1 - \lambda) \left( k_e \frac{T_E^P - T_A^P}{\Delta x} - k_w \frac{T_A^P - T_W^P}{\Delta x} \right) \right] \\ & + \Delta x \, \Delta t \left[ \lambda \left( k_n \frac{T_N^{P+1} - T_A^{P+1}}{\Delta y} - k_s \frac{T_A^{P+1} - T_S^{P+1}}{\Delta y} \right) \right. \\ & \left. + (1 - \lambda) \left( k_n \frac{T_N^P - T_A^P}{\Delta y} - k_s \frac{T_A^P - T_S^P}{\Delta y} \right) \right] + \bar{\dot{q}}''' \, \Delta x \, \Delta y \, \Delta t \end{aligned} \tag{11.83}$$

Solving eq. (11.83) for $T_A^{P+1}$ yields

$$\begin{aligned} T_A^{P+1} = {} & \left[ \frac{\rho c \, \Delta x \, \Delta y}{\Delta t} + \lambda \frac{\Delta y}{\Delta x}(k_e + k_w) + \lambda \frac{\Delta x}{\Delta y}(k_n + k_s) \right]^{-1} \\ & \left\{ k_e \frac{\Delta y}{\Delta x}[\lambda T_E^{P+1} + (1 - \lambda) T_E^P] + k_w \frac{\Delta y}{\Delta x}[\lambda T_W^{P+1} + (1 - \lambda) T_W^P] \right. \\ & + k_n \frac{\Delta x}{\Delta y}[\lambda T_N^{P+1} + (1 - \lambda) T_N^P] + k_s \frac{\Delta x}{\Delta y}[\lambda T_S^{P+1} + (1 - \lambda) T_S^P] \\ & \left. + \left[ \frac{\rho c \, \Delta x \, \Delta y}{\Delta t} - (1 - \lambda) \frac{\Delta y}{\Delta x}(k_e + k_w) - (1 - \lambda) \frac{\Delta x}{\Delta y}(k_n + k_s) \right] T_A^P + \bar{\dot{q}}''' \, \Delta x \, \Delta y \right\} \end{aligned} \tag{11.84}$$

Equation (11.84) gives the desired temperature of point $A$ at time $t + \Delta t$ as a function of the (known) temperature of this point at the previous time step (time $t$) and the temperature of points $W$, $E$, $S$, and $N$ at times $t$ (known) and $t + \Delta t$ (unknown). Note that in the special case where $\lambda = 0$, eq. (11.84) expresses the temperature of point $A$ as a function of the known temperature at points $A$, $W$, $E$, $S$, and $N$ at time $t$ only. The various solution methods of the set of algebraic equations resulting from the application of eq. (11.84) at the grid points of the computational domain will be discussed later in this chapter.

The discretization procedure of the boundary conditions in transient conduction follows along the lines of what was discussed earlier. For this reason and in the interest of brevity, this procedure is not presented herein. In what follows immediately, two illustrative examples of discretization of heat conduction models with the control volume method are presented.

**Example 11.3**

Discretize the mathematical model of Example 11.1 utilizing the control volume method.

**Solution**

(a) The mathematical model for this case ($h \to \infty$) consists of eqs. (11.28)–(11.32). In the special case $\Delta x = \Delta y$ (the same grid size in both directions), $\bar{\dot{q}}''' = \dot{q}''' =$ const (constant heat generation), $k_e = k_w = k_n = k_s = k =$ const (constant thermal conductivity), eq. (11.67) reduces to

$$\Theta_A = \frac{1}{4}\left[\Theta_E + \Theta_W + \Theta_N + \Theta_S + \frac{\dot{q}'''(\Delta x)^2}{k}\right] \tag{11.85}$$

Taking into account the notation correspondence shown in Fig. 11.6, $(A) \to (i, j)$, $(E) \to (i + 1, j)$, $(W) \to (i - 1, j)$, $(S) \to (i, j - 1)$, $(N) \to (i, j + 1)$, eq. (11.85) becomes

$$\Theta_{i,j} = \frac{1}{4}\left[\Theta_{i+1,j} + \Theta_{i-1,j} + \Theta_{i,j-1} + \Theta_{i,j+1} + \frac{\dot{q}'''(\Delta x)^2}{k}\right] \tag{11.86}$$

Equation (11.86) is identical to eq. (11.34) derived with the Taylor series expansion method. This is a direct result of the fact that a piecewise-linear temperature profile (shape) between grid points was assumed in the derivation of eq. (11.67). Choosing a different profile would not result in agreement between the discretized conduction equation of the control volume method and the Taylor series expansion method.

The discretization of the three boundary conditions that state that the surface temperature is constant is straightforward. The given value of temperature is assigned to the corresponding boundaries. The resulting expressions are identical to eqs. (11.35)–(11–37). The last boundary condition is obtained by utilizing expression (11.74) since it refers to a bottom boundary of the type shown in Fig. 11.7a. This boundary is insulated; hence $\dot{q}''_D = 0$. In addition, we use the correspondence $(E) \to (i + 1, 1)$, $(W) \to (i - 1, 1)$, $(N) \to (i, 2)$ and recall that $k =$ const, $\Delta x = \Delta y$, to obtain

$$\Theta_{i,1} = \frac{1}{2}\left[\frac{1}{2}\Theta_{i+1,1} + \frac{1}{2}\Theta_{i-1,1} + \Theta_{i,2} + \frac{(\Delta x)^2}{2k}\dot{q}'''\right] \qquad i = 1, \ldots, m \tag{11.87}$$

(b) Equations (11.86) and (11.87) are valid in this case as well. We need to discretize the additional boundary conditions (11.40)–(11.42). Boundary condition (11.40) pertains to a left-hand-side boundary condition. Therefore, eq. (11.77) will be used for its discretization. Combining Fourier's law and eq. (11.40) yields

$$\dot{q}_D'' = -h\Theta_D \tag{11.88}$$

Accounting for the correspondence $(D) \rightarrow (1, j)$, $(N) \rightarrow (1, j + 1)$, $(S) \rightarrow (1, j - 1)$, $(E) \rightarrow (2, j)$, we substitute eq. (11.88) into eq. (11.77) for $\Delta x = \Delta y$, $k = \text{const}$, $\dot{q}''' = \text{const}$, and solve for $\Theta_{1,j}$:

$$\Theta_{1,j} = \frac{1}{2 + (h\,\Delta x/k)}\left[\frac{1}{2}\Theta_{1,j+1} + \frac{1}{2}\Theta_{1,j-1} + \Theta_{2,j} + \frac{(\Delta x)^2}{2k}\dot{q}'''\right] \qquad j = 1, \ldots, m \tag{11.89}$$

Treating boundary conditions (11.41) and (11.42) in an analogous manner results in the following discretized expressions:

$$\Theta_{m,j} = \frac{1}{2 + (h\,\Delta x/k)}\left[\frac{1}{2}\Theta_{m,j+1} + \frac{1}{2}\Theta_{m,j-1} + \Theta_{m-1,j} + \frac{(\Delta x)^2}{2k}\dot{q}'''\right] \qquad j = 1, \ldots, m \tag{11.90}$$

$$\Theta_{i,m} = \frac{1}{2 + (h\,\Delta x/k)}\left[\frac{1}{2}\Theta_{i+1,m} + \frac{1}{2}\Theta_{i-1,m} + \Theta_{i,m-1} + \frac{(\Delta x)^2}{2k}\dot{q}'''\right] \qquad i = 1, \ldots, m \tag{11.91}$$

Clearly, the results on the discretization of the boundary conditions involving a heat flux in this example differ from those of Example 11.1. The differences are consistent with the principles of the two discretization approaches utilized (Taylor series expansions versus control volume formulation).

**Example 11.4**

Discretize the heat conduction model of Example 11.2 with the control volume method.

**Solution**

The mathematical model under consideration consists of eqs. (11.47)–(11.50). To discretize the energy equation (11.47), we utilize eq. (11.84) for the special case of unidirectional heat conduction and for $\dot{q}''' = 0$, $k = \text{const}$. The unidirectional ($y$-direction) discretization equation is obtained if we set $\Delta x = 1$ and eliminate all the terms with subscripts $e$, $w$, $E$, and $W$ in eq. (11.84). The result of such operation reads

$$\Theta_A^{P+1} = \left[\frac{\rho c\,\Delta y}{\Delta t} + \lambda\frac{2k}{\Delta y}\right]^{-1}\left\{\frac{k}{\Delta y}[\lambda\Theta_N^{P+1} + (1-\lambda)\Theta_N^P] + \frac{k}{\Delta y}[\lambda\Theta_S^{P+1} + (1-\lambda)\Theta_S^P] + \left[\frac{\rho c\,\Delta y}{\Delta t} - (1-\lambda)\frac{2k}{\Delta y}\right]\Theta_A^P\right\} \tag{11.92}$$

The discretization of boundary condition (11.48) and initial condition (11.50) is straightforward and yields equations identical to (11.54) and (11.57), respectively. To discretize the flux boundary condition (11.49), we integrate the heat conduction equation (11.47) over the half control volume of Fig. 11.7b and over the time period from $t$ to $t + \Delta t$:

$$\int_{s}^{D} \int_{t}^{t+\Delta t} \frac{\partial \Theta}{\partial t}\, dt\, dy = \alpha \int_{s}^{D} \int_{t}^{t+\Delta t} \frac{\partial^2 \Theta}{\partial y^2}\, dt\, dy \tag{11.93}$$

Proceeding as earlier, we obtain

$$(\Theta_D^{P+1} - \Theta_D^P)\frac{\Delta y}{2} = \alpha \int_{t}^{t+\Delta t} \left[\left(\frac{\partial \Theta}{\partial y}\right)_D - \left(\frac{\partial \Theta}{\partial y}\right)_s\right] dt \tag{11.94}$$

Taking into account boundary condition (11.49) yields

$$(\Theta_D^{P+1} - \Theta_D^P)\frac{\Delta y}{2} = \alpha \int_{t}^{t+\Delta t} \left(-\frac{h}{k}\Theta_D - \frac{\Theta_D - \Theta_S}{\Delta y}\right) dt \tag{11.95}$$

Applying assumption (11.82), we have

$$(\Theta_D^{P+1} - \Theta_D^P)\frac{\Delta y}{2} = \alpha\left\{-\left(\frac{h\,\Delta y}{k} + 1\right)[\lambda\Theta_D^{P+1} + (1 - \lambda)\Theta_D^P]\frac{\Delta t}{\Delta y} + [\lambda\Theta_S^{P+1} + (1 - \lambda)\Theta_S^P]\frac{\Delta t}{\Delta y}\right\} \tag{11.96}$$

Solving for $\theta_D^{P+1}$ yields

$$\Theta_D^{P+1} = \left[\Delta y + 2\alpha\lambda\left(\frac{h\,\Delta y}{k} + 1\right)\frac{\Delta t}{\Delta y}\right]^{-1}\left\{\left[-2\alpha\left(\frac{h\,\Delta y}{k} + 1\right)(1 - \lambda)\frac{\Delta t}{\Delta y} + \Delta y\right] \cdot \Theta_D^P + 2\alpha[\lambda\Theta_S^{P+1} + (1 - \lambda)\Theta_S^P]\frac{\Delta t}{\Delta y}\right\} \tag{11.97}$$

At this point, the discretization of the model with the control volume formulation is complete.

## 11.3 SOLUTION METHODS OF DISCRETIZED HEAT CONDUCTION MODELS

It was shown in Section 11.2 that the discretization process transforms a heat conduction model (conduction equation and boundary and initial conditions) to a set of algebraic equations for the temperature at the grid points. In this section, representative methods of solution of these algebraic equations are discussed. The discussion is centered around transient conduction problems, for the sake of generality. Methods for the solution of steady-state problems can be obtained by neglecting the dependence of the temperature field on time and following similar procedures.

The solution methods are grouped into two categories: explicit methods and implicit methods. The main difference between these methods can be explained utilizing the example of unidirectional transient conduction. In explicit methods, after discretization, the unknown temperature $T_j^{P+1}$ is expressed in terms of known quantities (the already calculated temperature of grid points at the previous time step, for example). Therefore, the algebraic equation for the temperature of each grid point can be solved directly to yield the desired temperature of that particular grid point. In implicit methods, on the other hand, after discretization, the unknown temperature, $T_j^{P+1}$, is expressed not only in terms of known temperatures of the previous time step ($p$) but also unknown temperatures of the time step ($p + 1$). This leads to a coupling of the algebraic equations at each grid point and creates the necessity to solve a set of algebraic equations often in an iterative manner. Explicit and implicit methods in transient conduction are discussed sequentially. In both cases only representative, commonly used methods are covered. Additional methods are outside the scope of this chapter. They can be found in the voluminous numerical heat transfer literature [1, 4, 6, 7].

### Explicit Methods

Three methods (schemes) are presented in this category: the FTCS (forward in time central in space) method, the Richardson method, and the Dufort–Frankel method.

**FTCS method.** In this method, the discretization of the conduction equation is performed by utilizing a first-order-accurate forward-difference approximation for the time derivative and a third-order-accurate centered-difference approximation for the space derivatives. To exemplify, we apply this scheme to the two-dimensional transient conduction equation for a body with constant thermophysical properties:

$$\frac{\partial T}{\partial t} = \alpha\left(\frac{\partial^2 T}{\partial x^2} + \frac{\partial^2 T}{\partial y^2}\right) \tag{11.98}$$

Note that this scheme was in essence already employed earlier, to discretize the unidirectional conduction equation of Example 11.2. Utilizing eqs. (11.8), (11.12), and (11.26) as needed, eq. (11.98) can be written as

$$\frac{T_{i,j}^{P+1} - T_{i,j}^{P}}{\Delta t} = \alpha\left[\frac{T_{i+1,j}^{P} + T_{i-1,j}^{P} - 2T_{i,j}^{P}}{(\Delta x)^2} + \frac{T_{i,j+1}^{P} + T_{i,j-1}^{P} - 2T_{i,j}^{P}}{(\Delta y)^2}\right] \tag{11.99}$$

Solving for the unknown temperature $T_{i,j}^{P+1}$, we obtain

$$T_{i,j}^{P+1} = T_{i,j}^{P} + \alpha\,\Delta t\left[\frac{T_{i+1,j}^{P} + T_{i-1,j}^{P} - 2T_{i,j}^{P}}{(\Delta x)^2} + \frac{T_{i,j+1}^{P} + T_{i,j-1}^{P} - 2T_{i,j}^{P}}{(\Delta y)^2}\right] \tag{11.100}$$

It is worth noting that eq. (11.100) can also be obtained directly from eq. (11.84) for $k$ = const, no heat generation, and by setting $\lambda = 0$. This derivation is left as an exercise for the reader.

***Stability Considerations.*** An important issue that needs to be examined before eq. (11.100) is utilized to obtain the temperature field numerically is that of stability. The issue of stability needs to be examined regardless of the discretization method used. Here the FTCS method is utilized to illustrate the discussion.

Although it is possible for any perturbations resulting from round-off errors in numerical solutions to decay, this is not always the case. These errors may grow at any stage of the computation. It is worth stressing once more that the errors relevant to the stability performance of the numerical method are the round-off errors stemming from the fact that the computer carries out the calculation to a finite number of significant figures. Other errors, such as logic or programming errors, are not relevant to this discussion. Based on the above, the numerical solution for the temperature at any grid point is not an exact solution but an approximate solution. An error can then be defined for the temperature calculations at any grid point as follows:

$$\epsilon_{i,j}^P = T_{i,j}^P - \hat{T}_{i,j}^P \tag{11.101}$$

where $T_{i,j}^P$ denotes the exact temperature at grid point $(i, j)$ at the $p$th time step and $\hat{T}_{i,j}^P$ the calculated temperature at the same location and time step. Note that since both $T_{i,j}^P$ and $\hat{T}_{i,j}^P$ satisfy the discretization equation (11.100), it can easily be shown that the error $\epsilon_{i,j}^P$ also satisfies the same equation, that is,

$$\epsilon_{i,j}^{P+1} = \epsilon_{i,j}^P + \alpha\,\Delta t\left[\frac{\epsilon_{i+1,j}^P + \epsilon_{i-1,j}^P - 2\epsilon_{i,j}^P}{(\Delta x)^2} + \frac{\epsilon_{i,j+1}^P + \epsilon_{i,j-1}^P - 2\epsilon_{i,j}^P}{(\Delta y)^2}\right] \tag{11.102}$$

Note further that if the initial and boundary conditions are specified (in a rectangular domain, for example, where $i = 1, 2, \ldots, M$ and $j = 1, 2, \ldots, N$), then the initial errors ($\epsilon_{i,j}^0$) and the boundary errors ($\epsilon_{1,j}^P$, $\epsilon_{M,j}^P$, $\epsilon_{i,1}^P$, $\epsilon_{i,N}^P$) are all equal to zero (unless some interior grid points are involved in the evaluation of the boundary conditions).

Several well-established methods of stability analysis that study the behavior of the round-off errors exist in the literature [13,14]. One of the most common methods is the von Neumann method, which is discussed next in the context of the FTCS approach and used later in the chapter to evaluate the stability of all the other methods of solution of heat conduction problems.

**Von Neumann method of stability analysis for the FTCS scheme.** The von Neumann method is based on the prediction of the growth or decay of the errors identified earlier as the difference between the exact solution and the numerical solution. A limitation of this method is that it yields necessary and sufficient conditions for the stability of mathematical models that can be described as linear, initial value problems with constant coefficients. Otherwise, this method provides only necessary conditions for stability. Even though the von Neumann method pertains to interior points of the domain of interest, it can provide useful approximate information on the effect of the boundary conditions on the stability of the solution [15] if it is applied to the boundaries of the domain of interest.

To begin the discussion we expand the errors at all grid points at a given time step in terms of Fourier series. The object is, then, to study whether the individual Fourier components will grow or decay as we proceed from one time step to the next. To this end, assume that the error of the $P$th time step at grid point $(i, j)$ is expanded as

$$\epsilon_{i,j}^{P} = \sum_{m=1}^{M} \sum_{n=1}^{N} A_{mn}^{P} e^{I(\lambda_m i + k_n j)} \tag{11.103}$$

where $\lambda_m = m\pi\,\Delta x$, $k_n = n\pi\,\Delta y$, $I = \sqrt{-1}$, $M$ is the number of vertical gridlines, $N$ is the number of horizontal gridlines, $\Delta x$ is the grid size in the $x$-direction, and $\Delta y$ is the grid size in the $y$-direction.

Based on the form of eq. (11.103), for a linear problem it is sufficient to investigate the error propagation by examining the behavior of one term (Fourier component) of the double series shown in eq. (11.103). To this end we consider the general term

$$A_{mn}^{P} e^{I(\lambda_m i + k_n j)} \tag{11.104}$$

where the superscript $P$ denotes the time step and where the time dependence is included in the term $A_{mn}^{P}$. Equation (11.104) can be used to obtain a solution of the error equation (11.102) by the method of separation of variables. Substituting eq. (11.104) into eq. (11.102) yields

$$\begin{aligned} A_{mn}^{P+1} e^{I(\lambda_m i + k_n j)} &= A_{mn}^{P}\left\{1 - 2\left[\frac{\alpha\,\Delta t}{(\Delta x)^2} + \frac{\alpha\,\Delta t}{(\Delta y)^2}\right]\right\} e^{I(\lambda_m i + k_n j)} \\ &\quad + A_{mn}^{P}\frac{\alpha\,\Delta t}{(\Delta x)^2} e^{I[\lambda_m(i+1) + k_n j]} + A_{mn}^{P}\frac{\alpha\,\Delta t}{(\Delta x)^2} e^{I[\lambda_m(i-1) + k_n j]} \\ &\quad + A_{mn}^{P}\frac{\alpha\,\Delta t}{(\Delta y)^2} e^{I[\lambda_m i + k_n(j-1)]} + A_{mn}^{P}\frac{\alpha\,\Delta t}{(\Delta y)^2} e^{I[\lambda_m i + k_n(j+1)]} \end{aligned} \tag{11.105}$$

Simplifying gives

$$\begin{aligned} A = \frac{A_{mn}^{P+1}}{A_{mn}^{P}} &= 1 - 2\left[\frac{\alpha\,\Delta t}{(\Delta x)^2} + \frac{\alpha\,\Delta t}{(\Delta y)^2}\right] + \frac{\alpha\,\Delta t}{(\Delta x)^2} e^{I\lambda_m} + \frac{\alpha\,\Delta t}{(\Delta x)^2} e^{-I\lambda_m} \\ &\quad + \frac{\alpha\,\Delta t}{(\Delta y)^2} e^{Ik_n} + \frac{\alpha\,\Delta t}{(\Delta y)^2} e^{-Ik_n} \end{aligned} \tag{11.106}$$

Note that the ratio $A$ in eq. (11.106) is indicative of the growth or decay of the error between two subsequent time steps and can be viewed as an error amplification factor. Utilizing the fact that for any real variable $\theta$

$$e^{\pm I\theta} = \cos\theta \pm I\sin\theta \qquad \cos\theta = 1 - 2\sin^2\frac{\theta}{2} \qquad (11.107, 11.108)$$

eq. (11.106) simplifies to

$$A = 1 - 4\frac{\alpha\,\Delta t}{(\Delta x)^2}\sin^2\left(\frac{\lambda_m}{2}\right) - 4\frac{\alpha\,\Delta t}{(\Delta y)^2}\sin^2\left(\frac{k_n}{2}\right) \qquad (11.109)$$

The errors will remain bounded and the solution will be stable if:

$$|A| \le 1 \qquad (11.110)$$

Combining eqs. (11.109) and (11.110) yields the criterion necessary for the stability of the solution:

$$\frac{\alpha\,\Delta t}{(\Delta x)^2} + \frac{\alpha\,\Delta t}{(\Delta y)^2} < \frac{1}{2} \qquad (11.111)$$

It is worth stressing that while criterion (11.111) provides the necessary and sufficient stability condition for linear problems with constant coefficients, for more complex problems with nonlinearities and variable coefficients it usually provides only the necessary (but not always sufficient) stability condition.

**Richardson and Dufort–Frankel schemes.** Since the FTCS scheme utilizes a first-order-accurate forward-difference approximation for the time derivative, one may naturally assume that the accuracy of the scheme will improve if a second-order-accurate centered-difference approximation is utilized for the time derivative. This is the main premise of the Richardson scheme. The discretized two-dimensional transient conduction equation for a body with constant thermophysical properties then becomes

$$\frac{T_{i,j}^{P+1} - T_{i,j}^{P-1}}{2\,\Delta t} = \alpha\left[\frac{T_{i+1,j}^P + T_{i-1,j}^P - 2T_{i,j}^P}{(\Delta x)^2} + \frac{T_{i,j+1}^P + T_{i,j-1}^P - 2T_{i,j}^P}{(\Delta y)^2}\right] \qquad (11.112)$$

Unfortunately, even though eq. (11.112) is second-order accurate in time, a von Neumann stability analysis can show that it is unconditionally unstable [9]. Therefore, it cannot be used to obtain the temperature field. To circumvent this difficulty $T_{i,j}^P$ in eq. (11.112) is replaced by $\frac{1}{2}(T_{i,j}^{P-1} + T_{i,j}^{P+1})$. The resulting scheme is the **Dufort–Frankel** scheme:

$$\begin{aligned} T_{i,j}^{P+1} &= \frac{1 - 2[\alpha\,\Delta t/(\Delta x)^2] - 2[\alpha\,\Delta t/(\Delta y)^2]}{1 + 2[\alpha\,\Delta t/(\Delta x)^2] + 2[\alpha\,\Delta t/(\Delta y)^2]}T_{i,j}^{P-1} \\ &+ \frac{2[\alpha\,\Delta t/(\Delta x)^2]}{1 + 2[\alpha\,\Delta t/(\Delta x)^2] + 2[\alpha\,\Delta t/(\Delta y)^2]}(T_{i+1,j}^P + T_{i-1,j}^P) \qquad (11.113) \\ &+ \frac{2[\alpha\,\Delta t/(\Delta y)^2]}{1 + 2[\alpha\,\Delta t/(\Delta x)^2] + 2[\alpha\,\Delta t/(\Delta y)^2]}(T_{i,j+1}^P + T_{i,j-1}^P) \end{aligned}$$

It can be shown by application of the von Neumann stability analysis that the Dufort–Frankel scheme of eq. (11.113) is stable for any value of $\Delta t$. However, the scheme should be expected to be inaccurate if $\alpha[\Delta t/(\Delta x)^2]$ *or* $\alpha[\Delta t/(\Delta y)^2]$ are large.

The Dufort–Frankel scheme utilizes three levels in time $(P - 1, P, P + 1)$. In the special case $\alpha\,\Delta t/(\Delta x)^2 = \alpha\,\Delta t/(\Delta y)^2 = \frac{1}{4}$, this scheme becomes identical to the FTCS scheme of eq. (11.100).

### Implicit Methods

Two common implicit schemes will be considered: the fully implicit scheme and the Crank–Nicolson scheme.

**Fully implicit scheme.** In this scheme the space derivatives are evaluated at the time step $(P + 1)$ while a forward-difference approximation is used for the time derivatives. In the case of a two-dimensional heat conduction process in a body with constant thermophysical properties, the fully implicit scheme becomes

$$\frac{T_{i,j}^{P+1} - T_{i,j}^{P}}{\Delta t} = \frac{\alpha}{(\Delta x)^2}(T_{i+1,j}^{P+1} - 2T_{i,j}^{P+1} + T_{i-1,j}^{P+1}) + \frac{\alpha}{(\Delta y)^2}(T_{i,j+1}^{P+1} - 2T_{i,j}^{P+1} + T_{i,j-1}^{P+1}) \tag{11.114}$$

Note that eq. (11.114) can also be obtained from eq. (11.84) for $k$ = const, no heat generation and by setting $\lambda = 1$. This derivation is left as an exercise for the reader.

Solving eq. (11.114) for $T_{i,j}^{P+1}$ yields

$$\begin{aligned} T_{i,j}^{P+1} = {} & \frac{\alpha\,\Delta t/(\Delta x)^2}{1 + 2[\alpha\,\Delta t/(\Delta x)^2] + 2[\alpha\,\Delta t/(\Delta y)^2]}(T_{i+1,j}^{P+1} + T_{i-1,j}^{P+1}) \\ & + \frac{\alpha\,\Delta t/(\Delta y)^2}{1 + 2[\alpha\,\Delta t/(\Delta x)^2] + 2[\alpha\,\Delta t/(\Delta y)^2]}(T_{i,j+1}^{P+1} + T_{i,j-1}^{P+1}) \\ & + \frac{1}{1 + 2[\alpha\,\Delta t/(\Delta x)^2] + 2[\alpha\,\Delta t/(\Delta y)^2]}T_{i,j}^{P} \end{aligned} \tag{11.115}$$

Based on a von Neumann stability analysis, it can be shown that the fully implicit scheme is unconditionally stable.

The solution methodology of the set of algebraic equations resulting from eq. (11.115) deserves special consideration. To this end, the discussion will begin by examining the special case of unidirectional transient conduction (say, in the $x$-direction). In this case, eq. (11.115) reduces to

$$T_i^{P+1} = \frac{\alpha\,\Delta t/(\Delta x)^2}{1 + 2[\alpha\,\Delta t/(\Delta x)^2]}(T_{i+1}^{P+1} + T_{i-1}^{P+1}) + \frac{1}{1 + 2[\alpha\,\Delta t/(\Delta x)^2]}T_i^P \tag{11.116}$$

Note that the subscript $j$ has been dropped, for it is not relevant to the unidirectional conduction discretization equation. To proceed, we rewrite eq. (11.116) as follows:

$$-\frac{\alpha\,\Delta t}{(\Delta x)^2}T_{i-1}^{P+1} + \left[1 + 2\frac{\alpha\,\Delta t}{(\Delta x)^2}\right]T_i^{P+1} - \frac{\alpha\,\Delta t}{(\Delta x)^2}T_{i+1}^{P+1} = T_i^P \tag{11.117}$$

If eq. (11.117) is applied to all interior points of interest $i = 2, \ldots, M - 1$, we obtain

$$
\begin{aligned}
-\gamma T_1^{P+1} + (1 + 2\gamma)T_2^{P+1} - \gamma T_3^{P+1} &= T_2^P \\
-\gamma T_2^{P+1} + (1 + 2\gamma)T_3^{P+1} - \gamma T_4^{P+1} &= T_3^P \\
\cdots\cdots\cdots\cdots\cdots \\
\cdots\cdots\cdots\cdots\cdots \\
-\gamma T_{i-1}^{P+1} + (1 + 2\gamma)T_i^{P+1} - \gamma T_{i+1}^{P+1} &= T_i^P \\
\cdots\cdots\cdots\cdots\cdots \\
\cdots\cdots\cdots\cdots\cdots \\
-\gamma T_{M-2}^{P+1} + (1 + 2\gamma)T_{M-1}^{P+1} - \gamma T_M^{P+1} &= T_{M-1}^P
\end{aligned}
\tag{11.118}
$$

where for convenience the notation

$$\gamma = \frac{\alpha\, \Delta t}{(\Delta x)^2} \tag{11.119}$$

was used. Assuming that the temperatures $T_1^{P+1}$ and $T_M^{P+1}$ in the equations above are known from the boundary conditions of the problem of interest, the following matrix of equations can be constructed utilizing eq. (11.118):

$$
\begin{bmatrix}
1 + 2\gamma & -\gamma & & & \\
-\gamma & 1 + 2\gamma & -\gamma & & \\
\vdots & \vdots & \vdots & \vdots & \vdots \\
 & -\gamma & 1 + 2\gamma & -\gamma & \\
\vdots & \vdots & \vdots & \vdots & \vdots \\
 & & & -\gamma & 1 + 2\gamma
\end{bmatrix}
\begin{bmatrix}
T_2^{P+1} \\ T_3^{P+1} \\ \vdots \\ T_i^{P+1} \\ \vdots \\ T_{M-1}^{P+1}
\end{bmatrix}
=
\begin{bmatrix}
T_2^P + \gamma T_1^{P+1} \\ T_3^P \\ \vdots \\ T_i^P \\ \vdots \\ T_{M-1}^P + \gamma T_M^{P+1}
\end{bmatrix}
\tag{11.120}
$$

Clearly, the first matrix on the left-hand side of eq. (11.120) is tridiagonal and all its elements are known. In addition, all the elements of the column matrix on the right-hand side of eq. (11.120) are known. Therefore, the unknown elements (temperatures) of the column matrix on the left-hand side of eq. (11.120) ($T_2^{P+1}, T_3^{P+1}, \ldots, T_{M-1}^{P+1}$) can be obtained by using conventional methods of linear algebra such as variations and extensions of the Gauss elimination method. A well-established algorithm that performs this task is the Thomas algorithm that is described in detail in the numerical analysis literature [1, 4–7, 16]. To this end, FORTRAN subroutines containing the Thomas algorithm are readily available [4, 6, 7, 16]. Therefore, for brevity, no further details on this algorithm are given herein, with the exception of a description contained in Appendix G.

Unlike the unidirectional transient heat conduction problem described earlier, the two-dimensional transient heat conduction discretization equation (11.115) cannot be solved utilizing the Thomas algorithm. The reason for this fact is that while eq. (11.115) can be rearranged so that three of the terms are at or adjacent to the main diagonal, the remaining two terms are displaced. Conventional Gauss elimination would be possible, albeit undesirably time consuming. Two methods are commonly used for the solution of transient multidimensional heat conduction problems and are discussed sequentially: the Gauss–Seidel point-by-point method and the line-by-line method [5].

***Gauss–Seidel Method.*** This is perhaps the simplest to implement of all iterative methods. It makes use of the fact that at each grid point, the most recently calculated value of the temperature is stored in the computer memory. An iterative process at each time step is executed before proceeding to the next time step. During this process, the temperature at each grid point for two-dimensional heat conduction in a body with constant thermophysical properties is obtained from

$$T_{i,j}^{P+1} = \frac{[\alpha\,\Delta t/(\Delta x)^2](\hat{T}_{i+1,j}^{P+1} + \hat{T}_{i-1,j}^{P+1}) + [\alpha\,\Delta t/(\Delta y)^2](\hat{T}_{i,j+1}^{P+1} + \hat{T}_{i,j-1}^{P+1}) + T_{i,j}^{P}}{1 + 2[\alpha\,\Delta t/(\Delta x)^2] + 2[\alpha\,\Delta t/(\Delta y)^2]} \tag{11.121}$$

In eq. (11.121) the temperatures $\hat{T}_{i-1,j}^{P+1}$, $\hat{T}_{i+1,j}^{P+1}$, $\hat{T}_{i,j-1}^{P+1}$, $\hat{T}_{i,j+1}^{P+1}$ represent the temperatures at the four grid points surrounding point $(i, j)$ that are currently stored in the computer memory. More specifically, Fig. 11.8 shows a group of grid points in which the computation proceeds from bottom to top in the $y$-direction and from left to right in the $x$-direction. Imagine that eq. (11.121) is being used to calculate the temperature. As mentioned earlier, an iterative process is adopted at each time step. During the $n$th iteration of this process, since the computation proceeds from bottom to top and from left to right, the current values of $\hat{T}_{i-1,j}^{P+1}$ and $\hat{T}_{i,j-1}^{P+1}$ in eq. (11.121) are the values just obtained during the same $n$th iteration. On the other hand, the current values of $\hat{T}_{i+1,j}^{P+1}$ and $\hat{T}_{i,j+1}^{P+1}$ in eq. (11.121) are the values of the previous iteration $(n - 1)$. Once all the points of the domain of interest are visited utilizing eq.

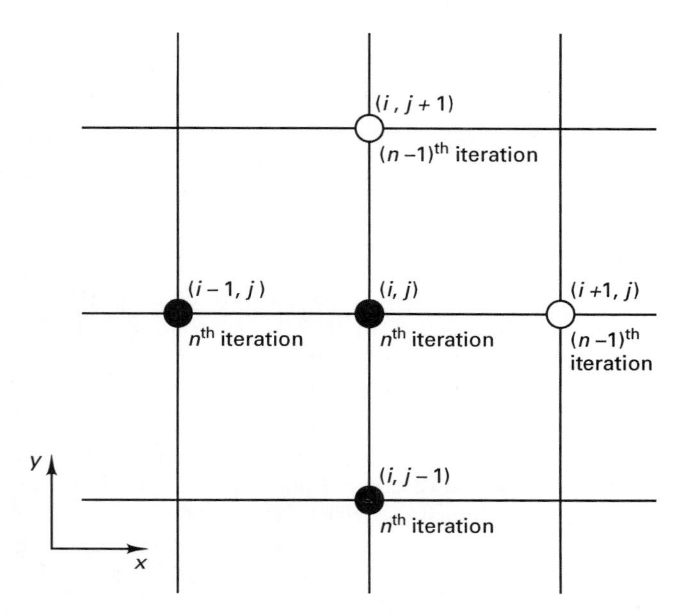


**Figure 11.8**

(11.121), one iteration is completed. The process is then repeated until convergence within acceptable error, at which point the solution is advanced to the next time step.

Regarding convergence, the Gauss–Seidel method does not always converge. A sufficient condition for its convergence is the Scarborough criterion [5,17]. This criterion requires that the sum of the absolute values of the coefficients of the temperatures of the neighboring points surrounding $(i, j)$, divided by the absolute value of the coefficient multiplying the unknown temperature of point $(i, j)$, is less or equal to unity for all grid points and less than unit for at least one grid point. With respect to eq. (11.121) the above-mentioned ratio of coefficients reads

$$\frac{|2[\alpha\, \Delta t/(\Delta x)^2] + 2[\alpha\, \Delta t/(\Delta y)^2]|}{|1 + 2[\alpha\, \Delta t/(\Delta x)^2] + 2[\alpha\, \Delta t/(\Delta y)^2]|} < 1 \tag{11.122}$$

Clearly, the Scarborough criterion is satisfied by eq. (11.121) since the ratio of coefficients of eq. (11.122) is less than unity for all grid points. The reader can show, however, that in the case of conduction in the presence of heat generation (for example), this criterion may not be satisfied. Overall, the Gauss–Seidel method is easy to use and is recommended except for the fact that its convergence is slow and it is not cost-efficient when a large number of grid points is involved.

***Line-By-Line Method.*** This method is recommended by Patankar [5] and it combines desirable features of the tridiagonal matrix algorithm method (Thomas's algorithm) used for unidirectional conduction problems and the Gauss–Seidel method. The basis for this method is that for a given grid line, the values of the temperature at its two neighboring grid lines are assumed known. For example, in Fig. 11.9 it is assumed that the values of the temperature at the left and right grid line (denoted by the open circles) are known. If the calculation proceeds from left to right, the values assigned to the temperatures of the left grid line are the new values of the current iteration. The values assigned to the right grid line are the values of the preceding iteration. The object is then to obtain the temperature of the central grid line (denoted by the black circles). This can be achieved by viewing the heat conduction in the central grid line as unidirectional (in the $y$-direction) and by making use of eq. (11.120) and the Thomas algorithm. The process is repeated to the next grid line until all the vertical grid lines are visited, at which point an iteration is completed. For the next iteration, one may choose to consider a horizontal grid line and repeat the process by visiting one horizontal grid line after the other. Changing directions from iteration to iteration may improve the communication of information between the boundaries and the interior points and facilitate convergence. After the temperature field reaches convergence at a given time step, the solution is advanced to the next time step. In this respect, the procedure resembles the one of the Gauss–Seidel method.

An advantage of the line-by-line method is that the information is transferred from an entire boundary grid line to an entire interior grid line, and so on, at once, speeding up the communication of information from the boundaries to the interior of the calculation domain. Deciding between alternative directions of calculation (be-

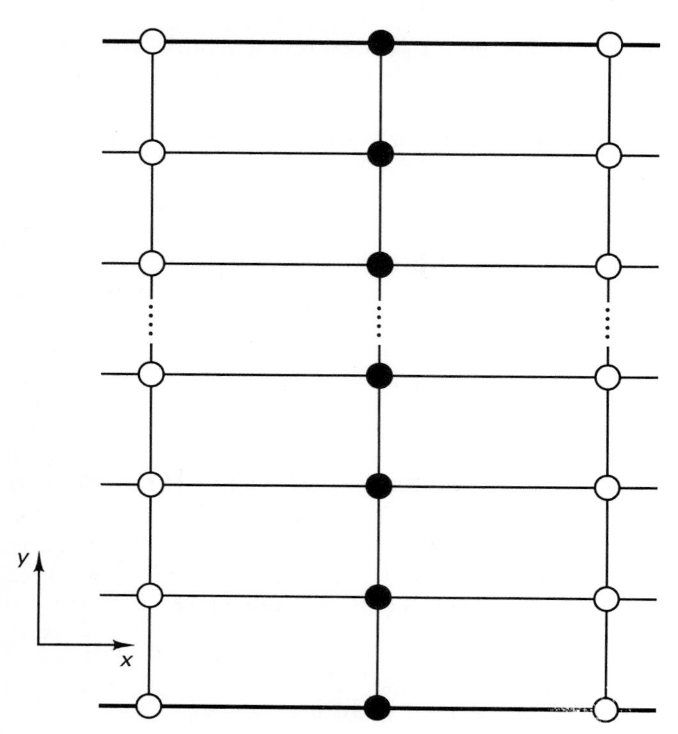


Figure 11.9

tween horizontal and vertical for two-dimensional conduction) or choosing a preferred calculation direction depends on the nature of the problem in hand (geometry and boundary conditions). As discussed in detail in Patankar [5], the correct choice will result in savings in computational time.

***The Issue of Relaxation.*** In iterative solutions, it is sometimes desirable to accelerate the changes in the temperature field from iteration to iteration. On the other hand, to achieve convergence it may actually be necessary to decelerate these changes. The former process (acceleration) is termed *overrelaxation* of the solution and the latter process (deceleration) *underrelaxation.*

Relaxation can be achieved by introducing a relaxation factor to the discretized energy equation. With reference to eq. (11.121), for example, a relaxation factor, $\zeta$, can be introduced as follows:

$$T_{i,j}^{P+1} = \tilde{T}_{i,j}^{P+1} + \zeta\left\{\frac{(\hat{T}_{i+1,j}^{P+1} + \hat{T}_{i-1,j}^{P+1})[\alpha\,\Delta t/(\Delta x)^2] + (\hat{T}_{i,j+1}^{P+1} + \hat{T}_{i,j-1}^{P+1})[\alpha\,\Delta t/(\Delta y)^2] + \hat{T}_{i,j}^{P}}{1 + 2\,[\alpha\,\Delta t/(\Delta x)^2] + 2\,[\alpha\,\Delta t/(\Delta y)^2]} - \tilde{T}_{i,j}^{P+1}\right\} \tag{11.123}$$

where $\tilde{T}_{i,j}^{P+1}$ is the value of $T_{i,j}^{P+1}$ at the end of the previous iteration. Clearly, when convergence is achieved after a number of iterations, $T_{i,j}^{P+1} = \tilde{T}_{i,j}^{P+1}$ and eq. (11.123) becomes identical to the original energy equation (11.121).

Underrelaxation is achieved if $\zeta$ is the range $0 < \zeta < 1$. Overrelaxation, on the other hand, takes place if $1 < \zeta < 2$. The optimum value of $\zeta$ is usually problem dependent and it is determined from experience and by a trial-and-error procedure. One may use different values of $\zeta$ at different stages of the computation or even at different regions of the computational domain.

**Crank–Nicolson scheme.** In this scheme, the space derivatives are averaged between the $P$th time step and the time step $(P + 1)$. A forward-difference approximation is used to discretize the time derivative. The resulting discretization equation for two-dimensional heat conduction in a body with constant thermophysical properties in the absence of heat generation is

$$\frac{T_{i,j}^{P+1} - T_{i,j}^{P}}{\Delta t} = \alpha\left[\frac{1}{2}\frac{T_{i-1,j}^{P+1} - 2T_{i,j}^{P+1} + T_{i+1,j}^{P+1}}{(\Delta x)^2} + \frac{1}{2}\frac{T_{i-1,j}^{P} - 2T_{i,j}^{P} + T_{i+1,j}^{P}}{(\Delta x)^2}\right] + \alpha\left[\frac{1}{2}\frac{T_{i,j-1}^{P+1} - 2T_{i,j}^{P+1} + T_{i,j+1}^{P+1}}{(\Delta y)^2} + \frac{1}{2}\frac{T_{i,j-1}^{P} - 2T_{i,j}^{P} + T_{i,j+1}^{P}}{(\Delta y)^2}\right] \tag{11.124}$$

It is worth noting that eq. (11.124) can be obtained from eq. (11.84) by setting $\lambda = 0.5$, for constant thermal conductivity and in the absence of heat generation. Solving eq. (11.124) for the temperature $T_{i,j}^{P+1}$, we obtain

$$\begin{aligned} T_{i,j}^{P+1} &= \frac{1 - [\alpha\,\Delta t/(\Delta x)^2] - [\alpha\,\Delta t/(\Delta y)^2}{1 + [\alpha\,\Delta t/(\Delta x)^2] + [\alpha\,\Delta t/(\Delta y)^2]}T_{i,j}^{P} \\ &+ \frac{[\alpha\,\Delta t/2(\Delta x)^2]}{1 + [\alpha\,\Delta t/(\Delta x)^2] + [\alpha\,\Delta t/(\Delta y)^2]}(T_{i-1,j}^{P+1} + T_{i+1,j}^{P+1} + T_{i-1,j}^{P} + T_{i+1,j}^{P}) \\ &+ \frac{[\alpha\,\Delta t/2(\Delta y)^2]}{1 + [\alpha\,\Delta t/(\Delta x)^2] + [\alpha\,\Delta t/(\Delta y)^2]}(T_{i,j-1}^{P+1} + T_{i,j+1}^{P+1} + T_{i,j-1}^{P} + T_{i,j+1}^{P}) \end{aligned} \tag{11.125}$$

A von Neumann stability analysis [4] shows that the Crank–Nicolson scheme is unconditionally stable. Regarding the solution of the set of algebraic equations resulting from the application of eq. (11.125) at all interior grid points, the methodologies utilized in the fully implicit scheme can be used in this case as well. For example, in the limit of unidirectional condition (say, in the $x$-direction) eq. (11.125) can be rearranged as follows:

$$\begin{aligned} -\frac{1}{2}\frac{\alpha\,\Delta t}{(\Delta x)^2}T_{i-1}^{P+1} + \left[1 + \frac{\alpha\,\Delta t}{(\Delta x)^2}\right]T_{i}^{P+1} - \frac{1}{2}\frac{\alpha\,\Delta t}{(\Delta x)^2}T_{i+1}^{P+1} \\ = \frac{1}{2}\frac{\alpha\,\Delta t}{(\Delta x)^2}T_{i-1}^{P} + \left[1 - \frac{\alpha\,\Delta t}{(\Delta x)^2}\right]T_{i}^{P} + \frac{1}{2}\frac{\alpha\,\Delta t}{(\Delta x)^2}T_{i+1}^{P} \end{aligned} \tag{11.126}$$

The reader can easily show that eq. (11.126) produces a tridiagonal system of equations that can be solved by the Thomas algorithm. For two- or three-dimensional conduction, the Gauss–Seidel method or the line-by-line method discussed in detail in connection with the fully implicit scheme is also appropriate for the Crank–Nicolson scheme.

### Multidimensional Splitting Methods

A group of methods that can be utilized to solve multidimensional heat conduction problems are the splitting methods. Here the advancement of the solution by one time step takes place as a sequence of two half-steps. At each half-step only the terms associated with a specific direction ($x$ or $y$, for example) are treated implicitly. Therefore, at each half-step a unidirectional heat conduction problem is considered. This problem can be solved efficiently by the Thomas algorithm. Perhaps the most popular splitting method is the alternating direction implicit (ADI) method. A description of this method, as well as other splitting methods, is outside the scope of this chapter. The interested reader can find a wealth of information on the ADI method and other splitting methods in the numerical heat transfer literature [1, 4, 6, 7].

## 11.4 STEADY-STATE CONDUCTION

The numerical solution of steady-state heat conduction problems can often be treated as a special case of the numerical solution of transient heat conduction problems by eliminating the time dependence. The implicit methods discussed in Section 11.3 can be used to solve steady-state heat conduction problems. Even though the procedure is self-explanatory to the reader already familiar with Section 11.3, to exemplify we present here the Gauss–Seidel method for the special case of steady-state conduction. To this end, eq. (11.121) in the limit $\Delta t \rightarrow \infty$ becomes

$$T^n_{i,j} = \frac{[1/(\Delta x)^2](\hat{T}^{n-1}_{i+1,j} + \hat{T}^n_{i-1,j}) + [1/(\Delta y)^2](\hat{T}^{n-1}_{i,j+1} + \hat{T}^n_{i,j-1})}{[2/(\Delta x)^2] + [2/(\Delta y)^2]} \qquad (11.127)$$

Note that the superscripts denoting the time step in eq. (11.121) have been replaced by superscripts denoting the iteration number in eq. (11.127). In writing these superscripts, it was assumed that the computational domain is swept from bottom to top and from left to right. Hence the temperature stored in the computer memory for grid points $(i-1, j)$ and $(i, j-1)$ is that of the current ($n$) iteration, while the temperature for grid points $(i, j+1)$, $(i+1, j)$ is that of the previous ($n-1$) iteration. An initial guess of the temperature field needs to be postulated to initiate the iterative process. An iteration is complete after all the interior grid points of the domain are visited by means of eq. (11.127). The process is completed if after a number of iterations a converged temperature field is obtained within acceptable error. The issues of convergence and relaxation discussed in Section 11.3 are valid here as well. Finally, as mentioned earlier, other implicit methods, such as the Crank–Nicolson method can be used to solve steady-state conduction problems numerically in a manner similar to what was shown in this section in connection with the Gauss–Seidel method.

## 11.5 SPECIAL CONSIDERATIONS IN NUMERICAL CONDUCTION

Two issues relevant to the successful numerical solution of heat conduction problems deserve special consideration. The first issue pertains to the treatment (modeling) of the conductivity at the interface between two different materials. The second issue is the proper linearization of source or sink terms in the energy equation. Both issues are discussed in detail in the numerical heat transfer literature [5–7] and are considered herein sequentially.

### The Issue of Interface Conductivity

The issue of nonuniform thermal conductivity arises in heat conduction problems in composite materials and structures, for example, where it may be needed to evaluate the conductivity at the interface of two different components of the composite material or structure. Figure 11.10 shows the interface of two materials that is assumed to be perpendicular to the $x$-direction of the computational domain. This interface is located between the $i$th and the $(i + 1)$th grid point. The conductivities at the interface, at grid point $(i)$ and at grid point $(i+1)$ and denoted by $k_{\text{int}}$, $k_i$, $k_{i+1}$, respectively. Perhaps the most obvious manner to express $k_{\text{int}}$ in terms of $k_i$ and $k_{i+1}$ is

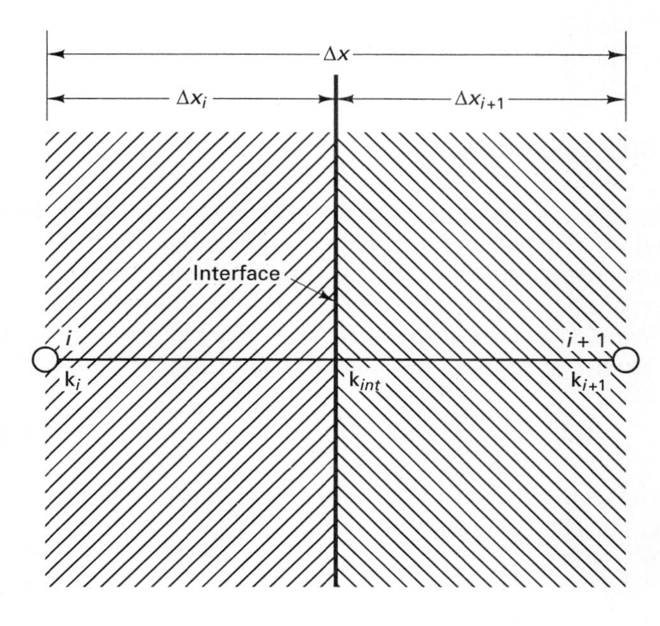


**Figure 11.10** Schematic for the discussion of the interface conductivity approximation.

linear interpolation, which yields (Fig. 11.9)

$$k_{\text{int}} = \left(1 - \frac{\Delta x_i}{\Delta x}\right)k_i + \frac{\Delta x_i}{\Delta x}k_{i+1} \tag{11.128}$$

Expression (11.128) is not always appropriate despite its simplicity. For example, in the limit where one of the materials is a perfect insulator ($k_i = 0$), eq. (11.128) still yields a finite value for the interface conductivity. This can lead to erroneous results featuring a nonzero heat flux at the interface between a material of finite thermal conductivity and a perfect insulator.

A better way to approximate the thermal conductivity is by utilizing the fact that the heat flux at the interface can be expressed in the following three equivalent ways:

$$\dot{q}''_{\text{int}} = k_{\text{int}}\frac{T_i - T_{i+1}}{\Delta x} = k_i\frac{T_i - T_{\text{int}}}{(\Delta x)_i} = k_{i+1}\frac{T_{\text{int}} - T_{i+1}}{(\Delta x)_{i+1}} \tag{11.129}$$

Utilizing eq. (11.129) first to obtain an expression for the interface temperature $T_{\text{int}}$, we can subsequently derive the following formula for the interface conductivity:

$$k_{\text{int}} = \frac{1}{\dfrac{1 - \Delta x_i/\Delta x}{k_{i+1}} + \dfrac{\Delta x_i/\Delta x}{k_i}} \tag{11.130}$$

Equation (11.130) is recommended for it conserves the heat flux at the interface. In addition, in the limit $k_i \to 0$ (one of the materials is a perfect insulation) it gives the realistic result that $k_{\text{int}} \to 0$ and therefore $\dot{q}''_{\text{int}} \to 0$.

Finally, it is worth noting that if the interface is placed halfway between the grid points ($i$) and ($i + 1$) ($\Delta x_i/\Delta x = \frac{1}{2}$), eq. (11.130) implies that $k_{\text{int}}$ is the harmonic mean of $k_i$ and $k_{i+1}$, while eq. (11.128) yields the arithmetic mean of $k_i$ and $k_{i+1}$ as the value of $k_{\text{int}}$.

## The Issue of Source or Sink Linearization

When a heat source or sink term is present in the mathematical model of a conduction problem of interest and it is temperature dependent, it needs to be linearized to facilitate discretization of the energy equation. The recommended linearization is by means of a Taylor series expansion accurate to the first order. To illustrate, assume that a temperature-dependent source term at point ($i$, $j$) is denoted by $Q_{i,j}(T)$. A first-order accurate Taylor series expansion of this source term yields

$$Q_{i,j}^{n+1} = Q_{i,j}^{n} + \left(\frac{\partial Q_{i,j}}{\partial T}\right)^n (T_{i,j}^{n+1} - T_{i,j}^{n}) \tag{11.131}$$

where the superscript $n$ denotes the previous (known) iteration and ($n+1$) the current iteration. Rearranging, eq. (11.131) can be rewritten as

$$Q_{i,j}^{n+1} = \left[Q_{i,j}^{n} - \left(\frac{\partial Q_{i,j}}{\partial T}\right)^n T_{i,j}^{n}\right] + \left(\frac{\partial Q_{i,j}}{\partial T}\right)^n T_{i,j}^{n+1} \tag{11.132}$$

Note that the right-hand side of eq. (11.132) is a linear function of the unknown temperature $T_{i,j}^{n+1}$, since the value of the bracketed term as well as the coefficient of $T_{i,j}^{n+1}$ are constants known from the previous ($n$) iteration. Expression (11.132) can be used to approximate the source term in the conduction equation and to obtain the final discretization equation for $T_{i,j}^{n+1}$ regardless of the numerical scheme utilized.

## PROBLEMS

**11.1.** Discretize the three-dimensional transient heat conduction equation in Cartesian coordinates [eq. (2.33)] utilizing the Taylor series expansion method.

**11.2.** Re-solve Problem 11.1 for the conduction equation in cylindrical coordinates [eq. (2.34)].

**11.3.** Discretize the three-dimensional transient heat conduction equation in Cartesian coordinates [eq. (2.33)] utilizing the control volume method.

**11.4.** Re-solve Problem 11.3 for the conduction equation in cylindrical coordinates [eq. (2.34)].

**11.5.** Discretize the three-dimensional bioheat equation for heat conduction in living tissues [eq. (2.43)] in Cartesian coordinates. Assume that $k_t$ = const.
**(a)** Utilize the Taylor series expansion method.
**(b)** Utilize the control volume method.

**11.6.** Apply the von Neumann stability analysis to the Dufort–Frankel scheme for the unidirectional transient heat conduction equation in Cartesian coordinates and prove that it is unconditionally stable.

**11.7.** Repeat Problem 11.6 for the two-dimensional Dufort–Frankel scheme of eq. (11.113).

**11.8.** Utilize the von Neumann stability analysis to show that the fully implicit scheme of eq. (11.116) is unconditionally stable.

**11.9.** Utilize the von Neumann stability analysis to show that the two-dimensional fully implicit scheme of eq. (11.115) is unconditionally stable.

**11.10.** Generalize the Gauss–Seidel expression (11.121) by including a variable known heat generation (absorption) term $\dot{q}'''(x, y, t)$ in the energy equation. Next, apply the Scarborough criterion and comment on the convergence of the resulting expression.

**11.11.** Prove that for $\lambda = 0.5$, constant thermophysical properties and in the absence of heat generation, eq. (11.84) becomes identical to the Crank–Nicolson scheme (11.124).

**11.12.** Generalize the Gauss–Seidel scheme for steady-state conduction [eq. (11.127)] by accounting for the presence of a variable heat generation term, $\dot{q}'''(x, y)$. You may assume that the thermal conductivity is constant.

**11.13.** Consider the problem described in Example 11.1, case (b) of this chapter. Assume that the bar is made out of stainless steel, $h = 20\ \text{W/m}^2 \cdot \text{K}$ and $s = 5$ mm.
**(a)** Solve the problem numerically utilizing the Gauss–Seidel method and obtain the temperature field in the bar for $\dot{q}''' = 10, 10^2$, and $10^3\ \text{W/m}^3$.
**(b)** Solve the problem analytically and compare the analytical to the numerical results by plotting the temperature distribution in the $y$-direction at the centerline of the bar, $\Theta(s/2, y)$. Comment on the accuracy of the numerical solution versus grid size.

**11.14.** Re-solve Problem 11.13 numerically utilizing the line-by-line method. Compare the efficiency of this method to that of the Gauss–Seidel method. How will the results for the temperature field change if the bar is made out of copper?

**11.15.** Reconsider the problem described and discretized in Example 11.2. The thickness of the steak is assumed to be 3.5 cm, its initial temperature $T_\infty = 20°$ C, and the heat transfer coeficient $h = 10$ W/m$^2 \cdot$ K.

**(a)** Solve the problem numerically utilizing the FTCS method and obtain the transient temperature distribution in the steak for $T_H = 150$°C and for $T_H = 200$°C. The cook is supposed to turn the steak over if the temperature at its center reaches the value of 120°C. After how much time should the cook turn the steak over? Plot the temperature distribution in the steak at characteristic times.

**(b)** Solve the problem analytically and compare your results to the findings of part (a).

**11.16.** Re-solve Problem 11.15 numerically utilizing

**(a)** The fully implicit scheme.

**(b)** The Crank–Nicolson scheme.

**11.17.** Assume that Example 4.1, case (c) is to be used as a benchmark problem for the evaluation of two numerical methods for the solution of steady-state heat conduction problems: the Gauss–Seidel method and the line-by-line method. Solve this example with the above-mentioned methods. Compare your results to the analytical solution offered in the text and comment on the required grid fineness. Rate the performance of the two numerical methods in question. Utilize reasonable values of your choice for the relevant parameters of the problem.

**11.18.** Assume that Example 5.1 is to be used as a benchmark problem for the evaluation of four numerical schemes for the solution of transient heat conduction problems. These are the FTCS, Dufort–Frankel, fully implicit, and Crank–Nicolson schemes. Utilizing realistic material properties and parameter values solve Example 5.1 with all four schemes mentioned above. Compare the performance of the various schemes. Also, compare the numerical results to the analytical solution for the temperature field offered in the text. How does the required grid fineness depend on the choice of numerical scheme?

**11.19.** A wall of a building is made out of concrete and has a thickness of 20 cm. Its outer surface is exposed to the daily temperature cycle and is measured to vary sinusoidally with time. During the summer, the temperature of the outer surface of this wall is approximated by $T(°\mathrm{C}) = 30 + 10 \sin[\pi(t - 6)/12]$, where $t$ is time in hours. The inner surface of the wall is cooled by natural convection. The heat transfer coefficient is assumed to be $h = 5$ W/m$^2 \cdot$ K. The temperature of the air in the building is kept at 25°C. Solve the problem numerically and obtain the steady-periodic temperature in the wall. Calculate the total amount of heat transferred into the building over a period of 24 hours. Solve this problem using **(a)** the FTCS scheme and **(b)** the Dufort–Frankel scheme.

**11.20.** An energy consultant proposes that adding an inner insulation layer of thickness 0.5 cm made out of cork to the building wall described in Problem 11.19 will result in significant energy savings. Neglecting the contact resistance between the wall and the cork layer obtain numerically the steady-periodic temperature distribution of the wall. Evaluate the total amount of energy transported into the building over a 24-hour period and evaluate the correctness of the assessment of the energy consultant.

**11.21.** On an Easter Sunday, prior to roasting the traditional Easter lamb on a spit (Fig. P11.21a), a debate is under way in a Greek village as to whether a stainless steel spit or a wooden spit should be used for the roasting process. It is believed that the presence and the material of the spit affects markedly the cooking time of the lamb as well as the temperature distribution in the lamb during roasting. To aid the avoidance of such arguments in the future you are asked to study a simplified heat transfer model of the lamb-roasting process based on the schematic of Fig. P11.21b. According to this model, the axial conduction (in the direction of the spit) and the transient nature of the imposed heat flux at the lamb surface are neglected. It is assumed that the lamb is heated radiatively with uniform heat transfer rate $\dot{q}$ = 15,000 W. Convection also takes place in the roasting process ($h$ = 40 W/m$^2$ · K, $T_\infty$ = 80°C). The length of the lamb is $L$ = 1.3 m. The radius of the lamb is $R_2$ = 14 cm and the radius of the spit

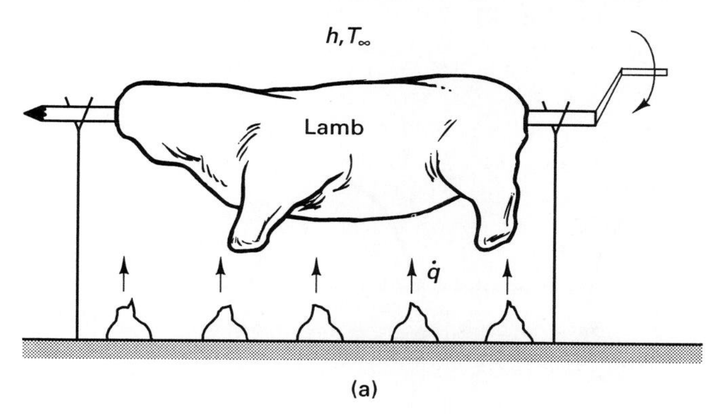


(a)

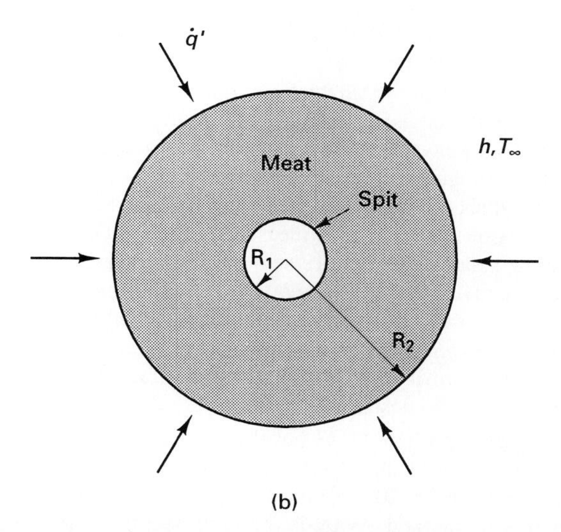


(b)

**Figure P11.21** (a) Schematic of a lamb roasting on a spit. (b) Cross section of the lamb modeled as a cylinder.

$R_1 = 2$ cm. The thermal conductivity, density, and thermal diffusivity of the lamb meat are $k_m = 0.5$ W/m · K, $\rho_m = 1070$ kg/m$^3$, and $\alpha_m = 0.14 \times 10^{-6}$ m$^2$/s. Obtain numerically the transient temperature distribution in the lamb if:

**(a)** A stainless steel spit is used.

**(b)** A wooden spit is used.

How does the spit material affect the temperature field in the lamb and the cooking time? The lamb is considered cooked if the temperature of the meat in contact when the spit surface reaches the value of 88°C or higher for 1 minute.

**11.22.** An improved model of the roasting process of Problem 11.21 can be obtained by accounting for the transient nature of the imposed radiative heat flux at the lamb surface. To this end, assuming that the lamb is roasted by rotating the spit at a speed of $N$ complete rotations per second, the heat flux at the lamb surface is approximated by the graph of Fig. P11.22. Re-solve Problem 11.21 with the outer surface heat transfer rate $\dot{q}(t)$ of Fig. P11.22. Determine the effect of $N$ on the temperature field.

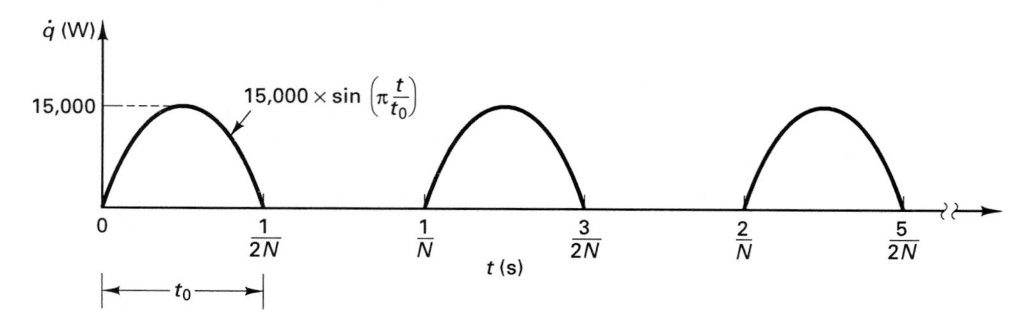


**Figure P11.22**

**11.23.** The thermal conductivity of a composite material is linearly dependent on temperature. This dependence is expressed by the following equation: $k$ (W/m · K) $= 1 + 0.02\Theta$. The excess of the material temperature above room temperature is denoted by $\Theta$. The units for the temperature are in degrees Kelvin. A long bar of this material is placed in a convection–radiation oven for curing purposes. The cross section of the bar is rectangular (5 cm × 10 cm). Both dimensions of the cross section are much smaller than the length of the bar. The bar is placed on a tray with one of its long sides whose temperature is measured to be 300°C during the curing process. The remaining sides of the bar are subjected to a radiative flux ($\dot{q}'' = 25$ kW/m$^2$) as well as to convective heat exchange with the oven ambient. The oven air temperature is 220°C and the convective heat transfer coefficient $h = 80$ W/m$^2$ · K. Obtain the steady-state temperature distribution in the bar using the Gauss–Seidel method. Weigh the contribution of radiative heat transfer versus convective heat transfer and comment on the design of the oven.

**11.24.** Reconsider Problem 11.23 but this time assume that a short block of the same composite material (5 cm × 10 cm × 20 cm) is placed on the oven tray with one of its largest (10 cm × 20 cm) sides. Obtain the temperature field in the block using the line-by-line method. What is the total heat transfer rate through the top and the four lateral sides of the block?

**11.25.** A long bar of steel at uniform temperature of 1300°C after undergoing heat treatment is placed to cool down upon a bar of ceramic as shown in Fig. P11.25. The system of the two bars is placed in a specially designed low-speed wind tunnel to aid the convective heat transfer mechanism. The convective heat transfer coefficient is $h = 80\ \text{W/m}^2 \cdot \text{K}$. The room temperature is 25°C. The relevant properties of steel are $k_s = 30\ \text{W/m} \cdot \text{K}$, $\rho_s = 7800\ \text{kg/m}^3$, $c_{Ps} = 0.45\ \text{kJ/kg} \cdot \text{K}$. The corresponding properties of the ceramic material are $k_c = 1\ \text{W/m} \cdot \text{K}$, $\rho_c = 1800\ \text{kg/m}^3$, and $c_{Ps} = 0.8\ \text{kJ/kg} \cdot \text{K}$. Utilizing finite differences, obtain the transient temperature field in the steel bar. Plot the time variation of the ratio of the total convective heat transfer rate through the three sides of the steel bar to the total conductive heat transfer rate through the side of the bar in contact with the ceramic. Plot the variation of the temperature of the center of the top side of the steel bar as well as the center of the bottom side of the steel bar versus time. Can you speculate whether significant thermal stresses are imposed on the bar during the cooling process?

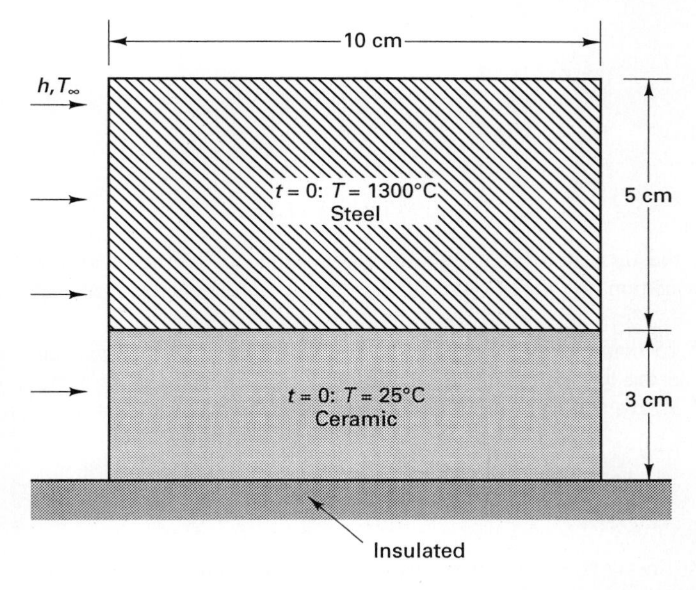


**Figure P11.25**

**11.26.** The cross section of an insulated electrical wire is shown in Fig. P11.26, where $R_1 = 3$ mm, $R_2 = 4$ mm. The outer surface of the wire is cooled convectively ($h = 10\ \text{W/m}^2 \cdot \text{K}$, $T_\infty = 25$°C). Initially, the wire is at room temperature, $T_\infty = 25$°C. Suddenly, the power is turned on generating heat in the wire $\dot{q}''' = 10^6\ \text{W/m}^3$. Utilizing finite differences, obtain the temperature field in the wire until steady state is reached. Plot the variation of the temperature of the copper–rubber interface with time. Is rubber an appropriate material for the insulation (i.e., is there danger for burnout of the rubber insulation)?

**11.27.** Reconsider the insulated electrical wire of Problem 11.26. A sinusoidal electric current is passing this time through the wire with a certain fequency. As a result, the heat generation in the wire is time dependent and it is described by $\dot{q}''' = A\,|\sin(\omega t)|$, where $A = 10^6\ \text{W/m}^3$ and $\omega = 50$ Hz. Determine the steady periodic temperature of

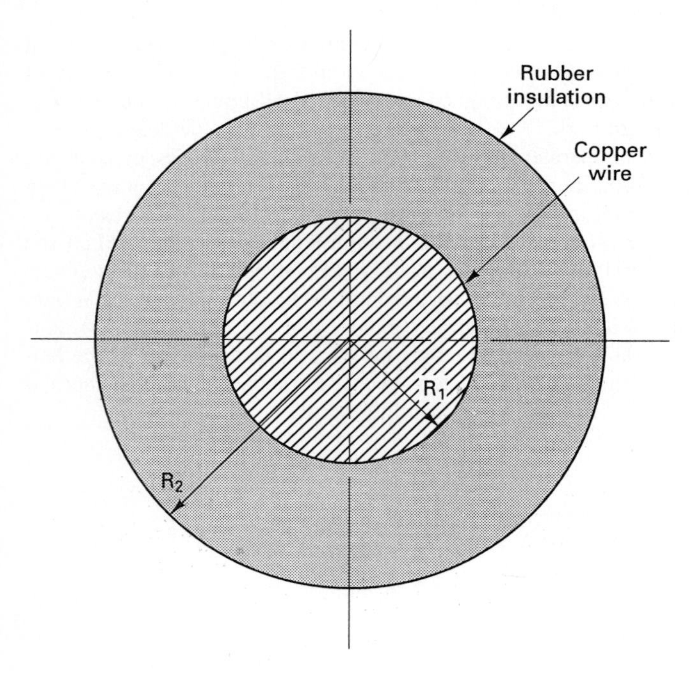


**Figure P11.26**

the copper wire and the insulation layer. What is the dependence of the maximum temperature of the insulation on time? Comment accordingly on the choice of insulating material.

**11.28.** Solve the Problem 5.25 numerically. Compare the results of the analytical and numerical solutions. Use the following values for the numerical solution. $R = 5$ cm, $T_i = 25°C$, $\dot{q}'' = 10^6$ W/m$^2$, and $T_m = 320°C$. The droplet material is lead.

## REFERENCES


1. W. J. Minkowycz, E. M. Sparrow, G. E. Schneider, and R. H. Pletcher, Eds., *Handbook of Numerical Heat Transfer*, Wiley, New York, 1988.
2. D. S. Burnett, *Finite Element Analysis*, Addison-Wesley, Reading, MA, 1987.
3. A. J. Baker, *Finite Element Computational Fluid Mechanics*, Hemisphere, New York, 1983.
4. C. A. J. Fletcher, *Computational Techniques for Fluid Dynamics*, Vol. I, 2nd ed., Springer-Verlag, Berlin, 1991.
5. S. V. Patankar, *Numerical Heat Transfer and Fluid Flow*, Hemisphere, New York, 1980.
6. D. A. Anderson, J. C. Tannehill, and R. H. Pletcher, *Computational Fluid Mechanics and Heat Transfer*, Hemisphere, New York, 1984.
7. Y. Jaluria and K. E. Torrance, *Computational Heat Transfer*, Hemisphere, New York, 1986.

8. FLETCHER, Ref. 4, p. 220.
9. B. J. NOYE, *Numerical Solution of Differential Equations*, North-Holland, Amsterdam, p. 138, 1983.
10. D. B. SPALDING, A novel finite-difference formulation for differential expressions involving both first and second derivatives, *Int. J. Numer. Methods Eng.*, Vol. 4, p. 551, 1972.
11. P. J. ROACHE, *Computational Fluid Dynamics*, Hermosa Publishers, Albuquerque, NM, p. 25, 1976.
12. S. V. PATANKAR and D. B. SPALDING, A calculation procedure for heat, mass and momentum transfer in three-dimensional parabolic flows, *Int. J. Heat Mass Transfer*, Vol. 15, pp. 1787–1806, 1972.
13. FLETCHER, Ref. 4, p. 79.
14. ROACHE, Ref. 11, pp. 36–53.
15. J. A. TRAPP and J. D. RAMSHAW, *J. Comput. Phys.*, Vol. 20, pp. 238–242, 1985.
16. ANDERSON et al., Ref. 6, pp. 128 and 549.
17. J. B. SCARBOROUGH, *Numerical Mathematical Analysis*, 4th ed., John Hopkins University Press, Baltimore, 1958.

CHAPTER **12**

# RECOMMENDED TERM PROJECTS IN HEAT CONDUCTION

This chapter contains a collection of 10 problems that are more involved than typical homework problems. As such, the problems of this chapter qualify to be used as term projects in a heat conduction course. All of these projects involve computer usage. The instructor may, of course, use different projects of his or her own preference or modify the projects of this chapter such that the individual needs of the class are best served.

## PROJECT 1

Figure 12.1 shows a board on which four identical electronic components have been mounted. Assuming that the components are generating heat at a rate $\dot{q}''$ (W/m$^2$), the surface of the board needs to be cooled to avoid undesirable overheating. Such cooling is performed by forced convection of air. The convective heat transfer coefficient is denoted by $h$ and the air temperature by $T_\infty$. The board is very thin compared to its surface dimensions ($\delta << H$ or $L$) such that the heat conduction phenomenon in the board can be assumed to be two-dimensional (i.e., temperature variations across the thickness of the board can be neglected). Further, the edges of the board can be modeled as adiabatic.

**(a)** Show that the mathematical model for the heat conduction process in the board, if the properties of the board and the electronic components do not vary

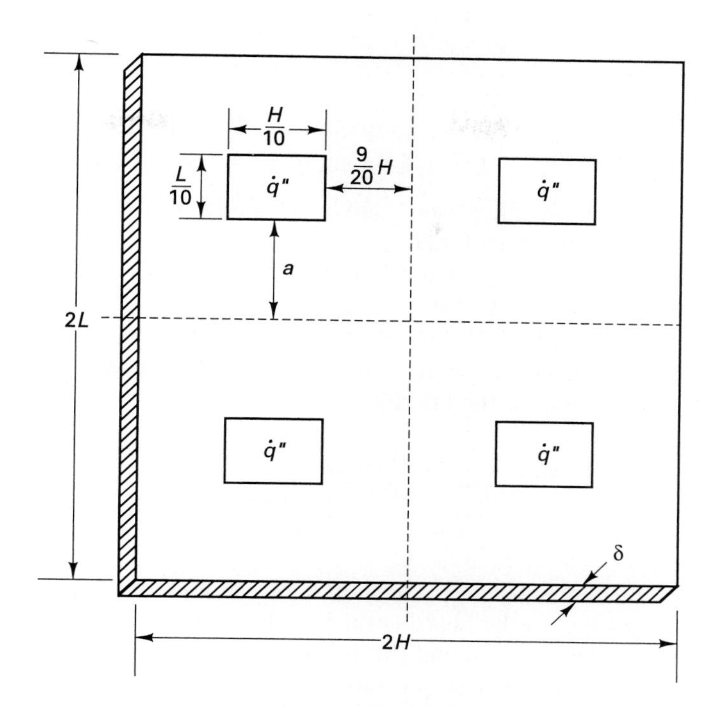


**Figure 12.1** Thin board with four electronic components mounted on it.

appreciably, is:

$$\frac{\partial^2 \hat{\Theta}}{\partial \hat{x}^2} + \frac{\partial^2 \hat{\Theta}}{\partial \hat{y}^2} + \lambda A - B\hat{\Theta} = 0$$

$$\hat{x} = 0: \qquad \frac{\partial \hat{\Theta}}{\partial \hat{x}} = 0$$

$$\hat{x} = 2\frac{H}{L}: \qquad \frac{\partial \hat{\Theta}}{\partial \hat{x}} = 0$$

$$\hat{y} = 0: \qquad \frac{\partial \hat{\Theta}}{\partial \hat{y}} = 0$$

$$\hat{y} = 2: \qquad \frac{\partial \hat{\Theta}}{\partial \hat{y}} = 0$$

The various dimensionless parameters in the equations above are defined as follows:

$$A = \frac{\dot{q}''L^2}{k\delta T_\infty} \qquad B = \frac{h}{k\delta}L^2 \qquad \hat{a} = \frac{a}{L}$$

$$\hat{x} = \frac{x}{L} \qquad \hat{y} = \frac{y}{L} \qquad \hat{\Theta} = \frac{T - T_\infty}{T_\infty}$$

$$\lambda = \begin{cases} 0 & \text{for a bare region of the board} \\ 1 & \text{for a component-covered region of the board} \end{cases}$$

**(b)** Using finite differences, solve the model above and obtain the temperature field in the board for $H = L$ and:
(1) $A = 1, 10, 100$ and $B = 1$, $\hat{a} = (9/20)\ H/L$.
(2) $B = 10, 100$ and $A = 1$, $\hat{a} = (9/20)\ H/L$.
(3) $\hat{a} = 0, (9/20)\ H/L, (9/10)\ H/L$ and $A = 10$, $B = 5$.

**(c)** Draw conclusions on the effect of the parameters $A$, $B$, and $\hat{a}$ on the maximum temperature on the board.

## PROJECT 2

Reconsider an electronic board similar to that of project 1 but with the following major differences: (a) the board is thick such that the two-dimensional heat conduction model is no longer valid and (b) the electronic components, although similar in size, are dissimilar in the heat generation: They produce heat at rates $\dot{q}''$, $2\dot{q}''$, $3\dot{q}''$, and $4\dot{q}''$, respectively, as shown in Fig. 12.2.

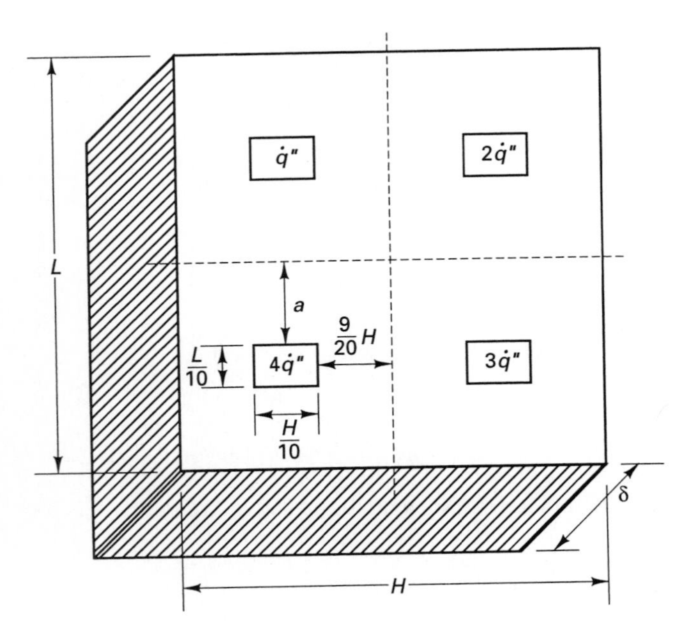


**Figure 12.2** Thick board with four electronic components mounted on it.

**(a)** Define and nondimensionalize the mathematical model for the heat conduction process along the lines of project 1.

**(b)** Using finite differences, solve this model for $A = 10$, $B = 1$, $H/L = 1$, $\hat{a} = (9/20)\,H/L$, and $\delta/L = 0.5$.

**(c)** Define the domain of $\delta/L$ in which a two-dimensional model neglecting heat conduction across the board thickness is appropriate.

You are urged to consider different combinations of the parameters in (b) and determine the effect of these parameters on the temperature field in the board.

## PROJECT 3

In the process of reflow soldering of electronic components, a board with the solder paste and the electronic components placed on it is traversed through the various chambers of a reflow soldering oven. The heating of the board and the components is achieved primarily by radiation. The solder paste, sandwiched between each component and the surface of the board melts and, subsequently, resolidifies, thus mounting (soldering) each component on the board. Figure 12.3a shows a simple arrangement of this type. Only three components are to be mounted on the board of Fig. 12.3a. For simplicity, the board is assumed to be thin (i.e., $\delta << H$ or $L$) and square in shape (i.e., $H = L$). The placement of the components of the board is defined by the following parameters (Fig. 12.3a): $a = L/10$, $b = L/5$.

The temperature of the heater elements of the oven heating up the board varies from chamber to chamber in the oven. Since the board is traversed through the oven, it can be assumed that it is heated radiatively by an emitter whose temperature ($T_0$) varies with time and is defined in Fig. 12.3b [1, 2]. Convection effects are negligible. The presence of the solder paste between the components and the board is also neglected to facilitate the present modeling. In reality the melting of the paste may have a serious effect on the heat conduction process. The edges of the board are assumed adiabatic. The radiation interaction between the bare surface of the board and the oven can be modeled as a source term of the type

$$Q_b(x, y, t) = \sigma \epsilon F(T^4 - T_0^4)$$

where $\sigma$ is the Stefan–Boltzmann constant, $\epsilon$ the emissivity of the board material, $F$ the view factor, $T$ the board surface temperature, and $T_0$ the oven (emitter) temperature defined in Fig. 11.3b. For simplicity, we decouple the radiative heating of the components from that of the board. To this end, the presence of the components is modeled as a source term.

$$Q_c^i(x, y, t) = \frac{M^i c^i}{A^i} \frac{dT^i}{dt} = \sigma \epsilon^i F^i (T_0^4 - T^{i4})$$

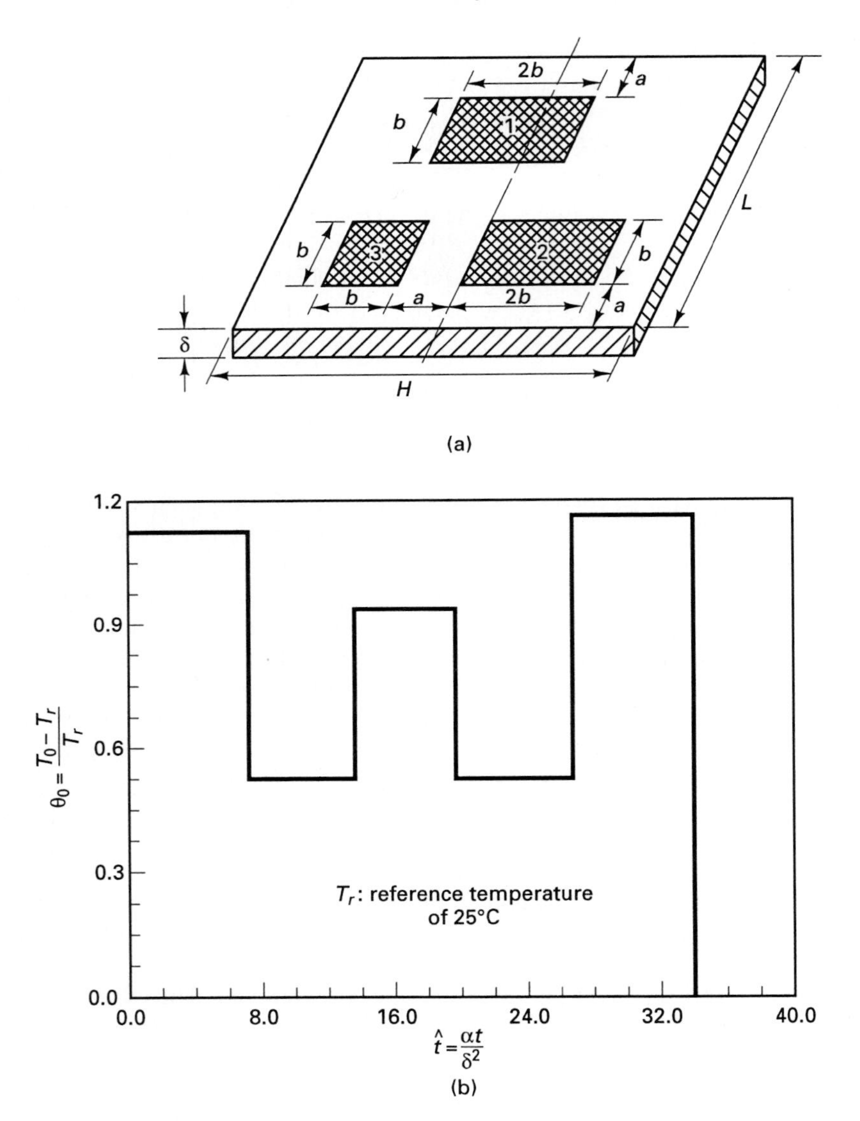


**Figure 12.3** (a) Board undergoing the process of reflow soldering. (b) Emitter temperature.

where superscript $i = 1, 2, 3$ denotes the individual components, $t$ the time, and $M^i$, $c^i$, $A^i$, $T^i$, $\epsilon^i$, and $F^i$ the mass, specific heat, surface area, temperature, emissivity, and configuration factor of the $i$th component.

**(a)** Define and nondimensionalize the conduction heat transfer model that describes the process of interest.

**(b)** Solve this model utilizing finite differences and obtain the temperature field in the board. Locate the region of maximum temperature. The nondimensionalization of the problem will yield the appearance of several dimensionless groups. Discuss the effect of these groups on the temperature field of the board and on the magnitude and the location of the maximum temperature. Perform a literature survey and obtain realistic values of the parameters involved ($M^i$, $c^i$, $\epsilon^i$, $F^i$, $A^i$, etc.). Utilize these values to define realistic ranges for the relevant dimensionless groups of the problem the effect of which is to be investigated.

## PROJECT 4

In the manufacturing process of splat cooling and solidification, a liquid metal droplet impacts and solidifies upon a cold surface (substrate), as shown in Fig. 12.4a. Fluid mechanics considerations indicate that the droplet spreads first and cools down

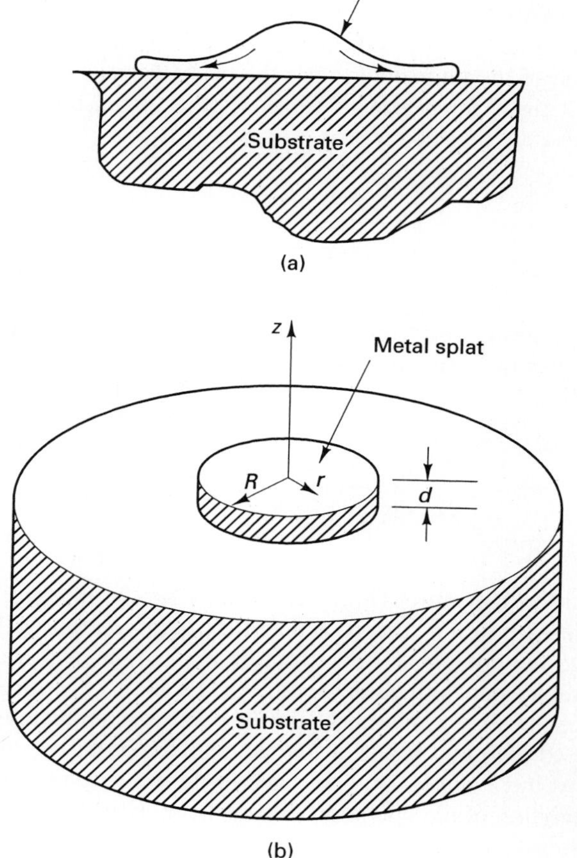


**Figure 12.4** (a) Liquid metal droplet spreading on a substrate after impact. (b) Idealized schematic used for heat conduction modeling.

and solidifies subsequently [3–7]. As a result, the splat cooling process can be approximated as heat conduction between a thin disk and a very thick substrate initially at different temperatures and brought suddenly into contact. A schematic of this approximate model is shown in Fig. 12.4b. The initial temperature of the splat is denoted by $T_0$ and the initial temperature of the substrate by $T_i$. The substrate and the splat are made out of different materials with known thermophysical properties. Two spread ratios of the splat are to be examined $d/R = \frac{1}{20}$ and $d/R = \frac{1}{10}$. As a first approximation, the phenomenon of phase change in the splat can be neglected.

**(a)** Define and nondimensionalize the mathematical model of the heat conduction process described above.

**(b)** Show, based on order-of-magnitude arguments, that the heat conduction process in the splat is adequately described by a one-dimensional heat conduction model.

**(c)** Solve the heat conduction model of (a) and (b) numerically and obtain the transient temperature field in the splat and the substrate. Show a descriptive map of isotherms.

**(d)** Determine the effect of the thermophysical properties of the substrate on the temperature field and on the heat transfer rate through the splat–substrate interface. Plot the variation of the temperature and the heat transfer rate at the bottom of the splat with time, for a host of values of the dimensionless groups characterizing the substrate thermophysical properties.

## PROJECT 5

Re-solve project 4 by accounting this time for the presence of one-dimensional phase change in the splat. A more general formulation and solution on this problem is contained in Refs. [4, 8].

## PROJECT 6

The miniature silicon microbridge sensor (Fig. 12.5a) used for the measurement of flow rate, is built by etching a pit on a silicon chip and installing on top of this pit a silicon nitride microbridge containing an appropriate combination of a heater and two sensors. The typical dimensions of this device are 0(1) mm. To measure the velocity of a gas flowing over the silicon microbridge sensor, the gas temperature prior and after the heater is sensed and, as a result, a voltage output is produced. After appropriate calibration, this voltage output can be related to the velocity of the flowing gas. A more detailed description of the operation of the silicon microbridge sensor is

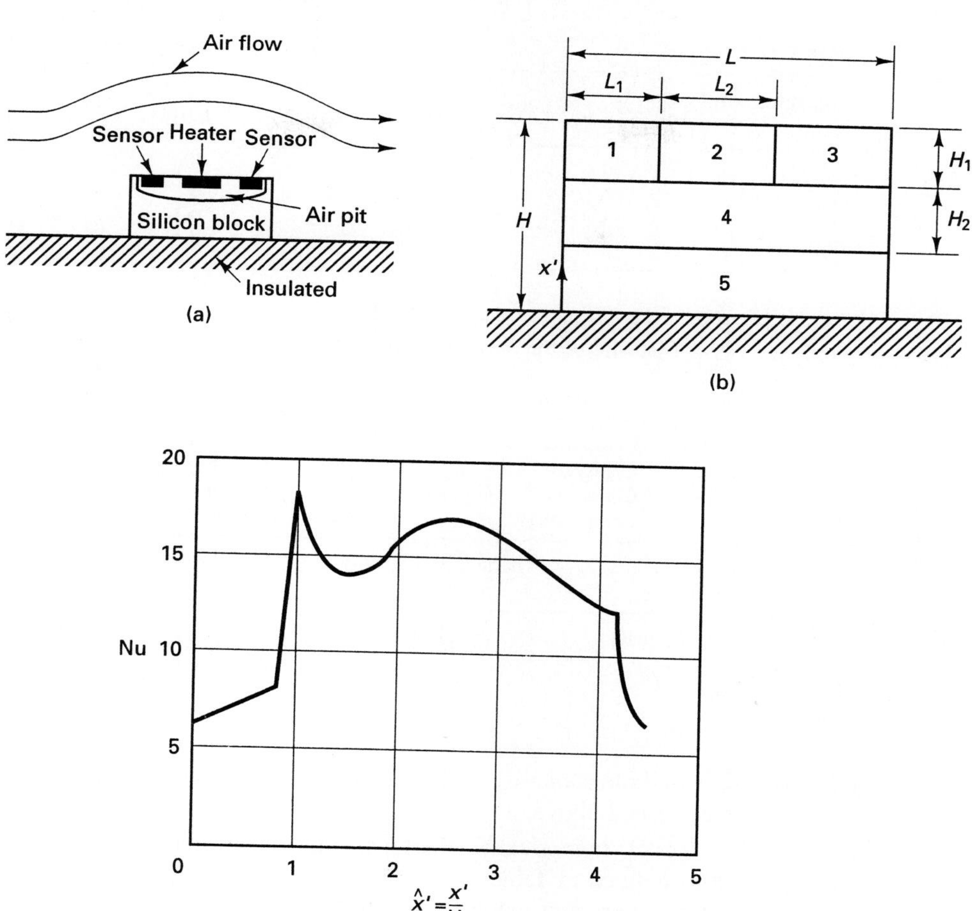


**Figure 12.5** (a) Silicon microbridge sensor placed in an airstream. (b) Details of the construction of the sensor. (c) Nusselt number around the sensor surface.

contained in Ref. [9]. A simple two-dimensional model for the silicon microbridge sensor is shown in Fig. 12.5b. The material of the various sections of the sensor is defined in Table 12.1. The local heat transfer coefficient ($h$) at the surface of the sensor (front face, top, and back face) can be approximated, for the purposes of this project, by the graph shown in Fig. 12.5c, where

$$\text{Nu} = \frac{hH}{k_w} = -\frac{(\partial\Theta/\partial n)_w}{\Theta_w} \qquad \Theta = \frac{T - T_\infty}{T_\infty}$$

The temperature of the fluid far away from the sensor is denoted by $T_\infty$ and subscript $w$ denotes quantities evaluated at the sensor wall surface. The direction normal to the sensor surface is denoted by $n$.

**TABLE 12.1**

| Configuration number | $H_1/H$ | $H_2/H$ | $L/H$ | $L_1/H$ | $L_2/H$ |
|---|---|---|---|---|---|
| 1 | 0.1 | 0.1 | 2.5 | 0 | 2.5 |
| 2 | 0.1 | 0.2 | 2.5 | 0 | 2.5 |
| 3 | 0.1 | 0.1 | 2.5 | 0.5 | 1.5 |
| 4 | 0.1 | 0.1 | 2.5 | 0.5 | 1.5 |

| | Material of the various block sections[a] | | | | | Heat-generating sections |
|---|---|---|---|---|---|---|
| | 1 | 2 | 3 | 4 | 5 | |
| 1 | Composite | Composite | Composite | Air | Silicon | 1, 2, 3 |
| 2 | Composite | Composite | Composite | Air | Silicon | 1, 2, 3 |
| 3 | Composite | Composite | Composite | Air | Silicon | 2 |
| 4 | Air | Composite | Air | Air | Silicon | 2 |

| [a] Material | Thermal conductivity (W/m · K) |
|---|---|
| Composite (silicon nitride) | 11.00 |
| Silicon | 120.00 |
| Plastic compound | 0.14 |
| Air | 0.03 |

**(a)** Define and nondimensionalize a two-dimensional heat conduction model for the sensor described above.

**(b)** Solve this model numerically. Perform the numerical runs for all four configurations defined in Table 12.1 and comment on the effect of the relevant geometric parameters and material properties on the temperature field and on the maximum temperature of the sensor.

## PROJECT 7

In a freeze-coating process, a rather thick cold metal slab is immersed with a constant velocity $U$ in a melt tank of coating material of temperature $T_\infty$. The coating material freezing temperature, $T_f$, is considerably higher than the initial temperature of the slab, $T_0$. As a result, the coating material freezes and forms a thin $[\delta(x) << x]$ layer on the surface of the slab (Fig. 12.6). The heat transfer coefficient at the coating–melt interface can be approximated, for simplicity, with the heat transfer coefficient for laminar boundary layer flow over a flat plate $h = 0.332(k/\tilde{x})\text{Re}_{\tilde{x}}^{1/2}\text{Pr}^{1/3}$ [10], where $\tilde{x}$ measures from the leading edge of the plate. The Reynolds and the Prandtl numbers are defined as $\text{Re}_{\tilde{x}} = U\tilde{x}/\nu$, $\text{Pr} = \nu/\alpha$. All

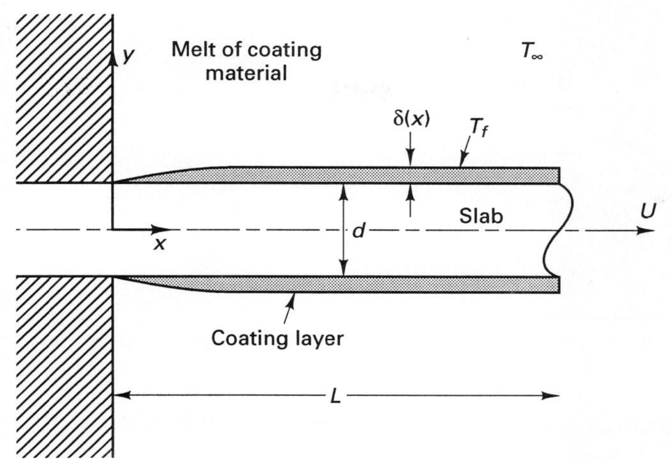


**Figure 12.6** Schematic of the freeze-coating process.

the properties (thermal conductivity, $k$, thermal diffusivity, $\alpha$, and viscosity $\nu$) are those of the fluid. A better approximation for the heat tranfer coefficient is the result for $h$ for laminar boundary layer flow from a cooled continuously moving flat sheet [11]. The reader is recommended to re-solve the problem by utilizing the expression of $h$ in Ref. [11] and comment on the differences between the two solutions.

**(a)** Define and nondimensionalize a heat conduction model for the above described process, accounting for two-dimensional conduction in the slab.

**(b)** Solve this model numerically and obtain the temperature field in the slab and the coating layer. Determine the effect of the relevant dimensionless groups of the problem characterizing the material properties, the temperature conditions, and the velocity of propagation of the slab, on the temperature field and on the thickness of the coating layer.

**(c)** Under what conditions would a one-dimensional heat conduction model for the slab be acceptably accurate?
*Note:* It is recommended that the reader consult Refs. [12–15], dealing with more general formulations of the freeze-coating process.

## PROJECT 8

In an experiment on the freezing of water flowing around a horizontal cold pipe of radius $R_1$, the steady-state ice–water interface was traced carefully from photographs and is reported to scale in Fig. 12.7. Only the upper half of the interface is shown since this interface is symmetric about the horizontal axis. The pipe wall temperature is denoted by $T_w$, the ice–water interface temperature by $T_f$, and the water temperature far away from the pipe by $T_\infty$. In this problem, the ice–water interface tempera-

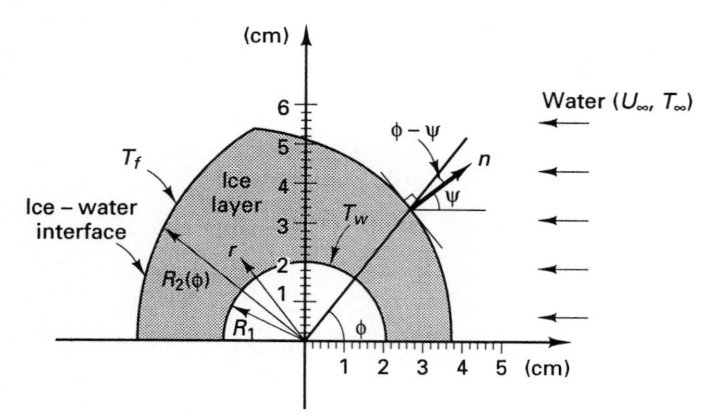


**Figure 12.7** Schematic of ice layer growing around a pipe in cross-flow.

ture is known, $T_f = 0°\text{C}$, and the pipe wall temperature can be measured easily with a thermocouple. A result of engineering importance is the local and overall heat transfer rates at the pipe surface or at the ice–water interface. This result determines the effect of the ice layer on the heat exchange between the water and the pipe. A point-by-point matching method for this problem was proposed by Cheng et al. [16]. The main goal of this project is to apply the Cheng et al. method [16] to the ice layer configuration of Fig. 12.7.

**(a)** Show that the mathematical model for the heat conduction process in the ice layer in cylindrical polar coordinates can be written as

$$\frac{\partial^2 \Theta}{\partial \hat{r}^2} + \frac{1}{\hat{r}}\frac{\partial \Theta}{\partial \hat{r}} + \frac{1}{\hat{r}^2}\frac{\partial^2 \Theta}{\partial \phi^2} = 0$$

$$\hat{r} = 1: \qquad \Theta = 0$$

$$\hat{r} = \frac{R_2(\phi)}{R_1}: \qquad \Theta = 1, \quad h\Theta_c = k\frac{\partial \Theta}{\partial n}$$

where

$$\hat{r} = \frac{r}{R_1} \qquad \Theta = \frac{T - T_w}{T_f - T_w} \qquad \Theta_c = \frac{T_\infty - T_f}{T_f - T_w}$$

**(b)** Even though the solution for the temperature field in the ice layer cannot be obtained with the classical separation-of-variables method, the point-by-point matching procedure can be initiated utilizing this method. Starting with the usual steps of the method of separation of variables, show that the expression for the temperature distribution in the ice layer, satisfying the boundary condition at $\hat{r} = 1$ reads

$$\Theta = b_0 \log \hat{r} + \sum_{j=1}^{\infty} [A_j(\hat{r}^j - \hat{r}^{-j}) \cos(j\phi)]$$

**(c)** The unknown coefficients $b_0, A_1, A_2, \ldots$ in the expression above can be obtained by satisfying the temperature boundary condition $\Theta(R_2/R_1, \phi) = 1$ at a sufficiently large number of equally spaced in the angular direction points, located at the ice–water interface. The exact location of each one of these points $(R_2/R_1, \phi)$ needs to be measured carefully from Fig. 12.7. Choosing 20 points, for example, implies that the series in the temperature expression should be expanded out to 19 terms yielding 19 unknown constants, $A_1, \ldots, A_{19}$. The twentieth constant is $b_0$. Applying the boundary condition $\Theta(R_2/R_1, \phi) = 1$ to all 20 points yields 20 equations for the 20 unknown constants. The number of points needed is determined by trial and error and depends on the desired accuracy of the solution. With the discussion above in mind, obtain the constants $b_0, A_1, A_2, \ldots$ in the expression for $\Theta$ utilizing 10, 20, and 30 equally spaced points in the angular direction on the ice–water interface of Fig. 12.7. Comment on the accuracy of the solution versus the number of points used.

**(d)** Show that the expression for the local heat transfer coefficient at the ice–water interface can be written as

$$h = \frac{k}{\Theta_c R_1}\left[\frac{\sin(\psi - \phi)}{R_2/R_1}\left(\frac{\partial\Theta}{\partial\phi}\right)_{\hat{r}=R_2/R_1} + \cos(\psi - \phi)\left(\frac{\partial\Theta}{\partial\hat{r}}\right)_{\hat{r}=R_2/R_1}\right]$$

With the temperature field from part (c), obtain the distribution of the local heat transfer coefficient around the ice–water interface for $\Theta_c = 50$ and present it graphically in terms of the distribution in the local Nusselt number $\mathrm{Nu} = hR_1/k$ around the interface.

**(e)** Obtain the total heat transfer rate by integrating the local heat transfer rate first around the cylinder surface and then around the ice–water interface. The two results should be identical since we are dealing with a steady-state process. Based on your findings, comment on the accuracy of your solution.

## PROJECT 9

Electromagnetic heating of tissues utilizing transmitting electromagnetic directional antennas (such as a 432-MHz antenna) is performed for therapeutic purposes. In the treatment method known as hyperthermia, tumor tissues are heated for a time period in a temperature range that is not harmful to normal cells but which destroys the cancerous cells, which are more sensitive to high temperatures [17–19]. The heat transfer analysis of the localized treatment of tissue using electromagnetic heating is the purpose of this project. More details on this topic can be found in the broader study of Kouremenos and Antonopoulos [19] and references therein.

Figure 12.8 shows the cross section (square, for simplicity) of a tissue with its top surface undergoing electromagnetic heating by a directional antenna. The heat

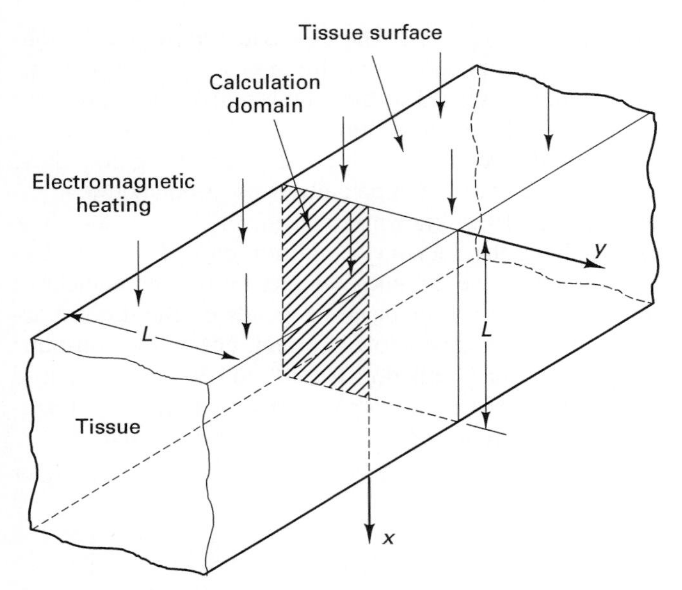


**Figure 12.8** Schematic of heated tissue showing the calculation domain.

conduction process in the tissue is adequately described by the bioheat equation discussed in Chapter 2:

$$\rho_t c_{pt} \frac{\partial T}{\partial t} = k_t \left( \frac{\partial^2 T}{\partial x^2} + \frac{\partial^2 T}{\partial y^2} \right) - \dot{m}_b''' c_{pb} (T - T_A) + \dot{q}_r'''$$

where the metabolic heat generation term was considered to be negligibly small. All the symbols in the equation above are as defined in Chapter 2. The heat generation term due to the electromagnetic radiation heating is approximated by [19]

$$\frac{\dot{q}_r'''}{\rho_t SP} = e^{A(x-0.01)} e^{By^2/(x+D)}$$

where $S$, $A$, $B$, and $D$ are the antenna constants and $P$ is the transmitted power, which depends on the therapeutic needs of the tissue under treatment. For the present application assume that the constants of a 432-MHz antenna are used [19] $S = 12.5\ \text{kg}^{-1}$, $A = -127\ \text{m}^{-1}$, $\text{B} = -129\ \text{m}^{-1}$, and $D = 0.0245$ m.

Making use of the symmetry of the problem, the calculation domain is shown in the crosshatched area of Fig. 12.8. From symmetry considerations, the centerline of the tissue is assumed to be adiabatic. The side and bottom surfaces are assumed to be sufficiently removed from the antenna to be maintained at constant temperature (37°C). The temperature of the top surface of the tissue, which is denoted by $T_s$, is controlled by cooling to avoid burning, and since it can easily be measured, it is considered known. The following tissue properties are also held constant [19]:

$$\rho = 10^3 \frac{\text{kg}}{\text{m}^3} \qquad c_{pt} = 4180 \frac{\text{J}}{\text{kg} \cdot \text{K}} \qquad c_{pb} = 3344 \frac{\text{J}}{\text{kg} \cdot \text{K}} \qquad T_A = 37°\text{C}$$

Note that, for simplicity, the arterial blood temperature $T_A$ is held constant. An improved model could account for the heating of the blood as it passes through the radiated tissue.

**(a)** Obtain the solution for the steady-state temperature field in the tissue in the special case where the conduction process is one-dimensional (it takes place in the $x$-direction only). Show the effect of $P$, $\dot{m}_b$, $T_s$, and $k_t$ on the temperature distribution by varying these parameters in the following ranges:

$$10 \text{ W} < P < 30 \text{ W}$$

$$5 \frac{\text{kg}}{\text{m}^3 \cdot \text{s}} < \dot{m}_b''' < 10 \frac{\text{kg}}{\text{m}^3 \cdot \text{s}}$$

$$20°\text{C} < T_s < 30°\text{C}$$

$$0.5 \frac{\text{W}}{\text{m} \cdot \text{K}} < k_t < 1 \frac{\text{W}}{\text{m} \cdot \text{K}}$$

**(b)** Obtain numerically the solution for the two-dimensional transient temperature field in the tissue if its initial temperature (prior to heating) is assumed to be 37°C. Show maps of isotherms at characteristic times. How much time does it take for the tissue temperature field to reach steady state? Comment on the validity of the unidirectional conduction model of part (a). What is the location and the magnitude of the maximum temperature in the tissue? Show the effect of $P$, $\dot{m}_b'''$, $T_s$, and $k_t$ on the temperature field, the maximum temperature, and the time it takes for the temperature field to reach steady state by varying the above-mentioned parameters within the ranges defined in part (a).

## PROJECT 10

When heat is conducted through the interface between two materials, the issue of contact resistance needs careful investigation (in simple heat conduction models, contact resistance is, for simplicity, often overlooked or neglected). Figure 12.9a shows a microscopic view of the interface between two materials, featuring imperfect contact. Assuming that the enclosed regions (pockets) of noncontact contain vacuum and that radiation is not important, heat flow has to be "squeezed" through the contact areas, as shown in the heat flow lines of Fig. 12.9a.

If two conducting metals (nickel and copper, for example) remain into contact over a long period of time, diffusion of one metal into the other will often take place. As a result, instead of dealing with a thermal conductivity discontinuity at the interface (from the conductivity of one metal, $k_1$, to the conductivity of the other metal, $k_2$) a continuous distribution in conductivity takes place in the diffusion zone [20]. A convenient approximation for the conductivity of the diffusion zone that

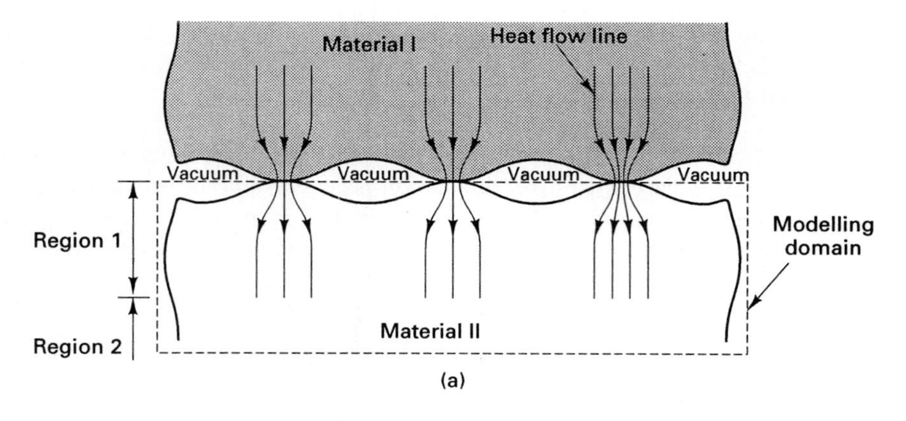


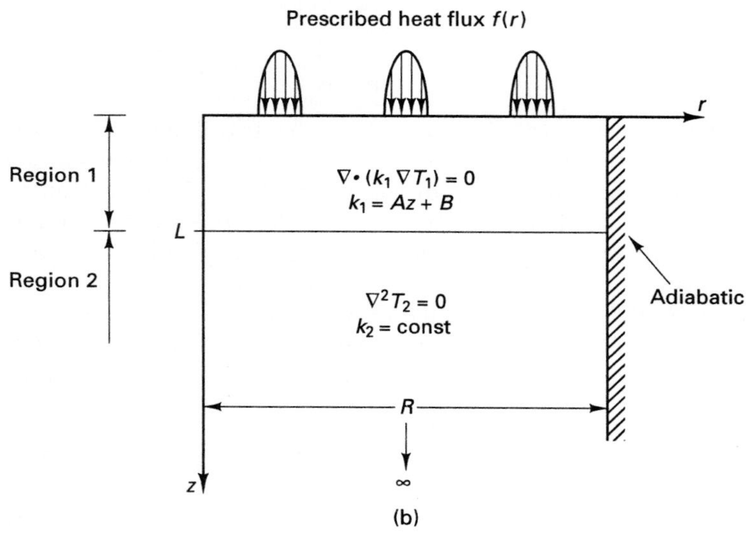


**Figure 12.9** (a) Contact resistance at the interface of two materials. (b) System used in mathematical modeling.

sometimes gives good results for engineering purposes is that the conductivity within this zone varies linearly with distance. Negus et al. [20] adopted this approximation in their study of heat conduction within a cylindrical body, such as the body termed "material II" in Fig. 12.9a, in contact with another body (material I).

Figure 12.9b shows a cylindrical body on the upper surface (the contact surface) of which a known heat flux $f(r)$ is prescribed. This heat flux accounts for the alternating presence of contact and noncontact areas at the interface between two conducting materials. The lateral surface is assumed adiabatic and the extend of the body, semi-infinite. Two distinct regions are identified: region 1, modeling the diffusion zone, in which the thermal conductivity is a linear function of the axial coordinate $k_1 = A(z + B)$, where $A$ and $B$ are known constants, and region 2, in which the thermal conductivity, $k_2$, is constant. This model was proposed by Negus et al. [20]. This project is based on the work reported in this reference.

**(a)** Solve the problem analytically and obtain an expression for the temperature distribution. Caution should be exercised during the solution process to determine the effect of the slope of the thermal conductivity $dk_1/dz$ on the nature of the solution for the temperature field. Determine the effect of the size of the diffusion zone on the temperature field.

**(b)** Solve the same problem numerically. Assign realistic values of your choice to parameters as needed to facilitate the numerical solution. Compare the analytical to the numerical results and discuss the advantages and disadvantages of each approach.

## REFERENCES


1. S. K. RASTOGI and D. POULIKAKOS, Modelling of heat transfer in the surface mounting of electronic components. *ASME J. Electron. Packag.*, 1993 (accepted for publication).
2. C. GARCIA, private communication, Motorola Inc., Arlington Heights, IL, 1991.
3. T. BENNETT and D. POULIKAKOS, Splat-quench solidification: estimating the maximum spreading of a droplet impacting a solid surface, *J. Mater. Sci.* Vol. 28, pp. 963–970, 1993.
4. T. BENNETT and D. POULIKAKOS, Heat transfer aspects of splat quench solidification: modelling and experiment, *J. Mater. Sci.*, 1993 (submitted for publication).
5. G.-X. WANG and E. F. MATTHYS, Numerical modelling of phase change and heat transfer during rapid solidification processes, *Int. J. Heat Mass Transfer,* Vol. 35, pp. 141–153, 1992.
6. E. J. LAVERNIA, E. M. GUTIERREZ, J. SZEKELY, and N. J. GRANT, *Int. J. Rapid Solidif.*, Vol. 4, pp. 89–124, 1988.
7. S. ANNAVARAPU, D. APELIAN, and A. LAWLEY, Spray casting of steel strip: process analysis, *Metall. Trans. Part A,* Vol. 21A, pp. 3237–3256, 1990.
8. T. BENNETT, Splat quench solidification: an investigation into influential parameters, *M.S. thesis,* University of Illinois at Chicago, 1992.
9. A. WIETRZAK and D. POULIKAKOS, Turbulent forced convective cooling of microelectronic devices, *Int. J. Heat Fluid Flow,* Vol. 11, pp. 105–113, 1990.
10. A. BEJAN, *Convection Heat Transfer,* Wiley, New York, 1984.
11. L. E. ERICKSON, L. C. CHA, and L. T. FAN, The cooling of a moving continuous flat sheet, *Chem. Eng. Prog. Symp. Ser. (Heat Tranfer),* AIChE, Vol. 62, pp. 157–165, 1966.
12. A. REZAIAN and D. POULIKAKOS, Heat and fluid flow processes during the coating of a moving surface, *AIAA J. Thermophys.* Vol. 5, pp. 192–198, 1991.
13. R. STEVENS and D. POULIKAKOS, Freeze coating of a moving substrate with the melt of an alloy, *Numer. Heat Transfer Part A,* Vol. 20A, pp. 409–432, 1991.

14. F. B. CHEUNG and S. W. CHA, Finite difference analysis of growth and decay of a freeze coat on a continuous moving cylinder, *Numer. Heat Transfer,* Vol. 12, pp. 41–56, 1987.
15. F. B. CHEUNG, Thermal boundary layer on a continuous moving plate with freezing, *AIAA J. Thermophys.,* Vol. 1, pp. 335–342, 1987.
16. K. C. CHENG, H. INABA, and R. R. GILPIN, An experimental investigation of ice formation around an isothermally cooled cylinder in crossflow, *J. Heat Transfer,* Vol. 103, pp. 733–738, 1981.
17. J. W. STROHBEHN and R. B. ROEMER, A survey of computer simulations of hyperthermia treatments, *IEEE Trans. Biomed. Eng.,* Vol. BME-31(1), pp. 136–149, 1984.
18. R. J. SPIEGEL, A review of numerical models for predicting the energy deposition and resultant thermal response of humans exposed to electromagnetic fields, *IEEE Trans. Microwave Theory Tech.,* Vol. MTT-32(8), pp. 730–746, 1984.
19. D. A. KOUREMENOS and K. A. ANTONOPOULOS, Heat transfer in tissues radiated by a 432 MHz directional antenna, *Int. J. Heat Mass Transfer,* Vol. 31, pp. 2005–2012, 1988.
20. K. J. NEGUS, C. A. VANOVERBEKE, and M. M. YOVANOVICH, Thermal constriction resistance with variable conductivity near the contact surface, *Fundamentals of Conduction and Recent Developments in Contact Resistance,* ASME HTD-Vol. 69, pp. 91–98, 1987.

# APPENDIX A

# BESSEL FUNCTIONS

This appendix contains material related to Bessel functions that is considered useful from the perspective of heat conduction as presented in this book. Consider the following differential equation:

$$x^2y'' + xy' + (x^2 - n^2)y = 0 \qquad n \geq 0 \tag{A.1}$$

where the primes denote differentiation with respect to $x$. It can be shown [1–3] that the solution of this equation is

$$y = C_1 J_n(x) + C_2 Y_n(x) \tag{A.2}$$

where $J_n(x)$ is the Bessel function of the first kind of order $n$ and $Y_n(x)$ is the Bessel function of the second kind of order $n$ and where $C_1$ and $C_2$ are constants. The above-mentioned Bessel functions are infinite series:

$$J_n(x) = \sum_{m=0}^{\infty} \frac{(-1)^m (x/2)^{2m+n}}{m!\,\Gamma(m + 1 + n)} \tag{A.3}$$

$$Y_n(x) = \begin{cases} \dfrac{J_n(x)\cos(n\pi) - J_{-n}(x)}{\sin(n\pi)} & n \neq 0, 1, 2, \ldots \\[2ex] \lim\limits_{m \to n} \dfrac{J_m(x)\cos(m\pi) - J_{-m}(x)}{\sin(m\pi)} & n = 0, 1, 2, \ldots \end{cases} \tag{A.4}$$

Note that the gamma function appears in definition (A.3):

$$\Gamma(k) = \begin{cases} \displaystyle\int_0^{\infty} x^{k-1} e^{-x}\, dx & k > 0 \\ \dfrac{\Gamma(k+1)}{k} & k < 0 \end{cases} \tag{A.5}$$

Some useful formulas involving the Bessel functions are listed below.

$$J_{-n}(x) = (-1)^n J_n(x) \tag{A.6}$$

$$Y_{-n}(x) = (-1)^n Y_n(x) \tag{A.7}$$

$$x \frac{dJ_n(x)}{dx} = xJ_{n-1}(x) - nJ_n(x) \tag{A.8}$$

$$x \frac{dJ_n(x)}{dx} = nJ_n(x) - xJ_{n+1}(x) \tag{A.9}$$

$$\frac{d}{dx}[x^n J_n(x)] = x^n J_{n-1}(x) \tag{A.10}$$

$$\frac{d}{dx}[x^{-n} J_n(x)] = -x^{-n} J_{n+1}(x) \tag{A.11}$$

The Bessel functions of the second kind satisfy relations identical with those of eqs. (A.6–A.11).

In addition to the Bessel functions, in heat conduction problems the modified Bessel functions often appear. These functions are solutions to the modified Bessel equation

$$x^2 y'' + xy' - (x^2 + n^2)y = 0 \qquad n \geq 0 \tag{A.12}$$

The general solution to eq. (A.12) is [1–3]

$$y = C_1 I_n(x) + C_2 K_n(x) \tag{A.13}$$

where $I_n(x)$ is the modified Bessel function of the first kind of order $n$, $K_n(x)$ is the modified Bessel function of the second kind of order $n$, and $C_1$ and $C_2$ are constants. The modified Bessel functions are given by

$$I_n(x) = \sum_{m=0}^{\infty} \frac{(x/2)^{n+2m}}{m!\,\Gamma(n+m+1)} \tag{A.14}$$

$$K_n(x) = \begin{cases} \dfrac{\pi}{2\sin(n\pi)}[I_{-n}(x) - I_n(x)] & \text{for } n \neq 0, 1, 2, \ldots \\ \displaystyle\lim_{m\to n} \dfrac{\pi}{2\sin(m\pi)}[I_{-m}(x) - I_m(x)] & \text{for } n = 0, 1, 2, \ldots \end{cases} \tag{A.15}$$

Some useful formulas involving the modified Bessel functions are listed below.

$$I_n(x) = i^{-n} J_n(ix) \tag{A.16}$$

$$I_{-n}(x) = i^n J_{-n}(ix) \tag{A.17}$$

$$x \frac{dI_n(x)}{dx} = xI_{n-1}(x) - nI_n(x) \tag{A.18}$$

$$x \frac{dI_n(x)}{dx} = xI_{n+1}(x) + nI_n(x) \tag{A.19}$$

$$\frac{d}{dx}[x^n I_n(x)] = x^n I_{n-1}(x) \tag{A.20}$$

$$\frac{d}{dx}[x^{-n} I_n(x)] = x^{-n} I_{n+1}(x) \tag{A.21}$$

$$K_{-n}(x) = K_n(x) \qquad n = 0, 1, 2, \ldots \tag{A.22}$$

$$x \frac{dK_n(x)}{dx} = nK_n(x) - xK_{n+1}(x) \tag{A.23}$$

$$x \frac{dK_n(x)}{dx} = -xK_{n-1}(x) - nK_n(x) \tag{A.24}$$

$$\frac{d}{dx}[x^n K_n(x)] = -x^n K_{n-1}(x) \tag{A.25}$$

$$\frac{d}{dx}[x^{-n} K_n(x)] = -x^{-n} K_{n+1}(x) \tag{A.26}$$

Figure A.1a shows a graphical representation of the Bessel functions of the first and second kind of both order zero and order one [$J_0(x)$, $Y_0(x)$, $J_1(x)$, $Y_1(x)$]. Similarly, Fig. A.1b shows a graphical representation of the modified Bessel functions of the first and second kind of both order one and order two. It is of interest to note that while the Bessel functions exhibit an "oscillatory" behavior of decaying amplitude with increasing $x$ (Fig. A.1a), the modified Bessel functions exhibit an "exponential" behavior. In addition, both the Bessel functions of the second kind [$Y_0(x)$, $Y_1(x)$ in Fig. A.1a] and the modified Bessel functions of the second kind [$K_0(x)$, $K_1(x)$ in Fig. A.1b] "blow up" at $x = 0$. To illustrate how the material contained in this chapter can be used to obtain the solution for the temperature distribution, we present the following examples.

**Example A.1**

Show that the solution of eq. (3.78) is given by eq. (3.81).

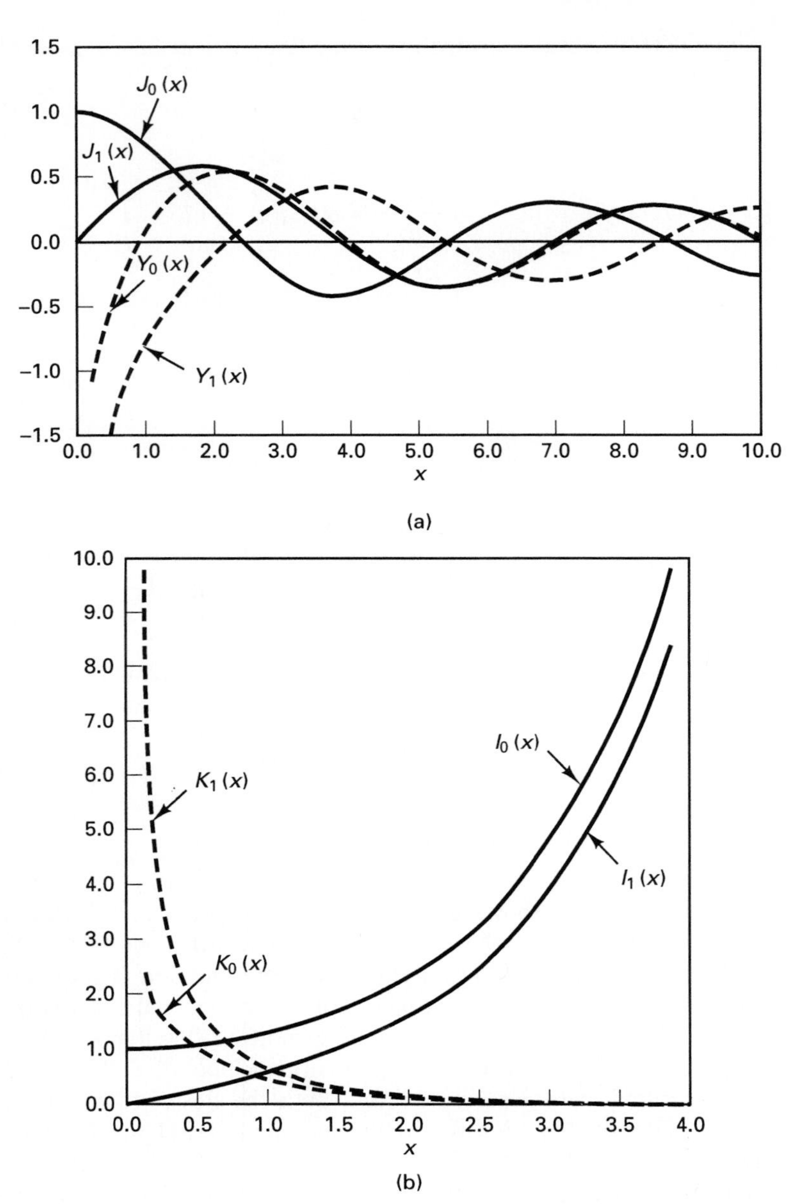


**Figure A1** (a) Variation of representative Bessel functions of $x$ with $x$. (b) Variation of representative modified Bessel functions of $x$ with $x$.

**Solution**

Equation (A.12) for $n = 0$ becomes

$$y'' + \frac{1}{x}y' - y = 0 \tag{A.27}$$

This equation is "similar" but not quite identical to eq. (3.78). The solution of eq. (A.27) is given by eq. (A.13) for $n = 0$:

$$Y = C_1 I_0(x) + C_2 K_0(x) \tag{A.28}$$

To be able to utilize eq. (A.28) to obtain the solution of eq. (3.78), we define

$$\xi = 2Bx_*^{1/2} \tag{A.29}$$

Clearly, then,

$$\frac{d\Theta}{dx_*} = \frac{d\Theta}{d\xi}\frac{d\xi}{dx_*} = \frac{d\Theta}{d\xi}\frac{B}{x_*^{1/2}} \tag{A.30}$$

$$\frac{d^2\Theta}{dx_*^2} = \frac{d}{dx_*}\left(\frac{d\Theta}{d\xi}\frac{B}{x_*^{1/2}}\right) = \frac{B^2}{x_*}\frac{d^2\Theta}{d\xi^2} - \frac{B}{2x_*^{3/2}}\frac{d\Theta}{d\xi} \tag{A.31}$$

Substituting eqs. (A.29)–(A.31) into eq. (3.78) and performing the algebra, we obtain

$$\frac{d^2\Theta}{d\xi^2} + \frac{1}{\xi}\frac{d\Theta}{d\xi} - \Theta = 0 \tag{A.32}$$

Equation (A.32) is identical in form to eq. (A.27) and the general solution (A.28) can be used directly to yield $\Theta$:

$$\Theta = C_1 I_0(\xi) + C_2 K_0(\xi) \tag{A.33}$$

Recalling the definition of $\xi$ [eq. (A.29)], the expression for $\Theta$ becomes

$$\Theta = C_1 I_0(2Bx_*^{1/2}) + C_2 K_0(2Bx_*^{1/2}) \tag{A.34}$$

The two constants $C_1$ and $C_2$ are obtained by straightforward application of the boundary conditions at the base and the tip of the fin [eqs. (3.71) and (3.75)]. It is left up to the student to perform this last step and show that eq. (A.34) is indeed identical to eq. (3.81). It closing, we should comment that the introduction of a rather "involved" new variable (such as $\xi$ in this example) is not always necessary to facilitate the utilization of eqs. (A.1), (A.2), (A.12), and (A.13) in obtaining the temperature distribution of heat conduction problems. It is common that the solution for the temperature is obtained simply by comparing the conduction equation to the Bessel equation (A.1) or to the modified Bessel equation (A.12) simply by inspection.

**Example A.2**

Prove that the solution of eq. (4.81) is given by eq. (4.83).

**Solution**

Equation (A.1) for $n = 0$ yields

$$y'' + \frac{1}{x}y' + y = 0 \tag{A.35}$$

The solution of this equation is obtained directly from eq. (A.2) for $n = 0$:

$$Y = CJ_0(x) + DY_0(x) \tag{A.36}$$

To be able to use eq. (A.36) to solve eq. (4.81), we note that we should rewrite it such that it has identical form to eq. (A.35). To this end, if instead of $r$ we use $\alpha r$ as the independent variable, in eq. (4.81) we obtain

$$\alpha^2 \frac{d^2G}{d(\alpha r)^2} + \alpha^2 \frac{1}{(\alpha r)} \frac{dG}{d(\alpha r)} + \alpha^2 G = 0 \tag{A.37}$$

or

$$G'' + \frac{1}{(\alpha r)}G' + G = 0 \tag{A.38}$$

where the primes now denote differentiation with respect to $\alpha r$. Clearly, eq. (A.38) is identical in form to eq. (A.35), and its solution is

$$G = CJ_0(\alpha r) + DY_0(\alpha r) \tag{A.39}$$

In closing this appendix we should comment that additional information on the Bessel functions and their applications can easily be found in the mathematics literature. For example, Refs. [1–3] are recommended for further reading. In addition, easily accessible subroutines exist in the IMSL library or other software packages available for mainframe or personal computers that evaluate a host of Bessel and modified Bessel functions for any value of their argument. These subroutines can be used to obtain specific "numerical" answers to heat conduction problems.

## REFERENCES

1. G. N. Watson, *A Treatise on the Theory of Bessel Functions,* 2nd ed., Cambridge University Press, London, 1966.
2. C. R. Wylie and L. C. Barret, *Advanced Engineering Mathematics,* 5th ed., McGraw-Hill, New York, 1982.
3. M. R. Spiegel, *Mathematical Handhook,* Schaum's Outline Series, McGraw-Hill, New York, 1968.

APPENDIX **B**

# SOLUTION OF SOME ORDINARY DIFFERENTIAL EQUATIONS

## B.1 Second-Order Ordinary Differential Equations with Constant Coefficients

Homogeneous second-order ordinary differential equations with constant coefficients appear frequently in the solution of heat conduction problems. These equations are of the type

$$a \frac{d^2y}{dx^2} + b \frac{dy}{dx} + cy = 0 \tag{B.1}$$

where $a$, $b$, and $c$ are constants. To solve eq. (B.1), consider the characteristic algebraic equation corresponding to eq. (B.1):

$$a\lambda^2 + b\lambda + c = 0 \tag{B.2}$$

Note that eq. (B.2) was obtained by replacing $d^2y/dx^2$ in eq. (B.1) by $\lambda^2$, $dy/dx$ by $\lambda$, and $y$ by $\lambda^0 = 1$. Let the two roots of the characteristic equation (B.2) be denoted by $\lambda_1$ and $\lambda_2$. The solution of the differential equation (B.1) depends on the nature of these roots in a manner summarized as follows:

1. If $\lambda_1$ and $\lambda_2$ are real and unequal ($\lambda_1 \neq \lambda_2$, $b^2 - 4ac > 0$), the solution to eq. (B.1) reads

$$y = c_1 e^{\lambda_1 x} + c_2 e^{\lambda_2 x} \tag{B.3}$$

2. If $\lambda_1$ and $\lambda_2$ are real and equal ($\lambda_1 = \lambda_2$, $b^2 - 4ac = 0$), the solution to eq. (B.1) reads

$$y = c_1 e^{\lambda_1 x} + c_2 x e^{\lambda_2 x} \tag{B.4}$$

3. If $\lambda_1$ and $\lambda_2$ are complex conjugates ($\lambda_1 = m + in$, $\lambda_2 = m - in$, $b^2 - 4ac < 0$), the solution to eq. (B.1) reads

$$y = e^{mx}[c_1 \sin(nx) + c_2 \cos(nx)] \tag{B.5}$$

## B.2 Second-Order Ordinary Differential Equation Accepting Reduction of Order

Another type of second-order differential equation that appears in conduction problems is that which becomes first-order separable by reducing its order. This equation is of the form

$$\frac{d^2y}{dx^2} + f(x)\frac{dy}{dx} = 0 \tag{B.6}$$

where $f(x)$ is a known function of $x$. The order reduction is obtained simply by setting

$$P = \frac{dy}{dx} \tag{B.7}$$

Combining eqs. (B.6) and (B.7), we obtain

$$\frac{dP}{dx} + f(x)P = 0 \tag{B.8}$$

Equation (B.8) is first-order separable and can be written as

$$\frac{dP}{P} = -f(x)\,dx \tag{B.9}$$

Integrating both sides of eq. (B.9) yields

$$\ln P = -\int f(x)\,dx + C \tag{B.10}$$

or

$$P = Ae^{-\int f(x)\,dx} \tag{B.11}$$

where $A$ and $C$ are constants of integration. Combining eqs. (B.7) and (B.11) yields the solution of eq. (B.6):

$$y = A\int e^{-\int f(x)\,dx}\,dx + B \tag{B.12}$$

### B.3 Homogeneous Euler Equation

The homogeneous Euler equation is of the form

$$a_n x^n \frac{d^n y}{dx^n} + a_{n-1} x^{n-1} \frac{d^{n-1} y}{dx^{n-1}} + \cdots + a_1 x \frac{dy}{dx} + a_0 y = 0 \tag{B.13}$$

Equation (B.13) accepts solutions of the form

$$y = x^P \tag{B.14}$$

Substituting eq. (B.14) into eq. (B.13), we obtain

$$\begin{aligned} a_n P(P-1)(P-2) \cdots (P-[n-1]) + a_{n-1} P(P-1)(P-2) \\ \cdots (P-[n-2]) + \cdots + a_1 P + a_0 = 0 \end{aligned} \tag{B.15}$$

Note that since the left-hand side of eq. (B.15) is an $n$th-order polynomial in $P$, this equation is an $n$th-order algebraic equation in $P$. In principle, solving this equation yields $n$ roots for $P$, namely, $P_1, P_2, P_3, \ldots, P_n$. The general solution of the Euler equation is then

$$y(x) = \sum_{k=1}^{n} A_k x^{P_k} \tag{B.16}$$

where the $A_k$ terms are constants of integration. Note further that in the special case where $n = 2$, eq. (B.15) is simply a quadratic equation for $P$:

$$a_2 P^2 + a_1 P + a_0 = 0 \tag{B.17}$$

whose solutions are

$$P_{1,2} = \frac{-a_1 \pm \sqrt{a_1^2 - 4a_0 a_2}}{2a_2} \tag{B.18}$$

Hence, in this case,

$$y = A_1 x^{P_1} + A_2 x^{P_2} \tag{B.19}$$

with $P_1$ and $P_2$ given by eq. (B.18).

### B.4 First-Order Linear Equation

The first-order linear equation is of the type

$$\frac{dy}{dx} + p(x) y = q(x) \tag{B.20}$$

To solve this equation, first the "integrating factor" must be calculated:

$$I(x) = e^{\int p(x)\, dx} \tag{B.21}$$

With $I(x)$ known, it can be shown [1, 2] that the solution to eq. (B.20) is

$$y(x) = \frac{1}{I(x)}\left[\int q(x)I(x)\,dx + C\right] \tag{B.22}$$

where $C$ is the integration constant.

The utilization of the material in this appendix in the solution of heat conduction problems is illustrated in the following examples.

### Example B.1

Show that the solution of eqs. (4.18) and (4.19) is given by eqs. (4.20) and (4.21), respectively.

#### Solution

The characteristic equation corresponding to eq. (4.18) is

$$\lambda^2 + \alpha^2 = 0 \tag{B.23}$$

This equation has two imaginary conjugate solutions $\lambda_1 = i\alpha$, $\lambda_2 = -i\alpha$. Therefore, according to eq. (B.5), for $m = 0$, $n = \alpha$,

$$X = c_1 \sin(\alpha x) + c_2 \cos(\alpha x) \tag{B.24}$$

Similarly, the characteristic equation corresponding to eq. (4.19) is

$$\lambda^2 - \alpha^2 = 0 \tag{B.25}$$

which has two real and unequal roots, $\lambda_1 = \alpha$ and $\lambda_2 = -\alpha$. Hence, based on eq. (B.3),

$$Y = c_1 e^{\alpha y} + c_2 e^{-\alpha y} \tag{B.26}$$

Note that it can be easily shown that eq. (B.26) may be cast in terms of hyperbolic sines and cosines instead of exponentials:

$$Y = A\cosh(\alpha y) + B\sinh(\alpha y) \tag{B.27}$$

where $A$ and $B$ are constant.

### Example B.2

Prove that the solution of the Euler equation (4.150) is given by eqs. (4.152) and (4.153).

#### Solution

Assuming that $G(r) = r^P$, as discussed in this appendix, and substituting this quantity into eq. (4.150) yields the following quadratic equation for $P$:

$$P^2 + P - \alpha^2 = 0 \tag{B.28}$$

According to eq. (B.18), the two solutions for $P$ are

$$P_1 = -\tfrac{1}{2} + \sqrt{\tfrac{1}{4} + \alpha^2} \tag{B.29}$$

$$P_2 = -\tfrac{1}{2} - \sqrt{\tfrac{1}{4} + \alpha^2} \tag{B.30}$$

For simplicity, we set

$$P_1 = n \tag{B.31}$$

Then it can easily be shown that

$$P_2 = -n - 1 = -(n + 1) \tag{B.32}$$

Therefore, based on eq. (B.19),

$$G(r) = Ar^n + \frac{B}{r^{n+1}} \tag{B.33}$$

Finally, multiplying eqs. (B.29) and (B.30) side by side yields

$$\alpha^2 = n(n + 1) \tag{B.34}$$

**Example B.3**

Show that the solution of eq. (5.172) is given by eq. (5.174) if $\lambda_n \neq 0$.

**Solution**

If $\lambda_n \neq 0$, eq. (5.172) is a first-order linear equation of the type represented by eq. (B.20). Comparing eqs. (5.172) and (B20), we conclude that

$$p = \alpha\lambda_n^2 \tag{B.35}$$

$$q = \frac{2\alpha\dot{q}''}{Rk}\frac{1}{J_0(\lambda_n R)} \tag{B.36}$$

The integrating factor is [eq. (B.21)]

$$I(t) = e^{\int \alpha\lambda_n^2\, dt} = e^{\alpha\lambda_n^2 t} \tag{B.37}$$

Hence, based on eq. (B.22),

$$\begin{aligned} A_n &= e^{-\alpha\lambda_n^2 t}\left[\int \frac{2\alpha\dot{q}''}{Rk}\frac{e^{\alpha\lambda_n^2 t}}{J_0(\lambda_n R)}\, dt + C\right] \\ &= \frac{2\dot{q}''}{Rk\lambda_n^2 J_0(\lambda_n R)} + Ce^{-\alpha\lambda_n^2 t} \end{aligned} \tag{B.38}$$

Next, applying initial condition (5.173) to evaluate the constant $C$ it is straightforward to obtain eq. (5.174).

**Example B.4**

Show that the solution of eq. (7.24) is given by expression (7.27).

**Solution**

Equation (7.24) is of the type represented by eq. (B.6) with $y = \Theta$, $x = \eta$, and $f(x) = (1 + \eta)/\eta$. Utilizing eq. (B.11) yields

$$\frac{d\Theta}{d\eta} = Ce^{-\int[(1+\eta)/\eta]\,d\eta} = C\frac{e^{-\eta}}{\eta} \tag{B.39}$$

Multiplying both sides by $d\eta$ and integrating the resulting equation from $\eta$ to $\infty$ yields

$$\Theta(\eta \to \infty) - \Theta = C\int_{\eta}^{\infty} \frac{e^{-\lambda}}{\lambda}\,d\lambda \tag{B.40}$$

where $\lambda$ is a dummy variable. This equation is recognized to be identical to eq. (7.27).

For additional reading on the solution of ordinary differential equations and their use in engineering problems the reader may consider Refs. [1–3].

## REFERENCES

1. C. R. Wylie and L. C. Barret, *Advanced Engineering Mathematics,* 5th ed. McGraw-Hill, New York, 1982.
2. M. Braun, *Differential Equations and Their Applications,* Springer-Verlag, New York, 1975.
3. C. M. Bender and S. A. Orszag, *Advanced Mathematical Methods for Scientists and Engineers,* McGraw-Hill, New York, 1978.

*APPENDIX* C

# ORTHOGONALITY AND ORTHOGONAL FUNCTIONS

The definition of orthogonality is as follows [1]: Consider two sequences of functions $\Theta_m(x)$, $\Theta_n(x)$, where $m = 1, 2, 3, \ldots, n = 1, 2, 3, \ldots$. If

$$\int_a^b \Theta_m(x)\, \Theta_n(x) p(x)\, dx = \begin{cases} 0 & m \neq n \\ \neq 0 & m = n \end{cases} \tag{C.1}$$

then the functions $\Theta_m(x)$ and $\Theta_n(x)$ are *orthogonal* in the region $[a, b]$ with respect to the weighting factor $p(x)$.

It is worth noting that the *sine* function, *cosine* function, *Bessel* functions of the *first* kind, and *Bessel* functions of the *second* kind are orthogonal under a certain condition. This condition is that both the boundary conditions at the ends of the interval of integration $[a, b]$ in eq. (C.1) need to be homogeneous of the type shown later in eqs. (C.3) and (C4). On the other hand, exponentials, hyperbolic sines and cosines, and modified Bessel functions of the first and second kind are not orthogonal.

To show that the orthogonality property of functions depends on the homogeneity of the boundary conditions in the direction in which the integration of eq. (C.1) is performed, we present the following theorem.

**Theorem.** Consider the differential equation

$$\frac{d}{dx}\left[f(x)\frac{d\Theta}{dx}\right] + [g(x) + \lambda^2 p(x)]\Theta = 0 \tag{C.2}$$

where $f(x)$, $g(x)$, and $p(x)$ are continuous in the interval $a \leq x \leq b$. If $\lambda_1$, $\lambda_2$, $\lambda_3, \ldots$ are distinct values of the parameter $\lambda$ for which nontrivial solutions of eq.

(C.2) $\Theta_1, \Theta_2, \Theta_3, \ldots$ exist having continuous derivatives and satisfying the boundary conditions

$$A\Theta(a) + B\left(\frac{d\Theta}{dx}\right)_{x=a} = 0 \tag{C.3}$$

$$C\Theta(b) + D\left(\frac{d\Theta}{dx}\right)_{x=b} = 0 \tag{C.4}$$

where $A$, $B$, $C$, and $D$ are constants such that $A$ and $B$ are not both zero and $C$ and $D$ are not both zero, the sequences of functions $\Theta_m(x)$, $m = 1, 2, 3, \ldots$ and $\Theta_n(x)$, $n = 1, 2, 3, \ldots$ are orthogonal with respect to the weighting factor $p(x)$ over the integral $(a, b)$.

*Proof*. Before beginning the proof it is worth stressing that eqs. (C.3) and (C.4) are *homogeneous* boundary conditions. With reference to heat conduction they are "convection" boundary conditions of the type $\pm h\Theta \pm k(d\Theta/dx) = 0$. In the special case where $A = 0$ or $C = 0$, these boundary conditions become "insulation" boundary conditions ($d\Theta/dx = 0$). Similarly, if $B = 0$ or $D = 0$, eqs. (C.3) and (C.4) become zero-temperature boundary conditions ($\Theta = 0$). Having said this, assume that $\Theta_m$ and $\Theta_n$ are two distinct solutions of eq. (C.2) corresponding to the *distinct* values $\lambda_m$ and $\lambda_n$ of parameter $\lambda$. Therefore, they satisfy eq. (C.2), that is,

$$\frac{d}{dx}\left(f\frac{d\Theta_m}{dx}\right) + (g + \lambda_m^2 p)\Theta = 0 \tag{C.5}$$

$$\frac{d}{dx}\left(f\frac{d\Theta_n}{dx}\right) + (g + \lambda_n^2 p)\Theta = 0 \tag{C.6}$$

Multiplying eq. (C.5) by $\Theta_n$, eq. (C.6) by $\Theta_m$, and subtracting the resulting equations yields

$$(\lambda_m^2 - \lambda_n^2)p\Theta_m\Theta_n = \Theta_m\frac{d}{dx}\left(f\frac{d\Theta_n}{dx}\right) - \Theta_n\frac{d}{dx}\left(f\frac{d\Theta_m}{dx}\right) \tag{C.7}$$

Adding and subtracting the quantity $f(d\Theta_n/dx)(d\Theta_m/dx)$ to the right-hand side of eq. (C.7), we obtain

$$(\lambda_m^2 - \lambda_n^2)p\Theta_m\Theta_n = \frac{d}{dx}\left[f\left(\Theta_m\frac{d\Theta_n}{dx} - \Theta_n\frac{d\Theta_m}{dx}\right)\right] \tag{C.8}$$

Integrating both sides of eq. (C.8) from $x = a$ to $x = b$ yields

$$\begin{aligned}(\lambda_m^2 - \lambda_n^2)\int_a^b p\Theta_m\Theta_n &= \left[f\left(\Theta_m\frac{d\Theta_n}{dx} - \Theta_n\frac{d\Theta_m}{dx}\right)\right]_a^b \\ &= f(b)\left[\Theta_m(b)\left(\frac{d\Theta_n}{dx}\right)_{x=b} - \Theta_n(b)\left(\frac{d\Theta_m}{dx}\right)_{x=b}\right] \\ &\quad - f(a)\left[\Theta_m(a)\left(\frac{d\Theta_n}{dx}\right)_{x=a} - \Theta_n(a)\left(\frac{d\Theta_m}{dx}\right)_{x=a}\right]\end{aligned} \tag{C.9}$$

To proceed, let us consider boundary condition (C.4), which is satisfied by both $\Theta_m$ and $\Theta_n$:

$$C\Theta_m(b) + D\left(\frac{d\Theta_m}{dx}\right)_{x=b} = 0 \tag{C.10}$$

$$C\Theta_n(b) + D\left(\frac{d\Theta_n}{dx}\right)_{x=b} = 0 \tag{C.11}$$

Multiplying eq. (C.10) by $\Theta_n$, eq. (C.11) by $\Theta_m$, and subtracting the resulting expressions yields

$$\Theta_m(b)\left(\frac{d\Theta_n}{dx}\right)_{x=b} - \Theta_n(b)\left(\frac{d\Theta_m}{dx}\right)_{x=b} = 0 \tag{C.12}$$

Equation (C.12) proves that the first term in brackets on the right-hand side of eq. (C.9) is equal to zero. Using boundary condition (C.3) and following an identical procedure, we can easily show that the second term in brackets on the right-hand side of eq. (C.9) is equal to zero. Based on the above, we find that

$$(\lambda_m^2 - \lambda_n^2)\int_a^b p(x)\Theta_m\Theta_n\,dx = 0 \tag{C.13}$$

Since $m \neq n$ and $\lambda_m \neq \lambda_n$, eq. (C.13) proves that

$$m \neq n: \qquad \int_a^b p(x)\Theta_m\Theta_n\,dx = 0 \tag{C.14}$$

and the orthogonality theorem is established.

### Comments

1. Equation (C.2), for $f(x) = 1$, $g(x) = 0$, $p(x) = 1$, becomes the second-order differential equation with constant coefficients of the type $X'' + \lambda^2 X = 0$ or $Y'' + \lambda^2 Y$ that appears often in heat conduction problems. This equation has a general solution in terms of the sine and cosine functions. Clearly, then, as shown in this theorem, these functions are orthogonal, provided that the corresponding boundary conditions are homogeneous of the type shown in eqs. (C.3) and (C.4). This is the reason why it was stated in the text that separation of variables should be performed in such a way that sines and cosines (potentially orthogonal functions) appear in the solution corresponding to the direction ($x$ or $y$) with the two homogeneous boundary conditions.

2. For $f(x) = x$, $p(x) = x$, and $g(x) = -n^2/x$, it can easily be shown that eq. (C.2) becomes the Bessel equation

$$(\lambda x)^2\frac{d^2\Theta}{d(\lambda x)^2} + (\lambda x)\frac{d\Theta}{d(\lambda x)} + [(\lambda x)^2 - n^2]\Theta = 0 \tag{C.15}$$

which has a general solution in terms of the Bessel functions of order $n$ and argument $(\lambda x)$. Hence, according to this theorem, the separation of variables in cylindrical coordinates should be performed such that if both the boundary conditions in the radial direction are homogeneous of the type shown in eqs. (C.3) and (C.4), the solution to the equation in the radial direction is in terms of the Bessel functions. These functions, accompanied by the homogeneous boundary conditions, possess the orthogonality property. On the other hand, if the direction with the two homogeneous boundary conditions is the axial direction in a heat conduction problem in cylindrical coordinates, the separation of variables should be performed such that orthogonal functions (sines and cosines) are obtained in that direction, while modified Bessel functions, which are not orthogonal, are obtained in the radial direction. Further discussion on the material above is presented in the following example.

### Example C.1

Apply the orthogonality property of the Bessel functions in eq. (4.90) and obtain eq. (4.93).

**Solution**

Multiplying both sides of eq. (4.90) by $rJ_0(\alpha_m r)$ and integrating both sides from $r = 0$ to $r = R$, we obtain

$$\int_0^R \phi(r) r J_0(\alpha_m r)\, dr = \sum_{n=1}^{\infty} K_n \sinh(\alpha_n L) \int_0^R J_0(\alpha_n r) J_0(\alpha_m r) r\, dr \tag{C.16}$$

The integral on the right hand-side of eq. (C.16) is evaluated using integral tables [2].

$$\int_0^R J_0(\alpha_n r) J_0(\alpha_m r) r\, dr =$$

$$\begin{cases} \dfrac{(\alpha_n R) J_0(\alpha_m R) J_0'(\alpha_n R) - (\alpha_m R) J_0(\alpha_n R) J_0'(\alpha_m R)}{\alpha_n^2 - \alpha_m^2} & \text{for } m \neq n \\[2ex] R^2\{\frac{1}{2}[J_1(\alpha_n R)]^2 + \frac{1}{2}[J_0(\alpha_n R)]^2\} & \text{for } m = n \end{cases} \tag{C.17}$$

Taking into acount the fact that eq. (4.88), obtained from the homogeneous boundary condition at $r = R$, states that $J_0(\alpha_m R) = 0$, $n = 1, 2, \ldots, m, \ldots$, we conclude that the upper branch of eq. (C.17) for $m \neq n$ equals zero and that the second term in the expression of the lower branch is also zero. Hence

$$\int_0^R J_0(\alpha_n r) J_0(\alpha_m r) r\, dr = \begin{cases} 0 & m \neq n \\ \dfrac{R^2}{2} J_1^2(\alpha_n R) & m = n \end{cases} \tag{C.18}$$

At this point the orthogonality of $J_0(\alpha_n r)$, $J_0(\alpha_m r)$ with respect to the weighting factor $r$ has been established. Utilizing eq. (C.18), it is straightforward to solve eq. (C.16) for $K_n$ and obtain eq. (4.93).

## REFERENCES

1. C. R. WYLIE and L. C. BARRET, *Advanced Engineering Mathematics*, 5th ed., McGraw-Hill, New York, 1982.
2. M. R. SPIEGEL, *Mathematical Handbook*, Schaum's Outline Series, McGraw-Hill, New York, 1968.

APPENDIX D

# LEGENDRE EQUATION

The differential equation

$$(1 - x^2)\frac{d^2y}{dx^2} - 2x\frac{dy}{dx} + n(n + 1)y = 0 \tag{D.1}$$

is called Legendre's differential equation. The general solution of eq. (D.1) is

$$y = C_1 P_n(x) + C_2 Q_n(x) \tag{D.2}$$

The functions $P_n(x)$ are termed *Legendre polynomials of the first kind* and the functions $Q_n(x)$ are termed the *Legendre polynomials of the second kind*. The functions $P_n(x)$ and $Q_n(x)$ are given by the following equations:

$$P_n(x) = \frac{1}{2^n n!}\frac{d^n}{dx^n}(x^2 - 1)^n \tag{D.3}$$

If $n$ is an even number and $|x| < 1$,

$$Q_n(x) = \frac{(-1)^{n/2}2^n[(n/2)!]^2}{n!}\left[x - \frac{(n-1)(n+2)}{3!}x^3 + \frac{(n-1)(n-3)(n+2)(n+4)}{5!}x^5 - \cdots\right] \tag{D.4}$$

If $n$ is odd and $|x| < 1$,

$$Q_n(x) = \frac{(-1)^{(n+1)/2}(2^{n-1})\{([(n-1)/2]!)^2\}}{1 \cdot 3 \cdot 5 \cdots n}\left[1 - \frac{n(n+1)}{2!}x^2 + \frac{n(n-2)(n+1)(n+3)}{4!}x^4 - \cdots\right] \tag{D.5}$$

Of particular interest to the solver of heat conduction problems is the *orthogonality property* of the Legendre polynomials in the region $-1 < x < 1$ with a weighting factor equal to unity:

$$\int_{-1}^{1} P_m(x)P_n(x)\,dx = \begin{cases} 0 & \text{if } m \neq n \\ \dfrac{2}{2n+1} & \text{if } m = n \end{cases} \tag{D.6}$$

In addition, the following formulas are often utilized

$$Q_n(x = \pm 1) \to \infty \tag{D.7}$$

$$P_{n+1}(x) = \frac{2n+1}{n+1} x P_n(x) - \frac{n}{n+1} P_{n-1}(x) \tag{D.8}$$

$$\frac{d}{dx}[P_{n+1}(x)] - \frac{d}{dx}[P_{n-1}(x)] = 2(n+1)P_n(x) \tag{D.9}$$

$$P_n(-x) = (-1)^n P_n(x) \tag{D.10}$$

### Associated Legendre Functions

The differential euqation

$$(1 - x^2)\frac{d^2y}{dx^2} - 2x\frac{dy}{dx} + \left[n(n+1) - \frac{m^2}{1-x^2}\right]y = 0 \tag{D.11}$$

is called the associated Legendre equation. Note that for $m = 0$ it reduces to eq. (D.1). The general solution of this equation is

$$y = C_1 P_n^m(x) + C_2 Q_n^m(x) \tag{D.12}$$

where $P_n^m(x)$ and $Q_n^m(x)$ are the associated Legendre functions of the first and the second kind, respectively, given by

$$P_n^m(x) = (1 - x^2)^{m/2} \frac{d^m}{dx^m}[P_n(x)] \tag{D.13}$$

$$Q_n^m(x) = (1 - x^2)^{m/2} \frac{d^m}{dx^m}[Q_n(x)] \tag{D.14}$$

Of use to the solution of heat conduction problems is the orthogonality property of $P_n^m(x)$:

$$\int_{-1}^{1} P_n^m(x)P_k^m(x)\,dx = \begin{cases} 0 & n \neq k \\ \dfrac{2}{2n+1}\dfrac{(n+m)!}{(n-m)!} & n = k \end{cases} \tag{D.15}$$

In addition, the following formulas are listed:

$$P_n^m(x) = 0 \qquad \text{for } m > n \tag{D.16}$$

$$Q_n^m(x \pm 1) \to \infty \tag{D.17}$$

More information on the Legendre equation and its associated functions can easily be found in the literature [1–3].

## REFERENCES

1. C. R. WYLIE and L. C. BARRET, *Advanced Engineering Mathematics*, 5th ed., McGraw-Hill, New York, 1982.
2. M. R. SPIEGEL, *Mathematical Handbook*, Schaum's Outline Series, McGraw-Hill, New York, 1968.
3. S. J. FARLOW, *Partial Differential Equations for Scientists and Engineers*, Wiley, New York, 1982.

APPENDIX **E**

# ERROR FUNCTION AND EXPONENTIAL INTEGRAL FUNCTION

The error function is defined as [1, 2]

$$\mathrm{erf}(\eta) = \frac{2}{\sqrt{\pi}} \int_0^{\eta} e^{-x^2}\, dx \tag{E.1}$$

It can be shown [1] that for small values of the argument $\eta$,

$$\mathrm{erf}(\eta) = \frac{2}{\sqrt{\pi}}\left(\eta - \frac{\eta^3}{3 \cdot 1!} + \frac{\eta^5}{5 \cdot 2!} - \frac{\eta^7}{7 \cdot 3!} + \cdots\right) \tag{E.2}$$

In addition,

$$\mathrm{erf}(-\eta) = -\mathrm{erf}(\eta) \tag{E.3}$$

$$\mathrm{erf}(0) = 0 \tag{E.4}$$

$$\mathrm{erf}(\infty) = 1 \tag{E.5}$$

The complementary error function is defined as

$$\mathrm{erfc}(\eta) = 1 - \mathrm{erf}(\eta) \tag{E.6}$$

Therefore, based on eqs. (E.2), (E.4), and (E.5),

$$\mathrm{erfc}(\eta) = 1 - \frac{2}{\sqrt{\pi}}\left(\eta - \frac{\eta^3}{3 \cdot 1!} + \frac{\eta^5}{5 \cdot 2!} - \frac{\eta^7}{7 \cdot 3!} + \cdots\right) \quad \text{(for small values of } \eta\text{)} \tag{E.7}$$

$$\mathrm{erfc}(0) = 1 \tag{E.8}$$

$$\mathrm{erfc}(\infty) = 0 \tag{E.9}$$

The definition of the exponential integral function is [1, 2]

$$\mathrm{Ei}(\eta) = \int_{\eta}^{\infty} \frac{e^{-\lambda}}{\lambda}\, d\lambda \tag{E.10}$$

For small values of $\eta$ the exponential integral function can be approximated by [1]

$$\mathrm{Ei}(\eta) = -\gamma - \ln(\eta) + \left( \frac{\eta}{1 \cdot 1!} - \frac{\eta^2}{2 \cdot 2!} + \frac{\eta^3}{3 \cdot 3!} - \cdots \right) \tag{E.11}$$

where $\gamma$ is the Euler constant, $\gamma = 0.5772 \ldots$. In addition, the following relations are useful:

$$\mathrm{Ei}(\eta) = -\mathrm{Ei}(-\eta) \tag{E.12}$$

$$\mathrm{Ei}(\infty) = 0 \tag{E.13}$$

$$\mathrm{Ei}(0) = \infty \tag{E.14}$$

The error function and the exponential integral can easily be evaluated numerically with the help of IMSL subroutines available in most mainframe computers or other subroutines available for most types of personal computers. A list of values of the error function for the range of its argument $0 < \eta < 3$ is given in Table E.1. Similarly, Table E.2 shows typical values of the exponential integral in the range $0 < \eta < 3$.

**TABLE E.1**

| $\eta$ | $\mathrm{erf}(\eta)$ |
|---|---|
| 0.0 | 0.00000 |
| 0.1 | 0.11246 |
| 0.2 | 0.22270 |
| 0.3 | 0.32863 |
| 0.4 | 0.42839 |
| 0.5 | 0.52050 |
| 0.6 | 0.60386 |
| 0.7 | 0.67780 |
| 0.8 | 0.74210 |
| 0.9 | 0.79691 |
| 1.0 | 0.84270 |
| 1.1 | 0.88021 |
| 1.2 | 0.91031 |
| 1.3 | 0.93401 |
| 1.4 | 0.95229 |
| 1.5 | 0.96611 |
| 1.6 | 0.97635 |
| 1.7 | 0.98379 |
| 1.8 | 0.98909 |
| 1.9 | 0.99279 |
| 2.0 | 0.99532 |
| 2.1 | 0.99702 |
| 2.2 | 0.99814 |
| 2.3 | 0.99886 |
| 2.4 | 0.99931 |
| 2.5 | 0.99959 |
| 2.6 | 0.99976 |
| 2.7 | 0.99987 |
| 2.8 | 0.99992 |
| 2.9 | 0.99996 |
| 3.0 | 0.99998 |

**TABLE E.2**

| $\eta$ | $\mathrm{Ei}(\eta)$ |
|---|---|
| 0.00 | $+\infty$ |
| 0.01 | 4.0379 |
| 0.02 | 3.3547 |
| 0.04 | 2.6813 |
| 0.07 | 2.1508 |
| 0.10 | 1.8229 |
| 0.15 | 1.4645 |
| 0.20 | 1.2227 |
| 0.30 | 0.9057 |
| 0.40 | 0.7024 |
| 0.50 | 0.5598 |
| 0.60 | 0.4544 |
| 0.70 | 0.3738 |
| 0.80 | 0.3106 |
| 0.90 | 0.2602 |
| 1.00 | 0.2194 |
| 1.10 | 0.18599 |
| 1.20 | 0.15841 |
| 1.30 | 0.13545 |
| 1.40 | 0.11622 |
| 1.50 | 0.10002 |
| 1.60 | 0.08631 |
| 1.70 | 0.07465 |
| 1.80 | 0.06471 |
| 1.90 | 0.05620 |
| 2.00 | 0.04890 |
| 2.20 | 0.03719 |
| 2.40 | 0.02844 |
| 2.60 | 0.02185 |
| 2.80 | 0.01686 |
| 3.00 | 0.01305 |

## REFERENCES

1. M. R. SPIEGEL, *Mathematical Handbook*, Schaum's Outline Series, McGraw-Hill, New York, 1968.
2. M. D. GREENBERG, *Foundations of Applied Mathematics*, Prentice Hall, Englewood Cliffs, NJ, 1978.

APPENDIX **F**

# LAPLACE TRANSFORM AND INVERSE LAPLACE TRANSFORM

The Laplace transform of a function $T(x, t)$, such as the temperature, with respect to a variable $t$, such as the time, is defined by the following integral:

$$\mathcal{L}_t \{T(x, t)\} = \int_0^\infty e^{-st} T(x, t)\, dt = \bar{T}(x, s) \tag{F.1}$$

Note that the remaining variable ($x$) is treated as constant in the integration (F.1). Note further that the subscript $t$ on the left-hand side of eq. (F.1) denotes the fact that the Laplace transform is taken with respect to the variable $t$. For simplicity and since only Laplace transforms with respect to $t$ will be discussed, this subscript will be dropped henceforth.

The inverse Laplace transform is defined as the algebraic operation which if performed on the Laplace transform of a function $\bar{T}(x, s)$ yields the original function $T(x, t)$ [i.e., if $\mathcal{L}\{T(x, t)\} = \bar{T}(x, s)$, then $\mathcal{L}^{-1}\{\bar{T}(x, s)\} = T(x, t)$].

The inverse Laplace transform of function can be calculated directly from the following integral utilizing complex variable theory [1–3]:

$$T(x, t) = \frac{1}{2\pi i} \int_{a-i\infty}^{a+i\infty} e^{st} \bar{T}(x, s)\, ds = \frac{1}{2\pi i} \lim_{b\to\infty} \int_{a-ib}^{a+ib} e^{st} \bar{T}(x, s)\, ds \tag{F.2}$$

where $a$ is a real constant and where all the singularities of $\bar{T}(x, s)$ are located to the left of the line $\text{Re}(s) = a$ in the complex plane $s$.

The following properties of the Laplace transforms are often used in the solution of heat conduction problems.

$$\mathcal{L}[aT_1(x, t) + bT_2(x, t)] = a\mathcal{L}[T_1(x, t)] + b\mathcal{L}[T_2(x, t)] = a\bar{T}_1(x, s) + b\bar{T}_2(x, s) \tag{F.3}$$

$$\mathcal{L}\left[\frac{\partial T(x, t)}{\partial t}\right] = s\mathcal{L}[T(x, t)] - T(x, 0) = s\bar{T}(x, s) - T(x, 0) \tag{F.4}$$

$$\begin{aligned}\mathcal{L}\left\{\frac{\partial^2 T(x, t)}{\partial t^2}\right\} &= s^2\mathcal{L}[T(x, t)] - sT(x, 0) - \left(\frac{\partial T}{\partial t}\right)_{x,0} \\ &= s^2\bar{T}(x, s) - sT(x, 0) - \left(\frac{\partial T}{\partial t}\right)_{x,0}\end{aligned} \tag{F.5}$$

$$\mathcal{L}\left[\frac{\partial T(x, t)}{\partial x}\right] = \frac{\partial}{\partial x}\{\mathcal{L}[T(x, t)]\} = \frac{d}{dx}[\bar{T}(x, s)] \tag{F.6}$$

$$\mathcal{L}\left[\frac{\partial^2 T(x, t)}{\partial x^2}\right] = \frac{d^2}{dx^2}\{\mathcal{L}[T(x, t)]\} = \frac{d^2}{dx^2}[\bar{T}(x, s)] \tag{F.7}$$

$$\mathcal{L}^{-1}\left[\frac{d^n T(x, s)}{ds^n}\right] = (-1)^n t^n T(x, t) \tag{F.8}$$

$$\mathcal{L}^{-1}\left[\int_s^\infty T(x, s)\, ds\right] = \frac{T(x, t)}{t} \tag{F.9}$$

$$\mathcal{L}^{-1}\left[\frac{T(x, s)}{s}\right] = \int_0^t T(x, \lambda)\, d\lambda \tag{F.10}$$

In addition to the foregoing properties, the following **convolution theorem** is often used to invert Laplace transforms with the help of tables: If the Laplace transform of $T_1(x, t)$ is denoted by $\bar{T}_1(x, s)$ and the Laplace transform of $T_2(x, t)$ by $\bar{T}_2(x, s)$, then,

$$\mathcal{L}^{-1}[\bar{T}_1(x, s)\bar{T}_2(x, s)] = \int_0^t T_1(x, \lambda)T_2(x, t - \lambda)\, d\lambda \tag{F.11}$$

Equations (F.3)–(F.11) were included in this appendix without proof. The proof of these properties is rather straightforward and can be found in a variety of mathematics textbooks, exemplified by Refs. [1–3], where several additional properties of the Laplace transforms are also discussed.

Of particular interest to the solver of heat conduction problems is the availability of the inverse Laplace transform of a large variety of functions. To this end, rather extensive tables of inverse Laplace transforms exist in the mathematics literature [1–3]. For convenience, a host of inverse Laplace transforms are included in this appendix (Table F.1).

**TABLE F.1**

| | $\bar{T}(s)$ | $T(t)$ |
|---|---|---|
| 1 | $\frac{1}{s}$ | $1$ |
| 2 | $\frac{1}{s^2}$ | $t$ |
| 3 | $\frac{1}{s^n} \quad n = 1, 2, 3, \ldots$ | $\frac{t^{n-1}}{(n-1)!}$ |
| 4 | $\frac{1}{(s-a)}$ | $e^{at}$ |
| 5 | $\frac{1}{(s-a)^n} \quad n = 1, 2, 3, \ldots$ | $\frac{t^{n-1}e^{at}}{(n-1)!}$ |
| 6 | $\frac{1}{s^2+a^2}$ | $\frac{\sin at}{a}$ |
| 7 | $\frac{s}{s^2+a^2}$ | $\cos at$ |
| 8 | $\frac{1}{(s-b)^2+a^2}$ | $\frac{e^{bt}\sin at}{a}$ |
| 9 | $\frac{s-b}{(s-b)^2+a^2}$ | $e^{bt}\cos at$ |
| 10 | $\frac{1}{s^2-a^2}$ | $\frac{\sinh at}{a}$ |
| 11 | $\frac{s}{s^2-a^2}$ | $\cosh at$ |
| 12 | $\frac{1}{(s-b)^2-a^2}$ | $\frac{e^{bt}\sinh at}{a}$ |
| 13 | $\frac{s-b}{(s-b)^2-a^2}$ | $e^{bt}\cosh at$ |
| 14 | $\frac{1}{(s-a)(s-b)} \quad a \neq b$ | $\frac{e^{bt}-e^{at}}{b-a}$ |
| 15 | $\frac{s}{(s-a)(s-b)} \quad a \neq b$ | $\frac{be^{bt}-ae^{at}}{b-a}$ |

**TABLE F.1 (*cont.*)**

| | $\bar{T}(s)$ | $T(t)$ |
|---|---|---|
| 16 | $\dfrac{1}{(s^2 + a^2)^2}$ | $\dfrac{\sin at - at \cos at}{2a^3}$ |
| 17 | $\dfrac{s}{(s^2 + a^2)^2}$ | $\dfrac{t \sin at}{2a}$ |
| 18 | $\dfrac{s^2}{(s^2 + a^2)^2}$ | $\dfrac{\sin at + at \cos at}{2a}$ |
| 19 | $\dfrac{s^3}{(s^2 + a^2)^2}$ | $\cos at - \dfrac{1}{2} at \sin at$ |
| 20 | $\dfrac{s^2 - a^2}{(s^2 + a^2)^2}$ | $t \cos at$ |
| 21 | $\dfrac{1}{(s^2 - a^2)^2}$ | $\dfrac{at \cosh at - \sinh at}{2a^3}$ |
| 22 | $\dfrac{s}{(s^2 - a^2)^2}$ | $\dfrac{t \sinh at}{2a}$ |
| 23 | $\dfrac{s^2}{(s^2 - a^2)^2}$ | $\dfrac{\sinh at + at \cosh at}{2a}$ |
| 24 | $\dfrac{s^3}{(s^2 - a^2)^2}$ | $\cosh at + \dfrac{1}{2} at \sinh at$ |
| 25 | $\dfrac{s^2 + a^2}{(s^2 - a^2)^2}$ | $t \cosh at$ |
| 26 | $\dfrac{1}{(s^2 + a^2)^3}$ | $\dfrac{(3 - a^2t^2) \sin at - 3\, at \cos at}{8a^5}$ |
| 27 | $\dfrac{s}{(s^2 + a^2)^3}$ | $\dfrac{t \sin at - at^2 \cos at}{8a^3}$ |
| 28 | $\dfrac{s^2}{(s^2 - a^2)^3}$ | $\dfrac{(1 + a^2t^2) \sin at - at \cos at}{8a^3}$ |
| 29 | $\dfrac{s^3}{(s^2 + a^2)^3}$ | $\dfrac{3t \sin at + at^2 \cos at}{8a}$ |
| 30 | $\dfrac{s^4}{(s^2 + a^2)^3}$ | $\dfrac{(3 - a^2t^2) \sinh at + 5\, at \cosh at}{8a}$ |

**TABLE F.1 (*cont.*)**

| | $\bar{T}(s)$ | $T(t)$ |
|---|---|---|
| 31 | $\dfrac{1}{(s^2 - a^2)^3}$ | $\dfrac{(3 + a^2t^2)\sinh at - 3\,at\cosh at}{8a^5}$ |
| 32 | $\dfrac{s}{(s^2 - a^2)^3}$ | $\dfrac{at^2\cosh at - t\sinh at}{8a^3}$ |
| 33 | $\dfrac{s^2}{(s^2 - a^2)^3}$ | $\dfrac{at\cosh at + (a^2t^2 - 1)\sinh at}{8a^3}$ |
| 34 | $\dfrac{s^3}{(s^2 - a^2)^3}$ | $\dfrac{3t\sinh at + at^2\cosh at}{8a}$ |
| 35 | $\dfrac{s^4}{(s^2 - a^2)^3}$ | $\dfrac{(3 + a^2t^2)\sinh at + 5at\cosh at}{8a}$ |
| 36 | $\dfrac{1}{s^3 + a^3}$ | $\dfrac{e^{at/2}}{3a^2}\left\{\sqrt{3}\sin\dfrac{\sqrt{3}\,at}{2} - \cos\dfrac{\sqrt{3}\,at}{2} + e^{-3at/2}\right\}$ |
| 37 | $\dfrac{s}{s^3 + a^3}$ | $\dfrac{e^{at/2}}{3a}\left\{\cos\dfrac{\sqrt{3}\,at}{2} + \sqrt{3}\sin\dfrac{\sqrt{3}\,at}{2} - e^{-3at/2}\right\}$ |
| 38 | $\dfrac{s^2}{s^3 + a^3}$ | $\dfrac{1}{3}\left(e^{-at} + 2e^{at/2}\cos\dfrac{\sqrt{3}\,at}{2}\right)$ |
| 39 | $\dfrac{1}{s^3 - a^3}$ | $\dfrac{e^{-at/2}}{3a^2}\left\{e^{3at/2} - \cos\dfrac{\sqrt{3}\,at}{2} - \sqrt{3}\sin\dfrac{\sqrt{3}\,at}{2}\right\}$ |
| 40 | $\dfrac{s}{s^3 - a^3}$ | $\dfrac{e^{-at/2}}{3a}\left\{\sqrt{3}\sin\dfrac{\sqrt{3}\,at}{2} - \cos\dfrac{\sqrt{3}\,at}{2} + e^{3at/2}\right\}$ |
| 41 | $\dfrac{s^2}{s^3 - a^3}$ | $\dfrac{1}{3}\left(e^{at} + 2e^{-at/2}\cos\dfrac{\sqrt{3}\,at}{2}\right)$ |
| 42 | $\dfrac{1}{s^4 - a^4}$ | $\dfrac{1}{2a^3}(\sinh at - \sin at)$ |
| 43 | $\dfrac{s}{s^4 - a^4}$ | $\dfrac{1}{2a^2}(\cosh at - \cos at)$ |
| 44 | $\dfrac{s^2}{s^4 - a^4}$ | $\dfrac{1}{2a}(\sinh at - \sin at)$ |
| 45 | $\dfrac{s^3}{s^4 - a^4}$ | $\dfrac{1}{2}(\cosh at - \cos at)$ |

**TABLE F.1 (cont.)**

| | $\overline{T}(s)$ | | $T(t)$ |
|---|---|---|---|
| 46 | $\dfrac{1}{\sqrt{s+a}+\sqrt{s+b}}$ | | $\dfrac{e^{-bt}-e^{-at}}{2(b-a)\sqrt{\pi t^3}}$ |
| 47 | $\dfrac{1}{s\sqrt{s+a}}$ | | $\dfrac{\operatorname{erf}\sqrt{at}}{\sqrt{a}}$ |
| 48 | $\dfrac{1}{\sqrt{s}(s-a)}$ | | $\dfrac{e^{at}\operatorname{erf}\sqrt{at}}{\sqrt{a}}$ |
| 49 | $\dfrac{1}{\sqrt{s-a}+b}$ | | $e^{at}\left\{\dfrac{1}{\sqrt{\pi t}}-be^{b^2t}\operatorname{erfc}(b\sqrt{t})\right\}$ |
| 50 | $\dfrac{1}{\sqrt{s^2+a^2}}$ | | $J_0(at)$ |
| 51 | $\dfrac{1}{\sqrt{s^2-a^2}}$ | | $I_0(at)$ |
| 52 | $\dfrac{(\sqrt{s^2+a^2}-s)^n}{\sqrt{s^2+a^2}}$ | $n>-1$ | $a^nJ_n(at)$ |
| 53 | $\dfrac{(s-\sqrt{s^2-a^2})^n}{\sqrt{s^2-a^2}}$ | $n>-1$ | $a^nI_n(at)$ |
| 54 | $\dfrac{e^{b(s-\sqrt{s^2+a^2})}}{\sqrt{s^2+a^2}}$ | | $J_0(a\sqrt{t(t+2b)})$ |
| 55 | $\dfrac{e^{-b\sqrt{s^2+a^2}}}{\sqrt{s^2+a^2}}$ | | $\begin{cases} J_0(a\sqrt{t^2-b^2}) & t>b \\ 0 & t<b \end{cases}$ |
| 56 | $\dfrac{1}{(s^2+a^2)^{3/2}}$ | | $\dfrac{tJ_1(at)}{a}$ |
| 57 | $\dfrac{s}{(s^2+a^2)^{3/2}}$ | | $tJ_0(at)$ |
| 58 | $\dfrac{s^2}{(s^2+a^2)^{3/2}}$ | | $J_0(at)-atJ_1(at)$ |
| 59 | $\dfrac{1}{(s^2-a^2)^{3/2}}$ | | $\dfrac{tI_1(at)}{a}$ |
| 60 | $\dfrac{s}{(s^2-a^2)^{3/2}}$ | | $tI_0(at)$ |

**TABLE F.1 (cont.)**

| | $\bar{T}(s)$ | $T(t)$ |
|---|---|---|
| 61 | $\dfrac{s^2}{(s^2 - a^2)^{3/2}}$ | $I_0(at) + atI_1(at)$ |
| 62 | $\dfrac{e^{-a/s}}{\sqrt{s}}$ | $\dfrac{\cos 2\sqrt{at}}{\sqrt{\pi t}}$ |
| 63 | $\dfrac{e^{-a/s}}{s^{3/2}}$ | $\dfrac{\sin 2\sqrt{at}}{\sqrt{\pi a}}$ |
| 64 | $\dfrac{e^{-a/s}}{s^{n+1}} \quad n > -1$ | $\left(\dfrac{t}{a}\right)^{n/2} J_n(2\sqrt{at})$ |
| 65 | $\dfrac{e^{-a\sqrt{s}}}{\sqrt{s}}$ | $\dfrac{e^{-a^2/4t}}{\sqrt{\pi t}}$ |
| 66 | $e^{-a\sqrt{s}}$ | $\dfrac{a}{2\sqrt{\pi t^3}} e^{-a^2/4t}$ |
| 67 | $\dfrac{1 - e^{-a\sqrt{s}}}{s}$ | $\operatorname{erf}(a/2\sqrt{t})$ |
| 68 | $\dfrac{e^{-a\sqrt{s}}}{s}$ | $\operatorname{erfc}(a/2\sqrt{t})$ |
| 69 | $\dfrac{e^{-a\sqrt{s}}}{\sqrt{s}(\sqrt{s} + b)}$ | $e^{b(bt+a)} \operatorname{erfc}\left(b\sqrt{t} + \dfrac{a}{2\sqrt{t}}\right)$ |
| 70 | $\ln\left(\dfrac{s + a}{s + b}\right)$ | $\dfrac{e^{-bt} - e^{-at}}{t}$ |
| 71 | $\dfrac{\ln[(s + a)/a]}{s}$ | $Ei\,(at)$ |
| 72 | $-\dfrac{(\gamma + \ln s)}{s}$ <br> $\gamma = .5772156\ldots$ | $\ln t$ |
| 73 | $\dfrac{e^{a/s}}{\sqrt{s}} \operatorname{erfc}(\sqrt{a/s})$ | $\dfrac{e^{-2\sqrt{at}}}{\sqrt{\pi t}}$ |
| 74 | $e^{s^2/4a^2} \operatorname{erfc}(s/2a)$ | $\dfrac{2a}{\sqrt{\pi}} e^{-a^2t^2}$ |

**TABLE F.1 (cont.)**

| | $\bar{T}(s)$ | $T(t)$ |
|---|---|---|
| 75 | $\dfrac{e^{s^2/4a^2}\,\mathrm{erfc}(s/2a)}{s}$ | $\mathrm{erf}(at)$ |
| 76 | $\dfrac{e^{as}\,\mathrm{erfc}\sqrt{as}}{\sqrt{s}}$ | $\dfrac{1}{\sqrt{\pi\,(t+a)}}$ |
| 77 | $e^{as}\,\mathrm{Ei}(as)$ | $\dfrac{1}{t+a}$ |
| 78 | $1$ | $\delta(t) =$ delta function |
| 79 | $e^{-as}$ | $\delta(t-a)$ |
| 80 | $\dfrac{\sinh sx}{s\sinh sa}$ | $\dfrac{x}{a}+\dfrac{2}{\pi}\sum_{n=1}^{\infty}\dfrac{(-1)^n}{n}\sin\dfrac{n\pi x}{a}\cos\dfrac{n\pi t}{a}$ |
| 81 | $\dfrac{\sinh sx}{s\cosh sa}$ | $\dfrac{4}{\pi}\sum_{n=1}^{\infty}\dfrac{(-1)^n}{(2n-1)}\sin\dfrac{(2n-1)\pi x}{2a}\sin\dfrac{(2n-1)\pi t}{2a}$ |
| 82 | $\dfrac{\cosh sx}{s\sinh as}$ | $\dfrac{t}{a}+\dfrac{2}{\pi}\sum_{n=1}^{\infty}\dfrac{(-1)^n}{n}\cos\dfrac{n\pi x}{a}\sin\dfrac{n\pi t}{a}$ |
| 83 | $\dfrac{\cosh sx}{s\cosh sa}$ | $1+\dfrac{4}{\pi}\sum_{n=1}^{\infty}\dfrac{(-1)^n}{(2n-1)}\cos\dfrac{(2n-1)\pi x}{2a}\cos\dfrac{(2n-1)\pi t}{2a}$ |
| 84 | $\dfrac{\sinh sx}{s^2\sinh sa}$ | $\dfrac{xt}{a}+\dfrac{2a}{\pi^2}\sum_{n=1}^{\infty}\dfrac{(-1)^n}{n^2}\sin\dfrac{n\pi x}{a}\sin\dfrac{n\pi t}{a}$ |
| 85 | $\dfrac{\sinh sx}{s^2\cosh sa}$ | $x+\dfrac{8a}{\pi^2}\sum_{n=1}^{\infty}\dfrac{(-1)^n}{(2n-1)^2}\sin\dfrac{(2n-1)\pi x}{2a}\cos\dfrac{(2n-1)\pi t}{2a}$ |
| 86 | $\dfrac{\cosh sx}{s^2\sinh sa}$ | $\dfrac{t^2}{2a}+\dfrac{2a}{\pi^2}\sum_{n=1}^{\infty}\dfrac{(-1)^n}{n^2}\cos\dfrac{n\pi x}{a}\left(1-\cos\dfrac{n\pi t}{a}\right)$ |
| 87 | $\dfrac{\cosh sx}{s^2\cosh sa}$ | $t+\dfrac{8a}{\pi^2}\sum_{n=1}^{\infty}\dfrac{(-1)^n}{(2n-1)^2}\cos\dfrac{(2n-1)\pi x}{2a}\sin\dfrac{(2n-1)\pi t}{2a}$ |
| 88 | $\dfrac{\cosh sx}{s^3\cosh sa}$ | $\dfrac{1}{2}(t^2+x^2-a^2)-\dfrac{16a^2}{\pi^3}\sum_{n=1}^{\infty}\dfrac{(-1)^n}{(2n-1)^3}\cos\dfrac{(2n-1)\pi x}{2a}\cos\dfrac{(2n-1)\pi t}{2a}$ |
| 89 | $\dfrac{\sinh x\sqrt{s}}{\cosh a\sqrt{s}}$ | $\dfrac{2\pi}{a^2}\sum_{n=1}^{\infty}(-1)^n n e^{-n^2\pi^2 t/a^2}\sin\dfrac{n\pi x}{a}$ |

TABLE F.1 **(*cont.*)**

| | $\bar{T}(s)$ | $T(t)$ |
|---|---|---|
| 90 | $\dfrac{\cosh x\sqrt{s}}{\sinh a\sqrt{s}}$ | $\dfrac{\pi}{a^2}\sum_{n=1}^{\infty}(-1)^{n-1}(2n-1)e^{-(2n-1)^2\pi^2 t/4a^2}\cos\dfrac{(2n-1)\pi x}{2a}$ |
| 91 | $\dfrac{\sinh x\sqrt{s}}{\sqrt{s}\cosh a\sqrt{s}}$ | $\dfrac{2}{a}\sum_{n=1}^{\infty}(-1)^{n-1}\, e^{-(2n-1)^2\pi^2 t/4a^2}\sin\dfrac{(2n-1)\pi x}{2a}$ |
| 92 | $\dfrac{\cosh x\sqrt{s}}{\sqrt{s}\sinh a\sqrt{s}}$ | $\dfrac{1}{a}+\dfrac{2}{a}\sum_{n=1}^{\infty}(-1)^n e^{-n^2\pi^2 t/a^2}\cos\dfrac{n\pi x}{a}$ |
| 93 | $\dfrac{\sinh x\sqrt{s}}{s\sinh a\sqrt{s}}$ | $\dfrac{x}{a}+\dfrac{2}{\pi}\sum_{n=1}^{\infty}\dfrac{(-1)^n}{n}e^{-n^2\pi^2 t/a^2}\sin\dfrac{n\pi x}{a}$ |
| 94 | $\dfrac{\cosh x\sqrt{s}}{s\cosh a\sqrt{s}}$ | $1+\dfrac{4}{\pi}\sum_{n=1}^{\infty}\dfrac{(-1)^n}{(2n-1)}e^{-(2n-1)^2\pi^2 t/4a^2}\cos\dfrac{(2n-1)\,\pi x}{2a}$ |
| 95 | $\dfrac{\sinh x\sqrt{s}}{s^2\sinh a\sqrt{s}}$ | $\dfrac{xt}{a}+\dfrac{2a^2}{\pi^3}\sum_{n=1}^{\infty}\dfrac{(-1)^n}{n^3}(1-e^{-n^2\pi^2 t/a^2})\sin\dfrac{n\pi x}{a}$ |
| 96 | $\dfrac{\cosh x\sqrt{s}}{s^2\cosh a\sqrt{s}}$ | $\dfrac{1}{2}(x^2-a^2)+t-\dfrac{16a^2}{\pi^3}\sum_{n=1}^{\infty}\dfrac{(-1)^n}{(2n-1)^3}e^{-(2n-1)^2\pi^2 t/4a^2}\cos\dfrac{(2n-1)\pi x}{2a}$ |
| 97 | $\dfrac{J_0(ix\sqrt{s})}{sJ_0(ia\sqrt{s})}$ | $1-2\sum_{n=1}^{\infty}\dfrac{e^{-\lambda_n^2 t/a^2}J_0(\lambda_n x/a)}{\lambda_n J_1(\lambda_n)}$ where $\lambda_1, \lambda_2, \ldots$ are the positive roots of $J_0(\lambda)=0$ |
| 98 | $\dfrac{J_0(ix\sqrt{s})}{s^2 J_0(ia\sqrt{s})}$ | $\dfrac{1}{4}(x^2-a^2)+t+2a^2\sum_{n=1}^{\infty}\dfrac{e^{-\lambda_n^2 t/a^2}J_0(\lambda_n x/a)}{\lambda_n^3 J_1(\lambda_n)}$ where $\lambda_1, \lambda_2, \ldots$ are the positive roots of $J_0(\lambda)=0$ |

## REFERENCES


1. C. R. WYLIE and L. C. BARRET *Advanced Engineering Mathematics*, McGraw-Hill, NY, 1982.
2. M. D. GREENBERG, *Foundations of Applied Mathematics*, Prentice Hall, Englewood Cliffs, NJ, 1978.
3. M. R. SPIEGEL, *Laplace Transforms*, Schaum's Outline Series, McGraw-Hill, NY, 1965.

APPENDIX **G**

# SOLUTION OF A SYSTEM OF LINEAR ALGEBRAIC EQUATIONS USING THE THOMAS ALGORITHM

A convenient and popular algorithm for the solution of the system of equations (11.117) or (11.120) is the Thomas algorithm, also called the tridiagonal matrix algorithm, discussed in this appendix.

First, we note that eq. (11.117) can be written in the following form:

$$A_i T_i^{P+1} = B_i T_{i+1}^{P+1} + C_i T_{i-1}^{P+1} + D_i \tag{G.1}$$

where the coefficients are all known and are defined as

$$A_i = 1 + 2\frac{\alpha \, \Delta t}{(\Delta x)^2} \tag{G.2}$$

$$B_i = \frac{\alpha \, \Delta t}{(\Delta x)^2} \tag{G.3}$$

$$C_i = \frac{\alpha \, \Delta t}{(\Delta x)^2} \tag{G.4}$$

$$D_i = T_i^P \tag{G.5}$$

Equation (G.1) can be applied to all interior points, $i = 2, \ldots, M - 1$. Recall that we assumed that the boundary temperatures $T_1^{P+1}$ and $T_M^{P+1}$ are either known from the boundary conditions or that the boundary conditions can provide the needed relations for the boundary temperatures, in terms of the temperatures of the grid points near the boundaries ($T_2^{P+1}$, or $T_2^{P+1}$ and $T_3^{P+1}$ at one boundary, $T_{M-1}^{P+1}$, or

$T_{M-1}^{P+1}$ and $T_{M-2}^{P+1}$ at the other boundary). Applying equation (G.1) for $i = 2$ (the first interior point) yields

$$T_2^{P+1} = \frac{B_2}{A_2}T_3^{P+1} + \frac{C_2}{A_2}T_1^{P+1} + \frac{D_2}{A_2} \tag{G.6}$$

Since $T_1^{P+1}$ is either known or can be expressed in terms of $T_2^{P+1}$ and $T_3^{P+1}$ from the discretized boundary condition, equation (G.6) is effectively a relation between $T_2^{P+1}$ and $T_3^{P+1}$. Next, applying eq. (G.1) for $i = 3$ yields a relation between $T_2^{P+1}$, $T_3^{P+1}$, and $T_4^{P+1}$. However, utilizing eq. (G.6) to eliminate $T_2^{P+1}$, this relation becomes effectively one between $T_3^{P+1}$ and $T_4^{P+1}$. This process continues all the way until $i - 1$. Applying eq. (G.1) for $i = M - 1$ yields

$$T_{M-1}^{P+1} = \frac{B_{M-1}}{A_{M-1}}T_M^{P+1} + \frac{C_{M-1}}{A_{M-1}}T_{M-2}^{P+1} + \frac{D_{M-1}}{A_{M-1}} \tag{G.7}$$

Since a relation between $T_{M-1}^{P+1}$ and $T_{M-2}^{P+1}$ exists from the earlier application of eq. (G.1) for $i = M - 2$, eq. (G.7) is, effectively, a relation between $T_{M-1}^{P+1}$ and $T_M^{P+1}$. Further, since information on $T_M^{P+1}$ is provided by the boundary condition (as explained earlier in connection with $T_1^{P+1}$), eq. (G.7) can be solved for the value of $T_{M-1}^{P+1}$. With $T_{M-1}^{P+1}$ known, a process of back substitution can be initiated to obtain $T_{M-2}^{P+1}$ from $T_{M-1}^{P+1}$, $T_{M-3}^{P+1}$ from $T_{M-2}^{P+1} \cdots T_3^{P+1}$ from $T_4^{P+1}$ and $T_2^{P+1}$ from $T_3^{P+1}$.

The discussion above describes the basic principles of the Thomas algorithm. In terms of programming, at each point a relation of the type

$$T_i^{P+1} = F_i T_{i+1}^{P+1} + R_i \tag{G.8}$$

needs to be constructed during the forward marching process from ($i = 2$ to $i = M - 1$). Equation (G.8) connects the temperatures of two sequential grid points. Written for point $i - 1$, eq. (G.8) reads

$$T_{i-1}^{P+1} = F_{i-1}T_i^{P+1} + R_{i-1} \tag{G.9}$$

Combining eqs. (G.9) and (G.1) to eliminate $T_{i-1}^{P+1}$ yields, after some rearranging,

$$T_i^{P+1} = \frac{B_i}{A_i - C_iF_{i-1}}T_{i+1}^{P+1} + \frac{D_i + C_iR_{i-1}}{A_i - C_iF_{i-1}} \qquad i = 2, 3, \ldots, M - 2 \tag{G.10}$$

Note that eq. (G.10) is identical to eq. (G.8) with

$$F_i = \frac{B_i}{A_i - C_iF_{i-1}} \tag{G.11}$$

$$R_i = \frac{D_i + C_iR_{i-1}}{A_i - C_iF_{i-1}} \tag{G.12}$$

To start the calculation process, first information on $T_i^{P+1}$ and $T_M^{P+1}$ needs to be imputed in eq. (G.10) for $i = 2$ and $i = M - 1$, respectively, utilizing the boundary conditions according to the eariler discussion. Next, eqs. (G.11) and (G.12) should be used to obtain $F_i$ and $R_i$ for $i = 2, 3, \ldots, M - 1$. Finally, eq. (G.8) needs to be employed for $i = M - 1, M - 2, \ldots, 3, 2$, to yield the temperature at all interior points.

# INDEX

## E



## F



## G



## H



## I



## L



## M